Neurodegenerative Diseases

This reference book provides a comprehensive overview of models and therapeutic approaches against neurodegenerative diseases, including Parkinson's disease, Alzheimer's disease, Huntington's disease, and amyotrophic lateral sclerosis. It explores models based on the chemical, induced, cellular, genetic, transgenic, and 3D organoid approaches in neurodegenerative diseases. The book also reviews advantages and limitations of these models in designing the treatment strategies. Additionally, the book covers the emerging field of bioinformatics and its application in modeling various neurodegenerative diseases. Towards the end, the book highlights the role of holistic management, precision medicine, OMICS, and gene therapy against neurodegenerative disorders. It examines the implications and significance of stem cells therapy in translational models of neurodegenerative diseases. This book is an invaluable resource for researchers, neuroscientists, and neurosurgeons for getting in-depth information on the neurodegenerative models and therapeutic approaches.

Key Features

- Provides a comprehensive overview of neurodegenerative diseases and their models
- Examines the limitations associated with modeling neurodegenerative diseases
- Presents novel treatment strategies for Alzheimer's disease using cellular models
- Reviews importance of 3D organoid models for therapeutic approaches in Parkinson's disease
- Covers modeling techniques in understanding prion diseases
- Explores the role of genetic models in understanding Huntington's disease

Neurodegenerative Diseases

Translational Models, Mechanisms, and Therapeutics

Edited by
Sachchida Nand Rai
Sandeep Singh
Santosh Kumar Singh

CRC Press
Taylor & Francis Group
Boca Raton London New York

CRC Press is an imprint of the
Taylor & Francis Group, an **informa** business

Designed cover image: © Shutterstock

First edition published 2025
by CRC Press
2385 NW Executive Center Drive, Suite 320, Boca Raton FL 33431

and by CRC Press
4 Park Square, Milton Park, Abingdon, Oxon, OX14 4RN

CRC Press is an imprint of Taylor & Francis Group, LLC

ISBN: 978-1-032-61324-6 (hbk)
ISBN: 978-1-032-62042-8 (pbk)
ISBN: 978-1-032-62044-2 (ebk)

DOI: 10.1201/9781032620442

Typeset in Times
by KnowledgeWorks Global Ltd.

To those who care, conserve, and protect but do not destroy the beauty and unique characteristics of

Kshit (Earth)

Jal (Water)

Pawak (Fire)

Gagan (Sky)

and

Sameera (Air)

Contents

Preface

Neurodegenerative diseases represent a formidable challenge to both medical science and society at large. Their complex etiology, progressive nature, and often devastating impact on individuals and families underscore the urgent need for comprehensive understanding and effective therapeutic interventions. This book, *Neurodegenerative Diseases: Translational Models, Mechanisms, and Therapeutics*, emerges as a response to this pressing need, aiming to provide a cohesive resource that delves into various facets of neurodegenerative disorders, from their molecular mechanisms to translational therapeutic strategies.

In this preface, we offer a brief overview of the contents encapsulated within this volume, outlining the diverse range of topics covered and their significance in the broader landscape of neurodegenerative disease research and management.

Chapter 1: General Introduction to Different Neurodegenerative Diseases

The journey begins with a comprehensive survey of various neurodegenerative diseases, laying down the foundational knowledge necessary to navigate the subsequent chapters. From Alzheimer's to Huntington's, this chapter sets the stage by elucidating the commonalities and distinctions among these disorders.

Chapters 2–5: Parkinson's Disease: Models and Therapeutics

Subsequent chapters delve into Parkinson's disease, one of the most prevalent neurodegenerative conditions. From chemical-induced models to cutting-edge 3D organoid approaches, these chapters explore the multifaceted landscape of Parkinson's disease research, offering insights into both its pathogenesis and therapeutic avenues.

Chapters 6–8: Alzheimer's Disease: Models and Therapeutics

Moving onto Alzheimer's disease, another major neurodegenerative challenge, these chapters delve into chemical and genetic models, as well as the promise held by organoid technologies. Through a critical examination of therapeutic strategies and their limitations, these chapters illuminate the ongoing efforts to combat Alzheimer's disease.

Chapter 9: Huntington's Disease: Role of Genetic Models

Huntington's disease receives special attention in this chapter, focusing on the significance of genetic models in unraveling its pathophysiology and guiding therapeutic development.

Chapter 10: Natural Sources of Drugs for Neurodegenerative Diseases

Amidst the quest for novel therapeutics, this chapter explores the potential of natural compounds in mitigating neurodegenerative processes, offering insights into alternative therapeutic avenues.

Chapters 11–12: Limitations and Significance of Modeling Approaches

Acknowledging the complexities and challenges inherent in disease modeling, these chapters critically examine the limitations and significance of various modeling approaches, including 3D organoids, in advancing our understanding and treatment of neurodegenerative diseases.

Chapters 13–17: Therapeutic Advancements and Holistic Management

The latter chapters of this book delve into recent therapeutic advancements, including the role of bioinformatics, gene therapy, and stem cell interventions. Through a holistic lens, these chapters aim to synthesize diverse approaches toward the management of neurodegenerative diseases, offering insights into the future directions of translational research.

As editors of this volume, we hope that *Neurodegenerative Diseases: Translational Models, Mechanisms, and Therapeutics* serves as a valuable resource for researchers, clinicians, and policy makers alike, fostering interdisciplinary dialogue and driving progress toward a future where effective treatments for neurodegenerative diseases are within reach.

Sachchida Nand Rai

Sandeep Singh

Santosh Kumar Singh

About the Editors

Dr. Sachchida Nand Rai is currently serving as a Research Associate (ICMR-RA) at the Centre of Experimental Medicine and Surgery, Institute of Medical Sciences, Banaras Hindu University (BHU), Varanasi, a position he has held since January 27, 2023. Dr. Rai earned his PhD from the Biochemistry Department at BHU in 2018. His research focuses on exploring the anti-Parkinsonian properties of various herbal plants such as *Mucuna pruriens, Tinospora cordifolia,* and *Withania somnifera*, as well as their bioactive components like ursolic acid, betunilic acid, and chlorogenic acid, particularly in toxin-induced Parkinsonian mouse models. Banaras Hindu University consistently ranks among India's top universities for life sciences research output and quality.

Prior to his current role, Dr. Rai served as a post-doctoral fellow at the Centre of Biotechnology, University of Allahabad, Prayagraj, India, from January 27, 2020, to January 26, 2023, accumulating three years of experience. Dr. Rai has made significant contributions to the field, with impressive publications in reputable journals. He serves as an academic editor for BMC Research Notes and associate editor for 3-Biotech (Springer Nature), Molecular Neurobiology, Oxidative Medicine and Cellular Longevity, Journal of Inflammation Research, and Degenerative Neurological and Neuromuscular Disease (Dove Press). Additionally, he sits on the advisory board of Heliyon Neuroscience.

Dr. Rai has edited over 100 manuscripts and reviewed more than 1050 manuscripts for high-impact journals. He has authored two books with renowned publishers Bentham. Dr. Rai's Google Scholar citations stand at 3488, with an h-index of 32. He is a member of the Indian Academy of Neurosciences. His current research involves investigating the neuroprotective properties of commonly used Indian medicinal plants in toxin-induced Parkinsonian mouse models.

Dr. Sandeep Singh received his PhD (2020) from the Department of Biochemistry, University of Allahabad, Prayagraj (formerly known as Allahabad), Uttar Pradesh, India. It is one of the oldest modern universities in India and also known as the 'Oxford of the East'. His main research interests are in the crosstalk between biological activity of natural products, biochemical and molecular biology of aging, and age-associated diseases. The Department of Science and Technology (DST) and Indian Council of Medical Research (ICMR) in India have provided funding for his doctoral research. During his doctoral studies, he had an opportunity to work at the National Institute of Advanced Industrial Science & Technology (AIST) Tsukuba, Japan, supported by the International Brain Research Organization (IBRO) exchange fellowship. Subsequently upon receiving a postdoc fellowship at Hadassah Biological Psychiatry Laboratory Hadassah – Hebrew University Medical Centre, Jerusalem, Israel, his research was broadly focused to study psychedelic drugs and their role in psychiatric disorders. Following the completion of this project, Dr. Singh continued his research journey as a postdoctoral fellow at the Salk Institute for Biological Studies in San Diego, California, USA. Here, his primary objective was to screen compounds capable of impeding the progression of neurodegenerative diseases, particularly targeting cellular aging and memory loss observed in Alzheimer's disease.

To date, Dr. Sandeep Kumar Singh has contributed to the scientific community with 32 research articles and 11 book chapters published in esteemed international journals, having nearly 1,000 citations with an impressive h-index of 19 on Google Scholar. He has also served as a reviewer for numerous reputable journals including *Phytotherapy Research*, *Archives of Physiology and Biochemistry*,

PLos One, *Rejuvenation Research*, *Journal of Biochemical and Molecular Toxicology*, *3 Biotech*, *Frontiers in Oncology*, *International Journal of Inflammation*, and *Neural Plasticity*, among others.

Dr. Singh's scholarly endeavors extend beyond publications; he has showcased his work in various international conferences such as the Society for Neurosciences 2018 (USA), Japan Neurosciences Society 2018 (JNS), and the International Society for Neurochemistry 2017 (ISN). He holds memberships in these esteemed societies, as well as being a life member of the Indian Science Congress Association and the Indian Academy of Neurosciences, India.

Dr. Sandeep Singh is a post-doctoral fellow at the Cellular Neurobiology Laboratory, Salk Institute for Biological Studies, San Diego, California, USA. The Salk Institute consistently ranks among the top institutions in the USA in terms of research output and quality in the life sciences. Before joining the Salk Institute, he was a post-doctoral fellow at Hadassah Biological Psychiatry Laboratory, Hadassah – Hebrew University Medical Center, Jerusalem, Israel. During his PhD studies, he presented his work in many world-renowned international neuroscience conferences such as Society for Neuroscience (USA), International Society for Neurochemistry (Paris), and Japan Neuroscience Society (Japan). He was a recipient of a IBRO-APRC Exchange Fellowship (2019) to work at the National Institute of Advanced Industrial Science & Technology (AIST), Japan and Senior Research fellowship (2018) from the Indian Council of Medical Research, India. He has impressive publications in reputed journals. He is investigating the potential role of natural compounds and how they modulate the different molecular pathways during aging and age-associated diseases. He is also a member of the Indian Science Congress Association (India), Indian Academy of Neurosciences (India).

Dr. Santosh Kumar Singh is a distinguished figure in the scientific community, serving as a senior scientist and administrative in charge at the Centre of Experimental Medicine and Surgery, Institute of Medical Sciences, Banaras Hindu University, Varanasi. He obtained his PhD in botany from the same institution in 1998 and further enriched his expertise as a visiting scientist at the University of South Bohemia, Czech Republic. With an extensive career spanning over 23 years in academia and research, complemented by over 14 years of leadership in university administration, Dr. Singh's primary area of expertise revolves around the discovery and development of plant-derived drugs.

Dr. Singh's contributions are profound, having supervised 12 PhD and 10 MD/MS theses, authored numerous publications in esteemed journals, and edited an international book. He has also held editorial roles in various journals, showcasing his commitment to advancing scientific discourse. His leadership is evident through successfully spearheading multiple research projects funded by prestigious organisations such as DST, DBT, and SERB, India.

Acknowledged for his remarkable achievements, Dr. Singh has received accolades including the National Award of Science and Technology Communication from DST and the Excellence in Science Communication Award from Scientific & Education Research Society. His contributions extend beyond academia, as he actively engages in scientific awareness initiatives targeting diverse audiences such as school students, farmers, and rural communities.

Dr. Singh's dedication to uplifting economically disadvantaged regions through science and technology interventions is noteworthy. Serving as a technical advisor for FAARD-Foundation, he mentors over 60 farmer–producer organizations, positively impacting the lives of more than 100,000 farmers.

Contributors

Atifa Haseeb Ansari
S.S. Khanna Girls' Degree College
A Constituent College of University of Allahabad
Prayagraj, India

Vinjavarapu Lakshmi Anusha
Department of Pharmacy
Koneru Lakshmaiah Education Foundation
Vaddeswaram, India

Samuel Junior Osei Asokwah
Department of Pharmacy
Koneru Lakshmaiah Education Foundation
Vaddeswaram, India

Mehak Bajaj
School of Innovation in Biodesign
Translational Health Science and Technology Institute
Faridabad, India

Ishita Bhardwaj
S.S. Khanna Girls' Degree College
A Constituent College of University of Allahabad
Prayagraj, India

Kunal Biswas
Centre for Nanoscience and Nanotechnology
International Research Centre
Sathyabama Institute of Science and Technology
Tamil Nadu, India

Guntupalli Chakaravarthi
Department of Pharmacy
Koneru Lakshmaiah Education Foundation
Vaddeswaram, India

Naresh Dumala
Department of Pharmacy
Koneru Lakshmaiah Education Foundation
Vaddeswaram, India

Mansi Gandhi
School of Innovation in Biodesign
Translational Health Science and Technology Institute
Faridabad, India
and
Institute of Chemistry
Hebrew University of Jerusalem
Jerusalem, Israel

Ashutosh Gupta
ICMR-National Institute of Child Health and Development Research
Sajdarjung Hospital Campus
New Delhi, India

Saqib Hassan
Department of Biotechnology
Sathyabama Institute of Science and Technology
Tamil Nadu, India

Amita Jain
Department of Microbiology
Faculty of Medicine
King George's Medical University
Lucknow, India

Farrukh Jamal
Department of Biochemistry
Dr. Rammanohar Lohia Avadh University
Ayodhya, India

Priti Kanik
Department of Biological Engineering
Indian Institute of Technology Gandhinagar
Palaj, India

Neeraj Kumar
ICMR-National Institute of Child Health and Development Research
Sajdarjung Hospital Campus
New Delhi, India

Pradeep Kumar
Department of Neurology
Faculty of Medicine
Hadassah Medical Center
Hebrew University of Jerusalem
Jerusalem, Israel

Rashi Kumar
Department of Neurology
Institute of Human Behaviour and Allied Sciences
New Delhi, India

V. Ramesh Kumar
Department of Biotechnology
Sathyabama Institute of Science and Technology
Tamil Nadu, India

Shashank Shekhar Mishra
Centre of Experimental Medicine and Surgery
Institute of Medical Sciences
Banaras Hindu University
Varanasi, India

Hayiru Mohammed
Department of Pharmacy
Koneru Lakshmaiah Education Foundation
Vaddeswaram, India

Mohit
Department of Prosthodontics
Faculty of Dental Sciences
King George's Medical University
Lucknow, India

Buchi N Nalluri
Department of Pharmacy
Koneru Lakshmaiah Education Foundation
Vaddeswaram, India

Grace Mwihaki Njuguna
Department of Pharmacy
Koneru Lakshmaiah Education Foundation
Vaddeswaram, India

Shashank Kumar Ojha
The Hebrew University of Jerusalem
Jerusalem, Israel

Devhuti Palan
School of Innovation in Biodesign
Translational Health Science and Technology Institute
Faridabad, India

Prabhash Kumar Pandey
Centre for Advanced Research
Faculty of Medicine
King George's Medical University
Lucknow, India

Aakansha Pathak
ICMR-National Institute of Child Health and Development Research
Sajdarjung Hospital Campus
New Delhi, India

Shivani Pathak
Department of Zoology
S.S. Khanna Girls' Degree College Prayagraj
A Constituent College of University of Allahabad
Prayagraj, India

Faiza Parvez
S.S. Khanna Girls' Degree College
A Constituent College of University of Allahabad)
Prayagraj, India

Chandrabhan Prajapati
Centre of Experimental Medicine and Surgery
Institute of Medical Sciences
Banaras Hindu University
Varanasi, India

Chandra Prakash
Neurobiology Laboratory
School of Life Sciences
Jawaharlal Nehru University
New Delhi, India

Raghvendra Pratap
Prof. Rajendra Singh (Rajju Bhaiya) University
Prayagraj, India

Shyam Sunder Rabidas
Neurobiology Laboratory
School of Life Sciences
Jawaharlal Nehru University
New Delhi, India

Sachchida Nand Rai
Centre of Experimental Medicine and Surgery
Institute of Medical Sciences
Banaras Hindu University
Varanasi, India

Swayam Prabha Rai
Department of Zoology
S.S. Khanna Girls' Degree College
A Constituent College of University of Allahabad
Prayagraj, India

Kakarla Ramakrishna
Department of Pharmacy
Koneru Lakshmaiah Education Foundation
Vaddeswaram, India

Aakriti Sharma
Endocrinology division
CDRI, Lucknow
Lucknow, India

Deepak Sharma
Neurobiology Laboratory
School of Life Sciences
Jawaharlal Nehru University
New Delhi, India

Ashutosh Shrivastava
Centre for Advanced Research
Faculty of Medicine
King George's Medical University
Lucknow, India

Shiva Shukla
Centre for Advanced Research
Faculty of Medicine
King George's Medical University
Lucknow, India

Anurag Kumar Singh
Department of Pharmaceutical Engineering & Technology
Indian Institute of Technology, BHU
Varanasi, India

Durgesh Singh
S.S. Khanna Girls' Degree College
A Constituent College of University of Allahabad
Prayagraj, India

Sandeep Singh
Centre of Biotechnology
University of Allahabad
Prayagraj, India

Santosh Kumar Singh
Centre of Experimental Medicine and Surgery
Institute of Medical Sciences
Banaras Hindu University
Varanasi, India

Payal Singh
Department of Zoology
Banaras Hindu University
Varanasi, India

Prabhakar Singh
Department of Biotechnology
Sathyabama Institute of Science and Technology
Tamil Nadu, India

Sippy Singh
S.S. Khanna Girls' Degree College
A Constituent College of University of Allahabad
Prayagraj, India

Vineeta Singh
Department of Geriatric Medicine
Faculty of Medicine
Institute of Medical Sciences
Banaras Hindu University
Varanasi, India

Anshuman Srivastava
Centre for Advanced Research
Faculty of Medicine
King George's Medical University
Lucknow, India

Soni Tiwari
Department of Microbiology
Dr. Rammanohar Lohia Avadh University
Ayodhya, India

Manish Kumar Tripathi
Institute for Drug Research
School of Pharmacy
Faculty of Medicine
The Hebrew University of Jerusalem
Jerusalem, Israel

Ritu Varshney
PD Patel Institute of Applied Sciences
Charotar University of Science and Technology
CHARUSAT Campus
Anand, India

Vimala Venkatesh
Department of Microbiology
Faculty of Medicine
King George's Medical University
Lucknow, India

Matte Kasi Viswanadh
Department of Pharmacy
Koneru Lakshmaiah Education Foundation
Vaddeswaram, India

Sandhya Yadav
Department of Biochemistry and Molecular Pharmacology
University of Massachusetts Medical School
Worcester, USA

Acknowledgements

First and foremost, we express our deepest gratitude to the Almighty for guiding us through this journey and bestowing upon us the strength and wisdom to undertake this challenging endeavor.

It gives us immense pleasure to acknowledge **Bharat Ratna Mahamana Pt. Madan Mohan Malviya Ji,** founder of the Banaras Hindu University, Varanasi, Uttar Pradesh, India.

Sachchida Nand Rai would like to acknowledge the Indian Council of Medical Research (ICMR), Department of Health Research, Ministry of Health and Family Welfare, Government of India for support through Research Associate (ICMR-RA) Award (3/1/3/196/Neuro/2021-NCD-I).

Santosh Kumar Singh would like to acknowledge funding from the Institution of Eminence Incentive Grant Scheme, Trans Disciplinary Research Scheme, Department of Science and Technology, SEED grant, New Delhi and funding from the Institute of Medical Sciences intramural grant, Banaras Hindu University, Varanasi, India.

It is also our great pleasure to acknowledge and express our enormous debt to all the contributors who have provided their quality material to prepare this book. We are grateful to our beloved family members, who joyfully supported and stood with us in many hours of our absence to finish this book project. I would also like to acknowledge the publishing team at CRC Press for their professionalism, enthusiasm, and belief in this project. Their expertise in design, production, and marketing has been essential in bringing this book to life. We extend our heartfelt gratitude to all those who have contributed to the creation of this seminal work, *Neurodegenerative Diseases: Translational Models, Mechanisms, and Therapeutics.* This endeavor has been a culmination of collective efforts, steadfast dedication, and unwavering support from numerous individuals and entities.

We owe a profound debt of gratitude to our esteemed teachers, mentors, and advisors whose invaluable guidance, encouragement, and scholarly insights have been instrumental in shaping our understanding and approach towards unraveling the complexities of neurodegenerative diseases.

To all the contributors from India and across the globe, your expertise, passion, and commitment have enriched the pages of this book and have been pivotal in advancing our understanding of neurodegenerative disorders. Your diverse perspectives, groundbreaking research, and tireless efforts have significantly contributed to the comprehensiveness and depth of this work.

Sachchida Nand Rai

Sandeep Singh

Santosh Kumar Singh

1 General Introduction to Different Neurodegenerative Diseases

Prabhakar Singh, Saqib Hassan, and V. Ramesh Kumar
Department of Biotechnology, Sathyabama Institute of Science and Technology, Tamil Nadu, Chennai, India

Kunal Biswas
Centre for Nanoscience and Nanotechnology, International Research Centre, Sathyabama Institute of Science and Technology, Tamil Nadu, Chennai, India

1.1 INTRODUCTION

The WHO Global Action Plan for the Prevention and Management of Non-Communicable Diseases and the WHO Intersectoral Global Action Plan on Epilepsy and Other Neurological Disorders (2022–2031) are initiatives to reduce neurological disorders and help to achieve Sustainable Development Goal 3.4. With nine million fatalities annually, neurological illnesses rank as the 2nd greatest reason of mortality worldwide and the primary reason of disability-adjusted life years. According to a ground-breaking Global Burden of Disease (GBD) study, by 2050, there will be twice as many people with brain disease. The GBD study offers estimates on the prevalence, years of life lost or with disability, and disability-adjusted life-years for each of the 36 most prevalent neurological illnesses and conditions in the world. Stroke, neonatal encephalopathy, migraine, dementia, meningitis, epilepsy, neurological difficulties linked with preterm delivery, nervous system malignancies, Parkinson's disease, and autism spectrum disorders currently cover approximately 90% of all neurological disability-adjusted life-years [1].

The primary pathophysiological alteration found in brain-related diseases has been determined to be neurodegeneration [2]. Neurons are fundamental to the human brain's healthy operation. Although neurons are found throughout the body, most of the nervous system originates in the brain [3]. The majority of the nervous systems is produced by neural stem cells throughout childhood; in contrast, adult neurons are much fewer in number [4]. Neurodegeneration, the progressive reduction and dysfunction of neurons, is a major health problem and a pioneer factor in the causes of various brain diseases [5]. Synapse and neural network dysfunction, with the accumulation of physiochemically changed protein variations in the CNS, is linked to neurodegeneration [6]. Alzheimer's disease (AD), amyotrophic lateral sclerosis (ALS), Parkinson's disease (PD), Huntington's disease (HD), motor neuron disease (MND), spinocerebellar ataxia (SA), and spinal muscular atrophy (SMA) are some of the most prevalent neurodegenerative diseases (NDs) [7].

Age remains the major risk factor for the inception of non-communicable diseases (NCDs), but new research discoveries explain that an individual's genetic composition and environmental factors can also increase the risk for NCDs. Furthermore, the duration and degree of neurodegeneration primarily depend on their present environment [8, 9], even while specific genes responsible for NDs are expressed [10]. Recent research indicates that a single neurodegenerative disease may be

DOI: 10.1201/9781032620442-1

underlaid by several disorders [7, 11–13]. As a result, NDs can be extremely dangerous or even fatal in certain cases. Because the brain regulates several bodily functions, neurodegenerative diseases impact a variety of aspects of human functioning and impair the capacity to carry out both simple (such as movement, speech, balance, and stability) and complex (cognitive functions and bowel and bladder movements) tasks. While in some cases treatments focus on improving symptoms, relieving pain if it is present, and/or restoring movement and balance, most NDs develop without remission.

1.2 ALZHEIMER'S DISEASE (AD)

According to the World Health Organization (2022) and the World Alzheimer Report (Alzheimer's Disease International, 2022), AD and associated dementia (ADRD) are the leading causes of death, dependency, and impairment in the elderly. In terms of dementia, AD is the most prevalent. This is a sickness that progresses, starting with modest memory loss and perhaps ending with loss of capacity to respond to surroundings and carry on a conversation. AD affects the brain-specific areas responsible for language, memory, and thinking processes. The pathogenesis of Alzheimer's disease has demonstrated that amyloid-beta (Aβ) and tau protein buildup are essential to the disease's progression [14, 15].

The characteristic of AD is the development of Aβ-containing plaques along with neurofibrillary tangles (NFTs) within the brain due to hyperphosphorylated tau [16–18]. Hippocampal circuitry is disrupted by plaque formation, which results in weak short-term memory association into long-term traces [19]. AD is characterised by a sever loss of neural cells, abnormal synaptic connections, and impairment of the key neurotransmitter signaling molecules required for memory and other brain functions. Thus, selective memory impairment is the most prevalent clinical sign in the primary stages of AD. Furthermore, frequently impacted are the medial temporal lobe and hippocampal region dependent processes like declarative episodic memory. Lastly, other clinical indications typically show up early, such as impairment in problem-solving, judgment, and executive function [20]. The fundamental goal of treatments for controlling AD is to target several metabolic and signaling pathways involved in the course of the disease. The US FDA has currently approved three medication classes for the treatment of AD; these are detailed below.

1.3 AMYLOID-BETA (Aβ) PLAQUES AND AD THERAPY

In June 2021, Aducanumab, also known as Aduhelm, received its first approval as a disease-modifying medication for people with AD [21]. Monoclonal antibody (IgG1) aducanumab is highly active against extracellular Aβ plaques in the brain; it interacts with the plaques and facilitates their removal [21, 22]. Clinical research on aducanumab, although being conditionally licensed, indicates a decrease in the load of Aβ plaques, however, there is no correlation with better cognitive function in patients.

Phase I clinical trials have been completed by AAB-003, MEDI1814, RO7126209, SAR228810, and other Aβ targeting antibodies. Aducanumab is likewise specific to Aβ aggregation, despite completing phase III clinical trials. Comparably, gosuranemab (clinical trial phase II), whereas tau or TREM 2 specific antibodies, such as BIIB076, bepranemab, and JNJ-63733657, have finished phase I of clinical trials [23]. Consequently, more antibody-based targeted medications may soon receive FDA approval against AD.

1.4 CHOLINESTERASE INHIBITORS

For AD, cholinesterase inhibitors are currently 1st line drugs. In therapeutic settings, rivastigmine, galantamine, and donepezil are the three primary cholinesterase inhibitors. Acetylcholine levels in the brain's cortical regions decline and cholinergic neurons are lost in AD. It has been reported that elevated levels of acetylcholine in individuals with dementia mitigated their cognitive deterioration.

Increased cholinergic activity favors the patient when acetylcholine breakdown is limited by cholinesterase inhibitors [24].

1.5 GLUTAMATE REGULATORS

A neurotransmitter found in the brain, glutamate causes neuronal injury and neurodegeneration when it is overexpressed in postsynaptic neurons like NMDA receptors. However, dense adverse effects are associated with total blockage of NMDA receptors. As a result, memantine (an uncompetitive, NMDA receptor antagonist) was created. It shields patients from the inhibitory effects of overactivation while simultaneously offering pathological advantages with activation of NMDA receptor [25]. Compared to a placebo, memantine monotherapy enhanced cognition in AD patients. Clinical trials with acetylcholinesterase inhibitors demonstrated better efficacy compared to monotherapy after a year [26, 27].

1.6 PARKINSON'S DISEASE (PD)

PD is a multifaceted diverse disease that can range in severity from less complicated to advanced stages. It reflects a broad spectrum of motor/non-motor symptoms, including rigidity, bradykinesia, dysautonomia, sleep difficulties, and cognitive impairment [28]. In terms of frequency, after AD, PD is recognised as the 2nd neurodegenerative disease, and some recent research predicts a sharply increase in the total number of afflicted individuals in years to come. PD cannot be cured, and it will always progress. In harmony with medical practices, pharmaceutical treatment is typically initiated following the initial diagnosis to enhance the affected individual X's quality of life (QoL) [29].

PD is a degenerative condition inducing muscle rigidity, a shaky gait, tremors, and issues with balance and coordination. PD is brought on by both hereditary and non-genetic factors. PD is also mostly associated with age [30, 31]. Furthermore, many other factors are known to modify the chance of developing PD, including excessive coffee consumption, smoking, and exposure to environmental pollutants [32]. Ventricular enlargement and frontal cortical atrophy are the main pathogenesis of Parkinson's disease. Pigmentation loss in the substantia nigra pars compacta (SNpc) and locus coeruleus due to demise of neurons congaing dopaminergic (DA) neuromelanin is the most notable morphological change seen in the PD brain [33]. When a large number of cells are lost in PD, the nigrostriatal pathway becomes dysfunctional. This lowers the concentration of the signaling molecule dopamine in the striatum and causes cardinal motor symptoms [33]. The non-motor symptoms of PD are induced by cell loss in several locations, such vagus nerve nucleus, locus coeruleus, olfactory bulb, raphe nuclei, the pedunculopontine nucleus, and the hypothalamus [33, 34].

The advancement of PD is linked to several mechanisms, including misfolding and aggregation of α-synuclein, neuroinflammation, malfunctioning in protein clearance systems, mitochondrial dysfunction, ubiquitin-mediated proteasome system, and autophagy-mediated lysosome system [31, 33, 35]. Under a microscope, PD is identified by the presence of dystrophic neurites, also known as Lewy neurites, together with unregulated cytoplasmic deposits in neuronal cell bodies, an immunoreactive for α-synuclein [30, 36]. The CNS may seep Lewy bodies into other areas if they are phosphorylated. Protein misfolding happens in Parkinson's disease like in AD [36]. The tau protein is the protein that is frequently misfolded. NFT development is caused by aberrant tau protein hyperphosphorylation. Widespread NFT and Aβ plaques have been observed in a subset of PD patients [37].

1.7 PARKINSON'S DISEASE AND THERAPEUTIC APPROACHES

Restoring substantia nigra dopamine levels is the main strategy for treating PD. Several strategies are used to restore dopamine levels; however, levodopa with carbidopa is the most popular treatment. When dopamine is lost, levodopa, an instant precursor of dopamine, helps restore motor functioning. To prevent levodopa from being broken down peripherally before it reaches the brain, carbidopa and levodopa are combined. Furthermore, levodopa methylation by catechol-O-methyl transferase (COMT) is inhibited by entacapone and tolcapone, which prevents levodopa loss due to methylation. There are also dopaminergic agonists against Parkinson's disease (PD), including apomorphine hydrochloride, pramipexole dihydrochloride, pergolide, and rotigotine and ropinirole hydrochloride. These agonists have the same action as dopamine. Next in the line of accessible medications are monoamine oxidase inhibitors. They stop dopamine from being lost in the brain by inhibiting its oxidative deamination (e.g. rasagiline and selegiline) [38]. Like all NDs, PD is linked to major non-motor symptoms (MNMS), which are treated symptomatically and include psychosis, depression, constipation, sleep disturbances, dementia, and olfactory impairment [39]. Galantamine, donepezil, and rivastigmine are used to treat symptoms associated with dementia. Pimavanserin, quetiapine, and clozapine are authorised drugs to treat the psychosis. For problems related to sleep, clonazepam or melatonin is recommended [40].

1.8 AMYOTROPHIC LATERAL SCLEROSIS (ALS)

ALS is associated with nerve cells and spinal cord degeneration, inducing paralysis and muscle weakness. It is a motor neuron disease also called Lou Gehrig's disease [41–43]. As the motor neuron disease known as ALS progresses, patients may experience behavioral abnormalities, cognitive impairment, and frontotemporal dementia symptoms.

ALS people die due to respiratory failure three to five years after initial diagnosis. Motor neurons in ALS eventually degenerate, causing death [44]. Signals that ought to reach the brain are not sent when motor neurons are destroyed or injured.

More than 70% of ALS cases are caused by gene mutations, especially TARDBP, C9orf72, SOD1, and FUS, although over 30 other genes are also linked to the disease [44]. Major components of motor function, including homeostasis, repair of DNA, activity of mitochondria, and functions of glial cell, are encoded by these four genes. It is thought that a combination of these compromised processes is major cause of ALS. The pathogenic feature that distinguishes ALS is the deposition of intraneuronal protein aggregates. Superoxide dismutase-1 and neurofilament are proteins that can cluster; however, TAR DNA-binding protein is the most prevalent protein seen in the most of ALS people [45, 46]. However, it is not apparent if protein complexes or aggregates emerge first or if neuron injury comes first.

Gene-environment interaction could be involved in ALS in addition to solely genetic factors. Yoshida examined genetic-environmental research conducted in Guam, Kii, and Oshida regions on ALS and Parkinsonism-dementia complex (PDC). A significant result evidence supports for the idea that metallic ambient elements like Al, Ca, Fe, Mg, Cu, and Zn are important players in the oxidative reactions that lead to neuronal degeneration. Therefore, to stop the inception and progression of neurodegenerative diseases, the author emphasised the significance of controlling ionic homeostasis [47].

For ALS patients, edaravone and riluzole are the two approved treatments. The glutamate-receptor antagonist riluzole was first authorised in 1995 as a dose of 100 mg oral tablet. According to medical investigations, riluzole users live three to four months longer than those in a placebo group with ALS. The 2017 approval of edaravone, a free-radical scavenger, for intravenous infusion at a rate of 60 mg/day aids in delaying the course of the disease [48]. In addition to these drugs, patients receive symptomatic treatment to increases their life quality. As of right present, there are no effective disease-modifying treatments accessible.

Riluzole is a glutamate antagonist used to treat ALS. In some cases, individuals experience mild side effects; extremely infrequently, major adverse events occur. While some studies found that riluzole increases patient survival in population studies and clinical trials [49], other investigations concluded that none of these medications could significantly slow down the disease's course [48, 50]. Therefore, novel and creative ALS treatments are desperately needed. A biologically active protein called mouse nerve growth factor (mNGF) is taken out of the submaxillary gland of mice. Although no research has looked into its potential in treating ALS, it is legally approved in China to treat peripheral nerve injury.

1.9 HUNTINGTON'S DISEASE (HD)

Huntington's disease is an autosomal dominant disease characterised by cognitive impairment, abnormal movement (usually chorea), and psychiatric symptoms. According to Ghosh and Tabrizi (2018), the susceptible age of onset is 40 years, after which there is an unstoppable clinical decline and 15 to 20 years of average survival [51]. Before 21 years, juvenile HD starts, then gets worse quickly, frequently leading to death in ten years from the initial onset [52]. Worldwide variations in HD prevalence can range up to ten times [53]. According to prevalence studies, the prevalence is 2.7 per 100,000 worldwide, with Western nations having the greatest rates [53, 54]. As no medication has been reported to either stop or downregulate the HD progression, available treatments are restricted to symptomatic measures, such as quetiapine for behavioral symptoms of HD [55].

In Asia and Europe, HD prevalence is ≈ 0.4 per 100,000 people and 4–15 per 100,000 [56, 57]. Adults aged 30 to 50 years old are usually the ones who develop it, and the disease has a 20-year life expectancy following diagnosis. HD cannot be cured, but early intervention can reduce symptoms and enhance the quality of life [57]. The initial signs may manifest as mild psychosis, dysphoria, or cognitive impairments [56]. Psychotic symptoms are the initial signs in HD patients, which might make the clinical diagnosis more difficult. Here, we describe an early-stage HD patient who showed signs of psychosis and rapid cognitive impairment.

The pathogenesis of HD is typified by aberrant protein aggregation that spreads throughout the brain, a phenomenon also observed in PD, AD, and ALS [58]. An enlarged cytosine-adenine-guanine (CAG) repeat in exon 1 of chromosome 4 huntingtin gene (HTT) is responsible for this. An unusually lengthy polyglutamine (polyQ) region close to the huntingtin protein's (HTT) amino terminus is encoded by this genetic alteration. This protein exhibits toxic properties that cause neuronal malfunction and death. While HD has been identified as a condition affecting the nervous system and entire body, stroke-related GABAergic medium spiny neurons (MSNs) are more susceptible to mutant huntingtin-induced toxicity [59]. Testing for the disease's prognosis and diagnosis is made possible by its genetic pathogenesis. Additionally, because HD is genetically definite, we can create interesting genetic models to research disease causes and potential treatments.

A thorough investigation has yielded a significant understanding of the pathogenic processes underlying HD. Reducing mHTT has become a viable treatment strategy due to the monogenic character of HD, where the pathology is linked to the development of mutant huntingtin protein (mHTT). Novel treatments for HD are either being proposed or are in the process of clinical trials; these generally involve gene silencing or mHTT-lowering medications that work to diminish mHTT production. Delivering these HTT-lowering treatments – such as antisense oligonucleotides (ASOs) and tiny interference RNAs – has proven to be a significant challenge, though.

Viral vectors are required to transfer siRNA into nerve cellsvia striatum since naked siRNAs cannot cross the cell membrane or the blood-brain barrier (BBB) [42, 60]. Even still, given that the human brain is almost 3,000 times bigger than the brain of a mouse, any osiRNA sent into the brain would distribute throughout the parenchyma in a completely different way [61].

A possible treatment target for HD is G protein-coupled receptor 52 (GPCR52), an orphan GPCR that is expressed only in the brain, particularly in the striatum [42, 62–64]. It has been reported that deletion or reduction of GPR52 ameliorates HD-like symptoms [65, 66]. More recently, structure-activity relationship (SAR) research and high-throughput screening have led to the discovery of highly powerful, selective, and BBB-penetrating GPR52 antagonists. Moreover, these antagonists improve HD-associated abnormalities in HD mice in addition to lowering mHTT levels [42, 43].

1.10 FRONTOTEMPORAL DEMENTIA (FTD)

The frontotemporal lobar degeneration (FTLD) is progressive atrophy in the frontal and temporal lobes and varies in clinical presentation, genetic makeup, and neuropathological conditions [67]. The key clinical features are personality alterations, behavioral abnormalities, and social cognitive impairment in addition to language and executive dysfunction. It can result in progressive cognitive impairment.

A range of clinical disorders known as frontotemporal dementia are distinguished by neuronal degeneration affecting the brain's frontal and front temporal lobes [68, 69]. It is the 2nd most prevalent cause of early-onset dementia (age <65) and the third major cause of dementia in adults ≥65. Typically, patients in this age range are affected. In 1892, Arnold Pick was the first to describe clinical disorders related to frontotemporal association. In 1926, Hugo Spatz and Onari came up with the phrase "Pick's disease" to describe frontotemporal lobe atrophy accompanied by cytoplasmic inclusion inclusions or Pick bodies. Later, the term FTD took its place, determined by unique clinical and histological criteria in 1994, and was ascribed by a research team from the UK and Sweden [70].

The brain regions affected by FTD include those in charge of motivation, personality, behavior, language acquisition, abstract thought, and executive function. The clinical symptoms that manifest first are changes in behavior and/or language proficiency, which are followed by cognitive decline and executive function loss [71]. It is divided into behavior variants (bvFTD) and language type, two different clinical categories. FTD is further divided on to the basis of areas of neuronal loss viz semantic variant primary progressive aphasia (svPPA) and non-fluent variant primary progressive aphasia (nfvPPA) [72, 73]. The incidence ranges from 2.2/100,000 in the 40–49 age group to 3.3/100,000 in the 50–59 age group and peaks at 8.9/100,000 in the 60–69 age group [73, 74]. The average age of presentation is 58, while the age at onset varies. There are 15–22 cases per 100,000 people overall. Semantic variation (svPPA) and behavior variant (bvFTD) are more common in men, but nfvPPA is more common in women. It takes an average of 7.5 years to survive.

The underlying mechanism of frontotemporal dementia has been recognised by deposition of abnormal protein aggregates in the neural cells present in frontal and temporal lobes, inducing astrocytosis, neuronal degeneration, and microvacuole formation. Fused tumor-associated protein has been detected in 5–10% of cases; tau protein coupled with microtubules accounts for 45% of cases; and mutilated aggregates of TDP-43, a transactive DNA-binding protein, account for 50% of cases [72]. In patients with both familial and sporadic forms of FTD, aberrant tau protein aggregates have been reported to be the primary cause, except in the svPPA, where TDP-43 aggregates are more common [73].

Frontotemporal dementia is difficult to diagnose. It's possible to misdiagnose the initial clinical presentation as a stroke or a mental disease [11]. FTD offers a broad variety of expressions concerning the relevant fields. Functional abilities and daily living tasks are diminished in all kinds of FTD [68, 69, 75]. A delayed diagnosis of FTD due to early presentation could have a severe negative emotional, economic, and social impact on patients, their families, and caregivers. Compared to AD, which has a four-times more risk of death, front-end dementia (FTD) carries a six-fold risk [73, 76]. It takes an average of 7.5 years to survive. Sleep or nocturnal behavior disorder is reported by 33%–76% of FTD patients in various survey-based studies. PPA patients are more likely than bvFTD patients to report these. The situation with PPA sufferers' caretakers is comparable. Regular maintenance and good sleep hygiene are necessary for all sleep problems. Furthermore, there's a chance that FTD will coexist with overlapping diseases, like FTD-amyotrophic lateral sclerosis, FTD-motor neuron disease, or FTD-progressive supranuclear palsy. Treatment of the underlying condition and pharmacologic and non-pharmacologic methods are necessary for all of them.

1.11 MULTIPLE SYSTEM ATROPHY (MSA)

Multiple system atrophy (MSA) is a disorder of autonomic, cerebellar, and extrapyramidal dysfunction. It is necessary to have autonomic dysfunction to diagnose MSA. However, generally neurologists could find it difficult to identify autonomic dysfunction in MSA patients. In April

2022, the Movement Disorder Society (MDS) released the updated MSA diagnostic criteria [77]. "Unexplained voiding difficulties with post-void urinary residual volume" has been added to the key clinical symptoms. Within the parameters of clinically proven MSA, a limit of >100 ml post-void urine residual volume was established. Neurogenic orthostatic hypotension, in terms of cardiovascular autonomic dysfunction, is defined as a drop in systolic blood pressure (SBP) of at least 20 mmHg, typically accompanied by a drop in diastolic blood pressure (DBP) of at least 10 mmHg, and a ratio of Δheart rate (HR) to ΔSBP of less than 0.5 beats per minute (bpm)/mmHg within three minutes of standing or head-up tilt test (HUT) using oscillometric measurements. Moreover, REM sleep behavior disorder (RBD) is quite common in MSA, and the updated MDS diagnostic criteria for MSA include RBD as a necessary component of potential prodromal MSA based on polysomnography (PSG) evidence.

The most significant autonomic tests may be HUT and PVR measurement using ultrasonography, even though many other autonomic tests are helpful in the diagnosis of MSA. This is because the potential of orthostatic hypotension and PVR determines the criteria of autonomic dysfunctions [78–80]. Moreover, most institutions may be able to do HUT and PVR measurements with ultrasonography [79]. The diagnosis of MSA may be supported by normal cardiac sympathetic imaging by 123I-MIBG-scintigraphy and more than 100 pg/ml supine plasma noradrenaline level [78].

Kermorgant used overexpression of human alpha-synucle in transgenic mice under the control of PLP-α-syn, an oligodendrocyte-specific proteolipid promoter to investigate the impact of gender and age variations in MSA [81]. They found that there was an age-dependent alteration in baroreflex sensitivity in PLP-α-syn mice. At 12 months of age, PLP-α-syn female mice had reduced heart rate variability (HRV), whereas male mice did not exhibit this abnormality. According to Lazzeri et al., MSA-C patients performed worse on verbal and executive memory tests, and they found no statistically significant results in variation in MSA-P and MSA-C patients [82]. Additionally, they opined that their results do not corroborate the notion that dysautonomia plays a primary role in the cognitive disparities between MSA-P and MSA-C. RBD is one of the most reliable indicators of an underlying alpha-synucleinopathy, according to Giannini et al., and a substantial body of research has shown that RBD is highly prevalent in MSA. Research has thoroughly examined the frequency of RBD as a mechanism of disease beginning and its function in the development of the diseases. The authors concluded that RBD occurs in nearly all MSA patients at some time throughout the disease [83]. Between 10% and 60% of patients have RBD as their initial symptom at the beginning of their diseases. The scientists also noted that MSA patients with RBD at the beginning of the disease had a more severe and rapid disease progression, mostly because the autonomic nervous system was involved sooner, and early disease progression milestones were reached.

1.12 PROGRESSIVE SUPRANUCLEAR PALSY (PSP)

Atypical parkinsonism most commonly manifests as progressive supranuclear palsy (PSP). Tufted astrocytes are connected to the neuropathology of this tauopathy. PSP was initially explained by Steele and associates [84]. Numerous phenotypes were described as a result of observations of this disease. The most prevalent PSP is Richardson-Steele-Olszewski (PSP-RS), which is clinically associated with substantial postural instability and oculomotor dysfunction [85]. PSP-P, or the parkinsonism variant of PSP, has been reported as the 2nd most common form of the disease and is detected in 14%–35% of cases [86].

PSP-P diagnoses are frequently made after the fact. PSP-P, in contrast to PSP-RS, is not linked to midbrain tegmentum atrophy, particularly in its early phases [87]. PSP-P can change into PSP-RS as the disease progresses. The four fundamental features that characterise the PSP disease's phenotypes are postural instability (P), oculomotor dysfunction (O), cognitive deficits and linguistic problems (C), and akinesia (A). Within each factor, the severity is ranked from 1–3. It is found that PSP-P is linked to parkinsonism, akinetic-rigid mostly axial and levodopa resistive (A2), or parkinsonism with tremor and/or asymmetry and/or levodopa responsiveness (A3), as well as vertical supranuclear gaze palsy (O1) or delayed velocity of vertical saccades (O2). Cognitive or verbal impairments, as well as postural instability, are potential signs of PSP-P. It is required that A2 and A3 be present [85]. Despite various similarities in the clinical manifestation of PSP subgroups, there are also variations in the neuropathological examination linked to them. Differentiating between PSP-P and the postural instability gait disorder (PIGD) type of Parkinson's disease is made more difficult by the symptoms of PSP-P previously stated.

1.13 CORTICOBASAL DEGENERATION (CBD)

One rare neurodegenerative condition, corticobasal degeneration (CBD) is a collection of cortical and extrapyramidal symptoms and indications and is the most typical manifestation of CBD. Studies on the clinical-pathological presentation of CBD have shown that it can manifest in a variety of clinical phenotypes, such as behavioral, dysexecutive, progressive supranuclear palsy-like syndrome, and visuospatial syndromes. However, a patient with CBS may have more than one underlying pathology.

Three individuals with gradually worsening "clumsiness" and a range of asymmetrical extrapyramidal symptoms, such as stiffness, dystonia, and myoclonus, but otherwise intact cognitive abilities were initially reported [88]. To characterise this phenomenon, they used the phrase "corticonigral degeneration with neuronal achromasia." Before 1989, when the term "corticobasal degeneration" was first coined, this disorder was mainly ignored [89]. After years, the terms "corticobasal degeneration" (CBD) and "corticobasal syndrome" (CBS) have become synonymous. CBD describes the pathogenic entity, whereas CBS reflects phenotypic. It has been suggested, "All CBD patients appear with a CBS phenotype and vice versa," which in turn significantly increases the misunderstanding surrounding the topic. The pathological and clinical diagnostic criteria for CBD were established by Dickson et al. [90].

The clinical-pathological investigations explain the pathology and symptoms of these uncommon neurodegenerative diseases. It has been demonstrated that clinical variability of CBD manifests frontal-dysexecutive-spatial syndrome, progressive aphasia, and a Richardson-like syndrome [91]. Heterogeneity in pathology of CBS patients can also be manifested by FTD-TDP43, CBD, AD, and PSP [92].

Due to the disease's rarity, the disorder's high degree of clinical pathological heterogeneity, and the absence of standardised criteria until recently, there are insufficient robust epidemiological data on CBD. It is thought to be 10 times less common than PSP, or progressive supranuclear palsy [93]. It is expected to affect 0.6–0.9/100,000 people annually, or 4%–6% of parkinsonism patients. The prevalence estimated by a single study using data on incidence and mean survival was 4.9–7. 3 per 100,000. Nevertheless, no instance of CBD was reported in a cohort of 120,000 people [94]. The youngest patient verified pathologically was 43 years old; however, the common age at disease onset is found to be 64 years [91]. A minor female predominance might exist [95]. The projected mean survival is 6.5 years [91]. Over time, several sets of diagnostic standards that take into account only the CBS presentation of CBD [96] have been put forth. Introduced in 2013, the present set of recognised clinical diagnostic methods sought to contain a higher range of clinical symptoms to broaden the phenotypical spectrum of CBD [91]. These standards aimed to highlight the significant clinical-pathological variability of CBD and CBS and were based on clinical-pathological investigations of CBD.

The disease CBD is primarily sporadic and has been attributed to microtubule-associated tau protein mutation [97]. Although the family in this instance experienced early-onset dementia with severe behavioral symptoms, they also had the neuropathological traits of CBD. Furthermore, occasional CBS instances have been linked to the p.N410H, G389R, and P301S MAPT mutations [98, 99]. These standards allow CBD to be used in five clinical disorders.

No licensed CBD-related medication has been found to cure these diseases. However, a lot of CBS symptoms are treatable symptomatically. Levodopa is a treatment option for parkinsonism in

CBS, albeit its effects are only mild and temporary [100]. Botulinum toxin is a useful treatment for dystonia in CBS patients, as it can temporarily restore function to the afflicted limb [101]. The first-choice medication for myoclonus is clonazepam [100].

Aside from the conventional CBS, CBD can exhibit a wide range of phenotypes, such as behavioral, linguistic, or postural abnormalities. On the other hand, a diversity of diseases, containing CBD, PSP, AD, and FTD-TDP43, might be present in CBS. CSF biomarkers and imaging modalities (structural and functional) have been used to some extent. Given the particular protein-targeting, and disease-modifying therapies being evaluated in clinical trials, more study on biomarkers is needed.

1.14 VASCULAR DEMENTIA (VaD)

The common prevalent form of dementia in the older is vascular dementia (VaD). VaD affects 1%– 4% of subjects, compared to older people over age 65 years, and the prevalence seems to grow every 10–15 years after age 65. In the past, VaD was thought to be the second most prevalent cause of dementia in the elderly, behind AD. After 65 years, up to one-third of patients with clinically obvious ischemic stroke develop post-stroke dementia, which is incredibly prevalent. Three months following a stroke, dementia was found in 26.3% of patients in a recent series. Of those, 57.1% of the subtypes had VaD, 38.7% had stroke and AD, and 4.2% had other reasons [102].

According to Pohjasvaara and colleagues (2002), 61.7% of stroke patients between the ages of 55 and 85 had some form of cognitive deterioration. Although it is complicated to quantify the occurrence of mixed dementia because the idea is relatively new and most research has relied on pathology data, it is evident that AD and CVD commonly co-occur in the senior population. According to the autopsy series, 24%–28% of AD cases had co-existing vascular pathology, and 50% of individuals with vascular diseases had developed dementia with AD. The location, size, and type of brain injury influence the clinical manifestations of vasodilation, which might vary because of variety of pathogenic causes. The pathological aspects of this condition are typically characterised by an abrupt beginning, a progressive decline, a variable course, and localised sensory and motor abnormalities. Subcortical VaD, however, can exhibit a progressive onset and decline similar to AD. The clinical aspects can be further classified even within VaD [102].

Initial research on several drugs in VaD, such as thyrotropin-releasing hormone analogues, vasodilators, antithrombotics, nootropics, antioxidants, ergot alkaloids, and hyperbaric oxygen, has primarily shown unfavorable findings. The study designs, however, were not optimal due to their limited sample sizes, brief treatment periods, and inconsistent trial endpoints. The prevalence of vascular dementia is expected to rise, as it is a prevalent disorder. As medical professionals, we have a responsibility to identify risk factors and take aggressive action when necessary. Numerous clinical trials of cholinesterase inhibitors have shown encouraging outcomes, and no problems with safety or tolerability have been reported.

1.15 LEWY BODY DEMENTIA (LBD)

PD dementia and Lewy bodies dementia are the two clinical entities that makeup Lewy body dementia (LBD). It is a gradual degenerative brain disease with Parkinsonian symptoms, psychosis, and dementia. The symptoms change over time and vary from person to person. Since many of LBD's symptoms are similar to those of other dementia conditions, a comprehensive clinical examination is necessary for the diagnosis. After vascular dementia and AD, it is third most predominant kind of dementia. Alpha-synuclein and ubiquitin aggregates are seen in Lewy bodies, which are intraneuronal cytoplasmic inclusion bodies that are deposited in the nervous system [103–105].

The connection between PD and LBD is still up for dispute. To differentiate LBD from PD with dementia, the most recent LBD consortium established an arbitrary 12-month guideline. PD with dementia is the condition if the subject has the pathological situation for 12 months or more before any cognitive dysfunction is observed; if the period is less, the ailment is most likely LBD. The majority of LBD sufferers have dementia before motor symptoms, especially when it comes to sequence of reduced visual hallucinations and responsiveness.

The cause of LBD is a mystery. However, genetics, environmental factors, and aging-related changes may play a crucial role. Due to lack of understanding and clinical features, LBD is an underdiagnosed condition [106]. However, research indicates that it may account for as much as 20%–30% of dementia occurrences overall. It is common among European, Asian, and African racial groups. A patient is more at risk if they have a family history of PD or LBD. Acetylcholine insufficiency is a symptom of both LBD and AD; however, it is more noticeable in LBD. Visual hallucinations, a hallmark of LBD, are caused by decreased acetylcholine levels in the nervous system, whereas delusions are caused by an increase in muscarinic M1 receptors. Dopamine also starts to decline. Most parts of nervous system are impacted by LBD, according to postmortem investigations. The presynaptic protein alpha-synuclein, whose purpose is yet unclear, is a crucial component of LBD. Additional discoveries include the existence of neurofilament and ubiquitin proteins. The acetylcholine reduction in LBD is significantly more severe than in Parkinson's disease.

For LBD, the prognosis is fair to poor. Numerous difficulties, such as falls, immobility, heart issues, adverse drug reactions, pneumonia, difficulty swallowing, and depression leading to suicide, cause patients to pass away. After the first diagnosis, the typical life expectancy is only five to eight years. This may also be the result of a lack of awareness about LBD among the general public and medical professionals, as well as difficulties distinguishing it from other similar disorders. These factors might cause a delay in diagnosis and the start of a particular medication. Most patients are trapped in care facilities for a long term, where they may not always receive the best care possible for their most basic needs [40, 107, 108].

1.16 MECHANISM OF NEURODEGENERATION

The clinical appearance of neurodegenerative diseases is determined by the increasing malfunctioning of particular groups of neurons. Extracellular and intracellular misfolded protein accumulation is linked to neuronal death and is a characteristic feature of most neurodegenerative proteinopathies. The ubiquitin-proteosome-autophagy system deficiency that leads to abnormal protein dynamics, oxidative stress, and mitochondrial dysfunction is one of the major basic processes associated with neurodegeneration. Over a lengthy period, these interconnected processes result in programmed cell death. Although key components of protein deposits or established genetic processes are used to classify neurodegenerative diseases, recent research explains both intraindividual variations and overlap across various symptoms. Common pathogenic pathways may be suggested by synergistic mechanisms among diseased proteins. Further research and animal models have shed light on the fundamental processes behind neurodegeneration, opening up new avenues for treatment and prevention approaches in the future [109].

1.17 MISFOLDING AND AGGREGATION OF PROTEIN

Pathological characteristics of many, albeit varied, NDDs include abnormal protein interactions leading to aberrant intra- and extracellular misfolded protein deposition with the production of high-ordered insoluble fibrils. Generally speaking, each disease's clinical manifestation is determined by the identification of the underlying protein, which also dictates which neurons are afflicted. Neuronal networks disintegrate as a result of malfunction and death of glial and neuronal cells brought on by abnormal protein interactions, accumulation, precipitation and/or the lesions they cause [109].

Neurons that normally interact with each other in a very soluble way change their shape. Misfolding then turns these soluble proteins into insoluble polymers, which cluster either cross-pleated β-sheet structures as highly organised or amorphous forms as disordered. The diseases in question are collectively referred to as brain amyloidoses because deposits of amyloid features. When a normal, functional protein is folded incorrectly or misfolded, it can alter its structure and cause protein aggregation *in vivo*. Proteins that are aberrant due to production errors, genetic or acquired substitutions of amino acids, or damage frequently exhibit incorrect folding and become locked in misfolded protein conformations. The pathogenic mechanism entail hydrophobic interaction of prefibrillary amino-acids oligomers with other diversified cellular targets or channel mechanism-mediated membrane permeabilisation [109, 110].

The endoplasmic reticulum (ER) of living cells is home to a multitude of intracellular proteases that, in conjunction with chaperones, form a protein folding quality-control systems which eliminating misfolded polymer [111, 112]. Under physiological circumstances, a large number of proteins linked to non-diagnostic diseases (NDDs) are inherently disordered [113]. Recent research using transgenic (tg) mice models and cultured cells has shown that misfolded polymer aggregates can spread from one cell to various cells through the use of mutant huntingtin (mhtt), τ protein, and αSyn [113–115]. Misfolding protein conformations are linked to a significant number of NDDs because of their pivotal roles in cell biology [116].

1.18 MITOCHONDRIAL DYSFUNCTION AND OXIDATIVE STRESS

Oxidative stress (OS) is the result of the generation of free radicals or their byproducts in excess then the antioxidant defense capacity [117, 118]. When there is no hereditary predisposition, protein misfolding can be mediated by the production of excessive nitric oxide and reactive oxygen species (ROS) [119]. The most powerful free radicals produced by living things are hydroxyl radical and peroxynitrite, which may oxidise proteins, fatty acids, lipids, and nucleic acids without discrimination, causing cell damage [120]. One of the basic processes behind neurodegenerative diseases is the addition of Fe(II) to hydrogen peroxide, which subsequently transforms it into hydroxyl radicals. Protein precipitation and buildup may be exacerbated by OS, which could lead to increased oxidative damage by disrupting the proteasome's function and raising OS levels in proteins and other biomolecules [117].

ROS first targets proteins, and protein peroxides or radicals produced by ROS can oxidise other nearby biomolecules, causing reactive astrocytes, and ND-related abnormalities [121, 122]. Rapid apoptosis occurs in cells that are unable to adapt to oxidative imbalance (stress); conversely, long-term survival may be demonstrated by cells that have compensatory responses to ROS. RNA oxidation could be a major mechanism for non–disease-causing diseases (NDDs), yet precise and well-defined mechanisms still have to demonstrate the role of OS to regulates cell survival and death in NDDs [123]. Damage from oxidation to nuclear and mitochondrial DNA (mtDNA) can be found in both normal aging and the earliest stage of Alzheimer's and Parkinson's [124–127]. Increasing the redox-active transition metals in several areas of the brain most often damaged in NDs is consistent with their involvement in OS, even though the precise causes of enhanced oxidative damage remain unclear. Though it is currently unknown which forms of oxidative damage to biomolecules and cells happen early in the pathogenic cascade for neurons death, ROS chemistry seems to play a significant pathogenic role in all of these circumstances [128].

1.19 OXIDATIVE STRESS AND METALS

Changes in the metal homeostasis induce oxidative stress and related cellular consequences. In both the typical aging brain and other NDDs, increased redox-active Fe is accumulated from degenerating mitochondria [129]. Fe has the potential to catalyse the production of extremely reactive radicals from lipid peroxides and H_2O_2, which results in OS. In animal models of AD, enhanced lipid peroxidation occurs before the formation of Aβ plaque, and abnormal metal homeostasis may play a role in the generation of ROS and toxic Aβ oligomers in AD. The ND process

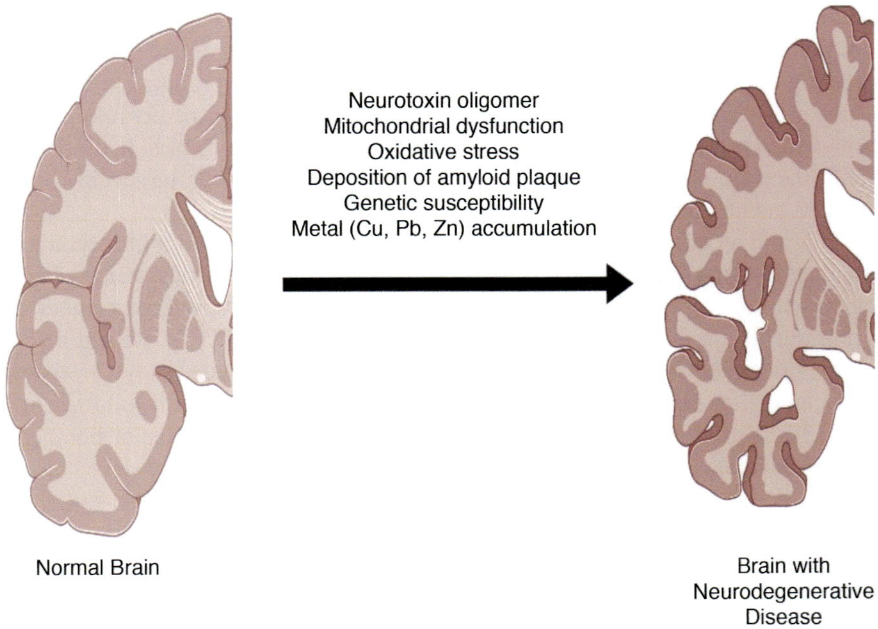

FIGURE 1.1 Factors affecting the neurodegeneration process.

that leads to AD is preceded by an imbalance in iron homeostasis. Mutations in genes encoding proteins related to iron metabolism [130] are linked to conditions including neuroferritinopathy [131] and Friedreich ataxia, which cause OS and cell death. Dopamine (DA) complexes of iron have a significant importance in the pathophysiology of PD [132]; accumulation of iron has been reported in PD and AD with brain aging [133]. **Figure 1.1** shows the different factors affecting neurodegeneration process.

1.20 MITOCHONDRIA AND NEURON DEGENERATION

Numerous biochemical, genetic, and clinical characteristics of sporadic AD may be explained by the recently revised "mitochondrial cascade hypothesis" [134, 135]. A notable rise in damage to mtDNA in the brain may result in aberrant dynamics and dysfunction of the mitochondria, which in turn harms AD neurons [126]. The development of the mitochondrial permeability transition pore (mPTP) is aided by Aβ [136]. Aβ-induced mitochondrial dysfunction initiates AD pathogenesis and is an early event in AD tg models [137]. Early Alzheimer's disease protein change patterns point to the activation of mitochondrial pathways, which involve proteins involved in ATP transport and consumption. Early AD may see the most noticeable alterations [138], but there is evidence of mitochondrial malfunction in AD, no causative mtDNA mutations have been found to date, and research on the significance of mtDNA haplogroups in AD has produced conflicting results, but it is indicated that the mitochondrial genome plays a significant role [139].

Important components of the complex pathogenic process of Parkinson's disease are mitochondrial changes and OS [140]. Abnormal mitochondrial paradigms include mtDNA damage, abnormalities in mitochondrial shape and dynamics, and compromised electron transport chain performance in both sPD and fPD. Increased free turbulin, which is triggered by mitochondrial failure, destabilises the microtubule network and encourages αSyn oligomerisation [134]. The UPS depends on ATP for multiple processes; hence a compromised mitochondrial ecosystem could hinder its functionality. More significantly, elevated ROS production results in misfolded, damaged proteins that the UPS must degrade.

1.21 NEUROINFLAMMATION

Prominent pathogenic aspects in the etiology and evolution of NDDs include chronic inflammatory biochemical reactions and evidence the immunological activation in the central nervous system, including MHC class II expression, glial reactivity, T-cell infiltration, and disruption of the blood-brain-barrier (BBB) [141]. It has been found that major contribution to inflammation and NDDs is inappropriate iron chelation. Iron has a variety of pro- and anti-inflammatory actions [142]. It has been suggested that AD "inflammation" is harmful or, conversely, beneficial [143]. However, a deeper comprehension of the interactions between the neurological system and the systemic and local immune systems will be necessary to comprehend the pathogenic processes of non-disruptive diseases (NDDs) [144], the function of innate "protective autoimmunity," and the reparative roles of autoimmunity in NDDs [145].

It has been reported that immunological changes could take place before neuronal loss and amyloid buildup. The pathophysiology of AD is made worse by complement activation, and it has been argued that this is a result of the brain's innate immune defense [146]. A key question is whether ND is promoted by immunological and inflammatory biochemical pathways that are hyperactivated with progressive aging, or if NDDs are caused by inadequate immune responses that are unable to handle age-related stress [147]. Neuronal function is severely impacted by age-related neuroinflammatory alterations [148]. According to certain theories, the initial processes of certain NDDs involve the breakdown of BBB and the influx of blood-borne molecules (plasma proteins), which can cause local harm [141]. In the "inflamed" AD brain, fibrinogen infiltration, leaky BBB, and microglial activation may produce neuronal injury [149]. Depending on several variables, such as age, activated microglia may have neurotoxic or neuroprotective effects; however, the reasons behind this transition between the two functions are still mostly unclear [150]. Despite being an innate immune cell type of the brain, microglia are also linked to the continuous phagocytotic removal of cell debris, serving as a crucial mediator between degeneration and regeneration.

An inadequate regeneration response is linked to unsatisfactory clearance by microglia, which are found in numerous NDDs and diminish with age [151]. In contrast, astrocytes generate neuroprotective and anti-inflammatory substances [152]. Both astrocytes and microglia in human brains and animal models can produce Aβ, which can function as a pro-inflammatory mediator by activating glia and other inflammatory components [153]. The involvement patterns of cytochemokines in AD support the inflammatory theory and point to a pathophysiological function for this network's selective modification [154].

1.22 MULTIFACETED NEURONAL DEATH

Different types of cellular death occur in the nervous system, based on the type and degree of stress. There is ongoing debate over the type, timing, and molecular reasons of cell death in NDDs as well as how these factors relate to fundamental processes. PCD is categorised into three main forms based on unique morphologic criteria and biochemical features: PCD type I is apoptosis; type II is autophagy; and type III is oncotic necrosis, which is a passive cell death [155–157]. Apoptosis is a particular type of neuronal death which is mediated by two main signaling biochemical pathways: mitochondria-based intrinsic system, and death-receptor–mediated extrinsic pathway, with multiple channels of communication between them. It is argued that there is no obvious distinction between apoptotic and autophagic cell death since nerve cells death can have morphologic aspects of autophagy or mortification, which differ from those of canonical apoptosis, or autophagic vacuolation can occur before apoptotic cell death [158].

Many non-apoptotic, controlled processes of cell death are known to exist (e.g. in tauopathies), and a few of them are exclusive or overlap with apoptosis [159]. More research indicates that nerve cells' death regulation is intricate, involving several biochemical pathways that depend on harmful insult, each of which demonstrates the unique function, modulation, and biochemical pathway involvement, all of which are regulated by minute variations in cell phenotypes [160].

1.23 CONCLUSIONS

The main hallmark of NDs, such as AD and Parkinson's, is neuronal death. Consequently, the most anticipated course of treatment for these conditions is neurogenesis. Nevertheless, there are still several important characteristics that make medication transport to the brain difficult, such as the BBB, molecular weight and lipophilicity of drugs, etc. These elements restrict the therapeutic ability of medications and increase the difficulty of treating NDs. In the fields of biomedicine and biopharmaceutics, developing new and potent anti-amyloidogenic agents to treat a range of human pathologies linked to protein aggregation is one of the main problems. The build-up and progressive accumulation of amyloid aggregation is thought to be a characteristic of several debilitating diseases that impact the brain or peripheral tissues. Researchers should put a lot of effort into understanding the protein aggregation mechanism and screening new anti-amyloid drugs that can considerably regulate the amyloid aggregation, given these remarkable discoveries and the implications for neurological diseases.

DECLARATION

Conflict of Interest: Authors declare there is no conflict of interest.

REFERENCES

1. Feigin, V.L., et al., *The global burden of neurological disorders: Translating evidence into policy.* The Lancet Neurology, 2020. **19**(3): p. 255–265.
2. Merelli, A., L. Czornyj, and A. Lazarowski, *Erythropoietin: A neuroprotective agent in cerebral hypoxia, neurodegeneration, and epilepsy.* Current Pharmaceutical Design, 2013. **19**(38): p. 6791–6801.
3. Pino, A., et al., *New neurons in adult brain: Distribution, molecular mechanisms and therapies.* Biochemical Pharmacology, 2017. **141**: p. 4–22.
4. Ganat, Y.M., et al., *Early postnatal astroglial cells produce multilineage precursors and neural stem cells in vivo.* Journal of Neuroscience, 2006. **26**(33): p. 8609–8621.
5. Przedborski, S., M. Vila, and V. Jackson-Lewis, *Series introduction: Neurodegeneration: What is it and where are we?* The Journal of Clinical Investigation, 2003. **111**(1): p. 3–10.
6. Kovacs, G.G., *Molecular pathology of neurodegenerative diseases: Principles and practice.* Journal of Clinical Pathology, 2019. 2(11): p. 725–735.
7. Esch, T., et al., *The role of stress in neurodegenerative diseases and mental disorders.* Neuroendocrinology Letters, 2002. **23**(3): p. 199–208.
8. Jain, N., and A.S. Chen-Plotkin, *Genetic modifiers in neurodegeneration.* Current Genetic Medicine Reports, 2018. **6**: p. 11–19.
9. Jain, V., et al., *Enriched environment prevents hypobaric hypoxia induced neurodegeneration and is independent of antioxidant signaling.* Cellular and Molecular Neurobiology, 2012. **32**: p. 599–611.
10. Liu, H., et al., *Mendelian randomization highlights significant difference and genetic heterogeneity in clinically diagnosed Alzheimer's disease GWAS and self-report proxy phenotype GWAX.* Alzheimer's Research & Therapy, 2022. **14**(1): p. 1–10.
11. Liu, Z., et al., *Oxidative stress in neurodegenerative diseases: From molecular mechanisms to clinical applications.* Oxidative Medicine and Cellular Longevity, 2017. **2017**: p. 1–11
12. Allan, S.M., and N.J. Rothwell, *Inflammation in central nervous system injury.* Philosophical Transactions of the Royal Society of London. Series B: Biological Sciences, 2003. **358**(1438): p. 1669–1677.
13. Brouwer-DudokdeWit, A.C., et al., *A hereditary disorder in the family and the family life cycle: Huntington disease as a paradigm.* Family Process, 2002. **41**(4): p. 677–692.
14. Bayer, T.A., and O. Wirths, *Intracellular accumulation of amyloid-beta-a predictor for synaptic dysfunction and neuron loss in Alzheimer's disease.* Frontiers in Aging Neuroscience, 2010. **2**: p. 1359.
15. Pickett, E.K., et al., *Amyloid beta and tau cooperate to cause reversible behavioral and transcriptional deficits in a model of Alzheimer's disease.* Cell Reports, 2019. **29**(11): p. 3592–3604. e5.
16. Paula, V.d.J.R.d., et al., *Trajetórias neurobiológicas para a doença de Alzheimer: Beta-amilóide, proteína TAU ou ambas?* Dementia & Neuropsychologia, 2009. **3**: p. 188–194.
17. Mondragón-Rodríguez, S., et al., *Amyloid beta and tau proteins as therapeutic targets for Alzheimer's disease treatment: Rethinking the current strategy.* International Journal of Alzheimer's Disease, 2012. **2012**: p. 1–7

18. Butterfield, D.A., et al., *Amyloid β-peptide and amyloid pathology are central to the oxidative stress and inflammatory cascades under which Alzheimer's disease brain exists.* Journal of Alzheimer's Disease, 2002. **4**(3): p. 193–201.
19. Spires-Jones, T.L., and B.T. Hyman, *The intersection of amyloid beta and tau at synapses in Alzheimer's disease.* Neuron, 2014. **82**(4): p. 756–771.
20. Akhondzadeh, S., and M. Noroozian, *Alzheimer's disease: Pathophysiology and pharmacotherapy.* IDrugs: the Investigational Drugs Journal, 2002. **5**(11): p. 1062–1069.
21. Dunn, B., P. Stein, and P. Cavazzoni, *Approval of aducanumab for Alzheimer disease—the FDA's perspective.* JAMA Internal Medicine, 2021. **181**(10): p. 1276–1278.
22. Walsh, S., et al., *Aducanumab for Alzheimer's disease?* British Medical Journal Publishing Group, 2021.
23. Wu, K. -M., et al., *The role of the immune system in Alzheimer's disease.* Ageing Research Reviews, 2021. **70**: p. 101409.
24. Hampel, H., et al., *Revisiting the cholinergic hypothesis in Alzheimer's disease: Emerging evidence from translational and clinical research.* The Journal of Prevention of Alzheimer's Disease, 2019. **6**: p. 2–15.
25. Johnson, J.W., and S.E. Kotermanski, *Mechanism of action of memantine.* Current Opinion in Pharmacology, 2006. **6**(1): p. 61–67.
26. Folch, J., et al., *Memantine for the treatment of dementia: A review on its current and future applications.* Journal of Alzheimer's Disease, 2018. **62**(3): p. 1223–1240.
27. Reisberg, B., et al., *Memantine in moderate-to-severe Alzheimer's disease.* New England Journal of Medicine, 2003. **348**(14): p. 1333–1341.
28. Fereshtehnejad, S.-M., et al., *Evolution of prodromal Parkinson's disease and dementia with Lewy bodies: A prospective study.* Brain, 2019. **142**(7): p. 2051–2067.
29. Fereshtehnejad, S.-M., et al., *Managing Parkinson's disease with a multidisciplinary perspective.* Frontiers in Neurology, 2021. **12**: p. 799017.
30. Kouli, A., K.M. Torsney, and W.-L. Kuan, *Parkinson's disease: Etiology, neuropathology, and pathogenesis.* Exon Publications, 2018: p. 3–26.
31. Spires-Jones, T.L., J. Attems, and D.R. Thal, *Interactions of pathological proteins in neurodegenerative diseases.* Acta Neuropathologica, 2017. **134**: p. 187–205.
32. Priyadarshi, A., et al., *Environmental risk factors and Parkinson's disease: A metaanalysis.* Environmental Research, 2001. **86**(2): p. 122–127.
33. Bartels, A.L., and K.L. Leenders, *Parkinson's disease: The syndrome, The pathogenesis and pathophysiology.* Cortex, 2009. **45**(8): p. 915–921.
34. Chaudhuri, K.R., and A.H. Schapira, *Non-motor symptoms of Parkinson's disease: Dopaminergic pathophysiology and treatment.* The Lancet Neurology, 2009. **8**(5): p. 464–474.
35. Moore, D.J., et al., *Molecular pathophysiology of Parkinson's disease.* Annual Review of Neuroscience, 2005. **28**: p. 57–87.
36. Schulz-Schaeffer, W.J., *The synaptic pathology of α-synuclein aggregation in dementia with Lewy bodies, Parkinson's disease and Parkinson's disease dementia.* Acta Neuropathologica, 2010. **120**: p. 131–143.
37. Gendron, T.F., and L. Petrucelli, *The role of tau in neurodegeneration.* Molecular Neurodegeneration, 2009. **4**: p. 1–19.
38. Emamzadeh, F.N., and A. Surguchov, *Parkinson's disease: Biomarkers, treatment, and risk factors.* Frontiers in Neuroscience, 2018. **12**: p. 612.
39. Alexander, G.C., et al., *Revisiting FDA approval of aducanumab.* New England Journal of Medicine, 2021. **385**(9): p. 769–771.
40. Armstrong, M.J., et al., *Cause of death and end-of-life experiences in individuals with dementia with Lewy bodies.* Journal of the American Geriatrics Society, 2019. **67**(1): p. 67–73.
41. Morris, J., *Amyotrophic lateral sclerosis (ALS) and related motor neuron diseases: An overview.* The Neurodiagnostic Journal, 2015. **55**(3): p. 180–194.
42. Wang, C., et al., *GPR52 antagonist reduces huntingtin levels and ameliorates Huntington's disease-related phenotypes.* Journal of Medicinal Chemistry, 2020. **64**(2): p. 941–957.
43. Wang, G., et al., *Advances in nanotechnology-based strategies for the treatments of amyotrophic lateral sclerosis.* Materials Today Bio, 2020. **6**: p. 100055.
44. Hardiman, O., et al., *Amyotrophic lateral sclerosis.* Nature Reviews Disease Primers, 2017. **3**(1): p. 1–19.
45. Rossi, F.H., M.C. Franco, and A.G. Estevez, *Pathophysiology of amyotrophic lateral sclerosis.* Current Advances in Amyotrophic Lateral Sclerosis, 2013. Chapter 1: P. 1–33.
46. Webster, C.P., et al., *The C9orf72 protein interacts with Rab1a and the ULK 1 complex to regulate initiation of autophagy.* The EMBO Journal, 2016. **35**(15): p. 1656–1676.

47. Yoshida, S., *Therapeutic strategies and metal-induced oxidative stress: Application of synchrotron radiation microbeam to amyotrophic lateral sclerosis in the Kii Peninsula of Japan.* Frontiers in Neurology, 2022. **13**: p. 884439.
48. Jaiswal, M.K., *Riluzole and edaravone: A tale of two amyotrophic lateral sclerosis drugs.* Medicinal Research Reviews, 2019. **39**(2): p. 733–748.
49. Andrews, J.A., et al., *Real-world evidence of riluzole effectiveness in treating amyotrophic lateral sclerosis.* Amyotrophic Lateral Sclerosis and Frontotemporal Degeneration, 2020. **21**(7–8): p. 509–518.
50. Witzel, S., et al., *Safety and effectiveness of long-term intravenous administration of edaravone for treatment of patients with amyotrophic lateral sclerosis.* JAMA Neurology, 2022. **79**(2): p. 121–130.
51. Ghosh, R., and S.J. Tabrizi, *Huntington disease.* Handbook of Clinical Neurology, 2018. **147**: p. 255–278.
52. Quigley, J., *Juvenile Huntington's disease: Diagnostic and treatment considerations for the psychiatrist.* Current Psychiatry Reports, 2017. **19**: p. 1–4.
53. Rawlins, M.D., et al., *The prevalence of Huntington's disease.* Neuroepidemiology, 2016. **46**(2): p. 144–153.
54. Pringsheim, T., et al., *The incidence and prevalence of Huntington's disease: A systematic review and meta-analysis.* Movement Disorders, 2012. **27**(9): p. 1083–1091.
55. Alpay, M., and W.J. Koroshetz, *Quetiapine in the treatment of behavioral disturbances in patients with Huntington's disease.* Psychosomatics, 2006. **47**(1): p. 70–72.
56. Dayalu, P., and R.L. Albin, *Huntington disease: Pathogenesis and treatment.* Neurologic Clinics, 2015. **33**(1): p. 101–114.
57. Frank, S., *Treatment of Huntington's disease.* Neurotherapeutics, 2014. **11**: p. 153–160.
58. Boland, B., et al., *Promoting the clearance of neurotoxic proteins in neurodegenerative disorders of ageing.* Nature Reviews Drug Discovery, 2018. **17**(9): p. 660–688.
59. Bates, G.P., et al., *Huntington disease (primer).* Nature Reviews: Disease Primers, 2015. **1**(1): p. 1–21.
60. Chen, W., Y. Hu, and D. Ju, *Gene Therapy for neurodegenerative disorders: Advances, insights and prospects.* Acta Pharmaceutica Sinica B, 2020. **10**(8): p. 1347–1359.
61. Pouladi, M.A., A.J. Morton, and M.R. Hayden, *Choosing an animal model for the study of Huntington's disease.* Nature Reviews Neuroscience, 2013. **14**(10): p. 708–721.
62. Komatsu, H., *Discovery of the first druggable GPR52 antagonist to treat Huntington's disease.* Journal of Medicinal Chemistry, 2021. **64**(2): p. 938–940.
63. Komatsu, H., *Innovative therapeutic approaches for Huntington's disease: From nucleic acids to GPCR-targeting small molecules.* Frontiers in Cellular Neuroscience, 2021. **15**: p. 785703.
64. Komatsu, H., et al., *Anatomical transcriptome of g protein-coupled receptors leads to the identification of a novel therapeutic candidate GPR52 for psychiatric disorders.* PLoS One, 2014. **9**(2): p. e90134.
65. Yao, Y., et al., *A striatal-enriched intronic GPCR modulates huntingtin levels and toxicity.* Elife, 2015. **4**: p. e05449.
66. Song, H., et al., *Targeting Gpr52 lowers mutant HTT levels and rescues Huntington's disease-associated phenotypes.* Brain, 2018. **141**(6): p. 1782–1798.
67. He, Q., M. Chu, and L. Wu, *Horizon in frontotemporal lobar degeneration related disorder.* Frontiers in Neurology, 2023. **14**: p. 1127073.
68. Mulkey, M., *Understanding frontotemporal disease progression and management strategies.* Nursing Clinics, 2019. **54**(3): p. 437–448.
69. Kurz, A., et al., *What is frontotemporal dementia?* Maturitas, 2014. **79** (2): p. 216–219.
70. Englund, B., et al., *Clinical and neuropathological criteria for frontotemporal dementia.* Journal of Neurology, Neurosurgery, and Psychiatry, 1994. **57**(4): p. 416–418.
71. Chan, D.K., et al., *Frontotemporal dementia: Features, diagnosis and management.* Australian Journal of General Practice, 2011. **40**(12): p. 968.
72. Olney, N.T., S. Spina, and B.L. Miller, *Frontotemporal dementia.* Neurologic Clinics, 2017. **35**(2): p. 339–374.
73. Bott, N.T., et al., *Frontotemporal dementia: Diagnosis, deficits and management.* Neurodegenerative Disease Management, 2014. **4**(6): p. 439–454.
74. Onyike, C.U., and J. Diehl-Schmid, *The epidemiology of frontotemporal dementia* International Review of Psychiatry, 2013. **25**(2): p. 130–137.
75. Sivasathiaseelan, H., et al. *Frontotemporal dementia: A clinical review.* In *Seminars in neurology.* 2019. Thieme Medical Publ Inc.
76. Bretag-Norris, R., L. Gallur, and P. Flynn, *Heterogeneity in the psychiatric presentation of behavioural variant frontotemporal dementia (bvFTD).* Australasian Psychiatry, 2019. **27**(5): p. 491–495.
77. Yamamoto, T., M.T. Pellecchia, and R. Sakakibara, *Autonomic dysfunction in multiple system atrophy.* Frontiers in Neurology, 2022. **13**: p. 1048895.

78. Wenning, G.K., et al., *The movement disorder society criteria for the diagnosis of multiple system atrophy.* Movement Disorders, 2022. **37**(6): p. 1131–1148.
79. Cheshire, W.P., et al., *Electrodiagnostic assessment of the autonomic nervous system: A consensus statement endorsed by the American Autonomic Society, American Academy of Neurology, and the International Federation of Clinical Neurophysiology.* Clinical Neurophysiology, 2021. **132**(2): p. 666–682.
80. Pellecchia, M.T., et al., *Can autonomic testing and imaging contribute to the early diagnosis of multiple system atrophy? A systematic review and recommendations by the Movement Disorder Society Multiple System Atrophy Study Group.* Movement Disorders Clinical Practice, 2020. **7**(7): p. 750–762.
81. Kermorgant, M., et al., *Age and gender differences in cardiovascular autonomic failure in the transgenic PLP-syn mouse, a model of multiple system atrophy.* Frontiers in Neurology, 2022. **13**: p. 874155.
82. Lazzeri, G., et al., *Cognitive and autonomic dysfunction in multiple system atrophy type p and c: A comparative study.* Frontiers in Neurology, 2022. **13**: p. 912820.
83. Giannini, G., et al., *REM sleep behaviour disorder in multiple system atrophy: From prodromal to progression of disease.* Frontiers in Neurology, 2021. **12**: p. 677213.
84. Steele, J.C., J.C. Richardson, and J. Olszewski, *Progressive supranuclear palsy: A heterogeneous degeneration involving the brain stem, basal ganglia and cerebellum with vertical gaze and pseudobulbar palsy, nuchal dystonia and dementia.* Archives of Neurology, 1964. **10**(4): p. 333–359.
85. Höglinger, G.U., et al., *Clinical diagnosis of progressive supranuclear palsy: The movement disorder society criteria.* Movement Disorders, 2017. **32**(6): p. 853–864.
86. Caso, F., et al., *Progression of white matter damage in progressive supranuclear palsy with predominant parkinsonism.* Parkinsonism & Related Disorders, 2018. **49**: p. 95–99.
87. Watanabe, H., et al., *The diagnosis of neurodegenerative disorders based on clinical and pathological findings using an MRI approach.* Rinsho Shinkeigaku= Clinical Neurology, 2011. **51**(11): p. 863–864.
88. Rebeiz, J.J., E.H. Kolodny, and E.P. Richardson, *Corticodentatonigral degeneration with neuronal achromasia.* Archives of Neurology, 1968. **18**(1): p. 20–33.
89. Gibb, W., P. Luthert, and C. Marsden, *Corticobasal degeneration.* Brain, 1989. **112**(5): p. 1171–1192.
90. Dickson, D.W., et al., *Office of rare diseases neuropathologic criteria for corticobasal degeneration.* Journal of Neuropathology & Experimental Neurology, 2002. **61**(11): p. 935–946.
91. Armstrong, M.J., et al., *Criteria for the diagnosis of corticobasal degeneration.* Neurology, 2013. **80**(5): p. 496–503.
92. Ling, H., et al., *Does corticobasal degeneration exist? A clinicopathological re-evaluation.* Brain, 2010. **133**(7): p. 2045–2057.
93. Mahapatra, R.K., et al., *Corticobasal degeneration.* The Lancet Neurology, 2004. **3**(12): p. 736–743.
94. Togasaki, D.M., and C.M. Tanner, *Epidemiologic aspects.* Advances in Neurology, 2000. **82**: p. 53–59.
95. Schneider, J., et al., *Corticobasal degeneration: Neuropathologic and clinical heterogeneity.* Neurology, 1997. **48**(4): p. 959–968.
96. Constantinides, V.C., et al., *Corticobasal degeneration and corticobasal syndrome: A review.* Clinical Parkinsonism & Related Disorders, 2019. **1**: p. 66–71.
97. Grazia Spillantini, M., et al., *A novel tau mutation (N296N) in familial dementia with swollen achromatic neurons and corticobasal inclusion bodies.* Annals of Neurology, 2000. **48**(6): p. 939–943.
98. Rossi, G., et al., *The G389R mutation in the MAPT gene presenting as sporadic corticobasal syndrome.* Movement Disorders, 2008. **23**(6): p. 892–895.
99. Kouri, N., et al., *Novel mutation in MAPT exon 13 (p. N410H) causes corticobasal degeneration.* Acta Neuropathologica, 2014. **127**: p. 271–282.
100. Kompoliti, K., et al., *Clinical presentation and pharmacological therapy in corticobasal degeneration.* Archives of Neurology, 1998. **55**(7): p. 957–961.
101. Cordivari, C., et al., *Treatment of dystonic clenched fist with botulinum toxin.* Movement Disorders, 2001. **16**(5): p. 907–913.
102. McVeigh, C., and P. Passmore, *Vascular dementia: Prevention and treatment.* Clinical Interventions in Aging, 2006. **1**(3): p. 229–235.
103. King, E., et al., *Peripheral inflammation in mild cognitive impairment with possible and probable Lewy body disease and Alzheimer's disease.* International Psychogeriatrics, 2019. **31**(4): p. 551–560.
104. Londos, E., et al., *Case report: Extreme sleep pattern in Lewy body dementia: A hypothalamic matter?* BMJ Case Reports, 2019. **12**(3): p. 1–4.
105. Chin, K.S., A. Teodorczuk, and R. Watson, *Dementia with Lewy bodies: Challenges in the diagnosis and management.* Australian & New Zealand Journal of Psychiatry, 2019. **53**(4): p. 291–303.
106. Brenowitz, W.D., et al., *Alzheimer's disease neuropathologic change, Lewy body disease, and vascular brain injury in clinic-and community-based samples.* Neurobiology of Aging, 2017. **53**: p. 83–92.

107. Siderowf, A., et al., *Biomarkers for cognitive impairment in Lewy body disorders: Status and relevance for clinical trials*. Movement Disorders, 2018. **33**(4): p. 528–536.
108. Morrin, H., et al., *Systematic review of the efficacy of non-pharmacological interventions in people with Lewy body dementia*. International Psychogeriatrics, 2018. **30**(3): p. 395–407.
109. Jellinger, K.A., *Basic mechanisms of neurodegeneration: A critical update*. Journal of Cellular and Molecular Medicine, 2010. **14**(3): p. 457–487.
110. Hartl, F.U., and M. Hayer-Hartl, *Converging concepts of protein folding in vitro and in vivo*. Nature Structural & Molecular Biology, 2009. **16**(6): p. 574–581.
111. Kubota, H., *Quality control against misfolded proteins in the cytosol: A network for cell survival*. Journal of Biochemistry, 2009. **146**(5): p. 609–616.
112. Scheper, W., and J.J. Hoozemans, *Endoplasmic reticulum protein quality control in neurodegenerative disease: The good, the bad and the therapy*. Current Medicinal Chemistry, 2009. **16**(5): p. 615–626.
113. Desplats, P., et al., *Inclusion formation and neuronal cell death through neuron-to-neuron transmission of α-synuclein*. Proceedings of the National Academy of Sciences, 2009. **106**(31): p. 13010–13015.
114. Frost, B., R.L. Jacks, and M.I. Diamond, *Propagation of tau misfolding from the outside to the inside of a cell*. Journal of Biological Chemistry, 2009. **284**(19): p. 12845–12852.
115. Ren, P.-H., et al., *Cytoplasmic penetration and persistent infection of mammalian cells by polyglutamine aggregates*. Nature Cell Biology, 2009. **11**(2): p. 219–225.
116. Mendoza-Espinosa, P., et al., *Disorder-to-order conformational transitions in protein structure and its relationship to disease*. Molecular and Cellular Biochemistry, 2009. **330**: p. 105–120.
117. Halliwell, B., *Oxidative stress and neurodegeneration: Where are we now?* Journal of Neurochemistry, 2006. **97**(6): p. 1634–1658.
118. Martínez, A., et al., *Protein targets of oxidative damage in human neurodegenerative diseases with abnormal protein aggregates*. Brain Pathology, 2010. **20**(2): p. 281–297.
119. Nakamura, T., and S.A. Lipton, *Cell death: Protein misfolding and neurodegenerative diseases*. Apoptosis, 2009. **14**: p. 455–468.
120. Nunomura, A., et al., *RNA oxidation in Alzheimer disease and related neurodegenerative disorders*. Acta Neuropathologica, 2009. **118**: p. 151–166.
121. Bond, C.E., and S.A. Greenfield, *Multiple cascade effects of oxidative stress on astroglia*. Glia, 2007. **55** (13): p. 1348–1361.
122. Ballatori, N., et al., *Glutathione dysregulation and the etiology and progression of human diseases*. 2009. **390** (3): p. 191–214.
123. Castellani, R.J., et al., *Sublethal RNA oxidation as a mechanism for neurodegenerative disease*. International Journal of Molecular Sciences, 2008. **9**(5): p. 789–806.
124. Browne, S.E., and M.F. Beal, *Oxidative damage in Huntington's disease pathogenesis*. Antioxidants & Redox Signaling, 2006. **8**(11–12): p. 2061–2073.
125. Wang, J., W.R. Markesbery, and M.A. Lovell, *Increased oxidative damage in nuclear and mitochondrial DNA in mild cognitive impairment*. Journal of Neurochemistry, 2006. **96**(3): p. 825–832.
126. Wang, X., et al., *The role of abnormal mitochondrial dynamics in the pathogenesis of Alzheimer's disease*. Journal of Neurochemistry, 2009. **109**: p. 153–159.
127. Praticò, D., and S. Sung, *Lipid peroxidation and oxidative imbalance: Early functional events in Alzheimer's disease*. Journal of Alzheimer's Disease, 2004. **6**(2): p. 171–175.
128. Sayre, L.M., G. Perry, and M.A. Smith, *Oxidative stress and neurotoxicity*. Chemical Research in Toxicology, 2008. **21**(1): p. 172–188.
129. Ward, R.J., and R.R. Crichton, *Iron and its role in neurodegenerative diseases*. Neurodegenerative Diseases and Metal Ions, 2006. **1**: p. 227–279.
130. Sparaco, M., et al., *Friedreich's ataxia: Oxidative stress and cytoskeletal abnormalities*. Journal of the Neurological Sciences, 2009. **287**(1–2): p. 111–118.
131. Mancuso, M., et al., *Hereditary ferritinopathy: A novel mutation, its cellular pathology, and pathogenetic insights*. Journal of Neuropathology & Experimental Neurology, 2005. **64**(4): p. 280–294.
132. Arreguin, S., et al., *Dopamine complexes of iron in the etiology and pathogenesis of Parkinson's disease*. Journal of Inorganic Biochemistry, 2009. **103**(1): p. 87–93.
133. Altamura, S., and M.U. Muckenthaler, *Iron toxicity in diseases of aging: Alzheimer's disease, Parkinson's disease and atherosclerosis*. Journal of Alzheimer's Disease, 2009. **16**(4): p. 879–895.
134. Esteves, A.R., et al., *Microtubule depolymerization potentiates alpha-synuclein oligomerization*. Frontiers in Aging Neuroscience, 2010. **1**: p. 1152.
135. Swerdlow, R.H., and S.M. Khan, *The Alzheimer's disease mitochondrial cascade hypothesis: An update*. Experimental Neurology, 2009. **218**(2): p. 308–315.

136. Du, H., and S. S. Yan, *Mitochondrial permeability transition pore in Alzheimer's disease: Cyclophilin d and amyloid beta.* Biochimica Et Biophysica Acta (BBA)-Molecular Basis of Disease, 2010. **1802**(1): p. 198–204.
137. Hauptmann, S., et al., *Mitochondrial dysfunction: An early event in Alzheimer pathology accumulates with age in AD transgenic mice.* Neurobiology of Aging, 2009. **30**(10): p. 1574–1586.
138. Lynn, B.C., et al., *Quantitative changes in the mitochondrial proteome from subjects with mild cognitive impairment, early stage, and late stage Alzheimer's disease.* Journal of Alzheimer's Disease, 2010. **19**(1): p. 325–339.
139. Mancuso, M., et al., *Is there a primary role of the mitochondrial genome in Alzheimer's disease?* Journal of Bioenergetics and Biomembranes, 2009. **41**: p. 411–416.
140. Vila, M., D. Ramonet, and C. Perier, *Mitochondrial alterations in Parkinson's disease: New clues.* Journal of Neurochemistry, 2008. **107**(2): p. 317–328.
141. Stolp, H., and K. Dziegielewska, *Role of developmental inflammation and blood–brain barrier dysfunction in neurodevelopmental and neurodegenerative diseases.* Neuropathology and Applied Neurobiology, 2009. **35** (2): p. 132–146.
142. Kell, D.B., *Iron behaving badly: Inappropriate iron chelation as a major contributor to the aetiology of vascular and other progressive inflammatory and degenerative diseases.* BMC Medical Genomics, 2009. **2**(1): p. 1–79.
143. Galimberti, D., C. Fenoglio, and E. Scarpini, *Inflammation in neurodegenerative disorders: Friend or foe?* Current Aging Science, 2008. **1**(1): p. 30–41.
144. Piehl, F., and T. Olsson, *Inflammation and susceptibility to neurodegeneration: The use of unbiased genetics to decipher critical regulatory pathways.* Neuroscience, 2009. **158**(3): p. 1143–1150.
145. Graber, J.J., and S. Dhib-Jalbut, *Biomarkers of disease activity in multiple sclerosis.* Journal of the Neurological Sciences, 2011. **305**(1–2): p. 1–10.
146. Salminen, A., et al., *Inflammation in Alzheimer's disease: Amyloid-β oligomers trigger innate immunity defence via pattern recognition receptors.* Progress in Neurobiology, 2009. **87**(3): p. 181–194.
147. Lucin, K.M., and T. Wyss-Coray, *Immune activation in brain aging and neurodegeneration: Too much or too little?* Neuron, 2009. **64**(1): p. 110–122.
148. Lynch, M.A., *Age-related neuroinflammatory changes negatively impact on neuronal function.* Frontiers in Aging Neuroscience, 2010. **1**: p. 1206.
149. Ryu, J.K., and J.G. McLarnon, *A leaky blood-brain barrier, fibrinogen infiltration and microglial reactivity in inflamed Alzheimer's disease brain.* Journal of Cellular and Molecular Medicine, 2009. **13**(9a): p. 2911–2925.
150. Sawada, M., H. Sawada, and T. Nagatsu, *Effects of aging on neuroprotective and neurotoxic properties of microglia in neurodegenerative diseases.* Neurodegenerative Diseases, 2008. **5**(3–4): p. 254–256.
151. Neumann, H., M.R. Kotter, and R.J. Franklin, *Debris clearance by microglia: An essential link between degeneration and regeneration.* Brain, 2009. **132**(2): p. 288–295.
152. Lee, M., et al., *Astrocytes produce the antiinflammatory and neuroprotective agent hydrogen sulfide.* Neurobiology of Aging, 2009. **30**(10): p. 1523–1534.
153. Sondag, C.M., G. Dhawan, and C.K. Combs, *Beta amyloid oligomers and fibrils stimulate differential activation of primary microglia.* Journal of Neuroinflammation, 2009. **6**(1): p. 1–13.
154. Reale, M., et al., *Peripheral chemokine receptors, their ligands, cytokines and Alzheimer's disease.* Journal of Alzheimer's Disease, 2008. **14**(2): p. 147–159.
155. Kroemer, G., et al., *Classification of cell death: Recommendations of the nomenclature committee on cell death 2009.* Cell Death & Differentiation, 2009. **16**(1): p. 3–11.
156. Bredesen, D.E., *Programmed cell death mechanisms in neurological disease* Current Molecular Medicine, 2008. **8**(3): p. 173–186.
157. Gorman, A.M., *Neuronal cell death in neurodegenerative diseases: Recurring themes around protein handling.* Journal of Cellular and Molecular Medicine, 2008. **12**(6a): p. 2263–2280.
158. Uchiyama, Y., et al., *Autophagic neuron death.* Methods in Enzymology, 2009. **453**: p. 33–51.
159. Spires-Jones, T.L., et al., *Tau pathophysiology in neurodegeneration: A tangled issue.* Trends in Neurosciences, 2009. **32**(3): p. 150–159.
160. Franco, R., et al., *Environmental toxicity, oxidative stress and apoptosis: Menage a trois.* Mutation Research/Genetic Toxicology and Environmental Mutagenesis, 2009. **674**(1–2): p. 3–22.

2 Chemical-Induced Parkinson's Disease Model and Associated Therapeutic Strategies

Sandhya Yadav
Department of Biochemistry and Molecular Pharmacology,
University of Massachusetts Medical School, Worcester, MA, USA

Manish Kumar Tripathi
Institute for Drug Research, School of Pharmacy, Faculty of Medicine,
The Hebrew University of Jerusalem, Jerusalem, Israel

2.1 INTRODUCTION

Parkinson's disease (PD) is an idiopathic neurodegenerative movement disorder. Death of dopamine (DA)-producing neuronal cells is responsible for the development of PD. Dopaminergic cell death in substantia nigra pars compacta (SNPc) of the midbrain, low levels of DA in the striatum, accumulation of intracellular debris, and α-synuclein overexpression/aggregation are the hallmarks of the disease. Although primary clinical symptoms such as tremors on rest, the presence of bradykinesia, unstable body posture, and rigidity show the presence of disease, this disorder is also associated with many other secondary symptoms that add to overall disability. The major molecular pathogenesis is α-synuclein protein dyshomeostasis, mitochondrial dysfunction, oxidative/nitrosative stress, calcium ion imbalance, axonal transport, aberrant autophagy, and neuroinflammation that involve multiple pathways and mechanisms [1–3].

2.1.1 Pathological Features

Many external and internal factors lead to the death of DA neurons in the midbrain. They are responsible for reduced stratal DA production which affects other motor functions. This leads to the accumulation of cytoplasmic inclusions in other neurons which are composed of the fibrillar structure of synuclein and are commonly called Lewy bodies and Lewy neuritis [4, 5]. Clinical symptoms appear when degeneration of neurons reaches up to 50%–60%. There are two main pathways, commonly known as direct and indirect, that eventually give signals to the brain's cortex region, which regulates and controls the body's movement. Two types of DA receptors mainly regulate these pathways in the striatum, which include the Dl-like subfamily and the D2-like subfamily, in which Dl regulates direct pathways and D2 regulates indirect pathways. Dl facilitates the synchronised movement, while D2 inhibits the movement in response to the DA [6]. When the dopamine level is depleted, this process favors the hypoactivation of direct pathways and hyperactivation of indirect pathways, which result in the impairment of the cortical motor system and lead to reduced motor movement, muscular rigidity, and resting tremor [6, 7]. Many molecular and biochemical changes are reported in PD. Some of the studies explained that mitochondrion is the center of PD pathogenesis. Disturbances in complex 1 and complex III activities and imbalance in energy metabolism have been reported in SNPc and muscle tissue of PD patients [8]. Altered calcium homeostasis and reduced levels of calcium-binding proteins are associated with aging and

DOI: 10.1201/9781032620442-2

PD [9]. Iron (Fe) accumulation in SNPc provides a favorable environment for DA auto-oxidation and lipid peroxidation, which further leads to an elevation in the levels of oxygen and hydrogen free radicals [8].

2.2 CHEMICAL-INDUCED PD MODELS

There are many chronic and acute chemical-induced rodent models used to mimic the early and later stages of PD. These chemicals/toxin models are described below:

2.2.1 6-Hydroxydopamine (6-OHDA)

6-OHDA is the hydroxylated analogue of the neurotransmitter DA, as well as catecholamines and noradrenaline. It is recognised for its lethal impact on catecholaminergic neurons [10]. The toxicity of 6-OHDA is observed in both the central nervous system (CNS) and peripheral nervous system (PNS). However, its lipophilic properties prevent it from entering into the brain, necessitating stereotaxic injection into the SNPc or stratum [11]. Since its introduction in 1970, 6-OHDA has emerged as a prominent neurotoxin utilised to model PD in experimental animals [12]. Following intra-nigral injection, 6-OHDA is observed to accumulate intracellularly, leading to an increase in the production of reactive oxygen species (ROS) and the creation of quinones through auto-oxidation [13]. The degradation of interneurons in the SNPc and ventral tegmental area (VTA) is attributed to the neurotoxin 6-OHDA [14]. The compound exhibits a greater affinity for catecholaminergic transporters, specifically DA transporters (DAT) and norepinephrine transporters, compared to other types of transporters. The toxicity of the substance is particularly targeted towards the catecholaminergic neurons as a result of the selective absorption of 6-OHDA by DAT [15, 16].

The entry of 6-OHDA into DA neurons is facilitated by DAT transporters. Once inside, it accumulates in the cytoplasm and undergoes oxidation, resulting in the formation of different types of free ROS such as superoxide, hydrogen peroxide, hydroxy radicals, and paraquinone [17–19]. Additionally, it elicits neurotoxic effects through mitochondrial complex 1 inhibition [20]. The infusion of 6-OHDA results in rapid degeneration of DA and norepinephrine as well as an elevation in striatal serotonin levels. This cascade of neurochemical changes induces various behavioral impairments in rats, such as alteration in sensorimotor function, deficits in learning and memory, and increased hyperactivity. These observed effects closely resemble the symptoms exhibited by individuals with PD [21–23]. The utilisation of 6-OHDA as a neurotoxin/chemical is prevalent in animal models for PD, rendering it one of the most extensively employed substances for this purpose. This model possesses several limitations in assessing the impact of antiparkinsonian medicines, as it solely induces degeneration in dopaminergic neurons. Furthermore, an additional drawback of 6-OHDA is its propensity to induce harm in other cerebral regions without exhibiting the formation of Lewy bodies, which are characteristic inclusions [24]. Motor deficits are predominantly evident in the limb of the opposite side of the hemisphere where the injection of 6-OHDA occurs [13]. The 6-OHDA model does not show a significant number of neuropathological characteristics observed in human PD, thereby making it a less effective model for investigating the molecular mechanisms behind dopamine neuron degeneration.

2.2.2 1-Methyl-4-Phenyl-1,2,3,6-TetrahydroPyridine (MPTP)

MPTP was accidentally discovered when drug addicts in California developed PD-like symptoms after injecting themselves with synthetic heroin [25]. Basically, MPTP is not reported to be a potent neurotoxic chemical. When MPTP makes entry into the brain, it gets converted into 1-methyl-4phenylpyridinium (MPP+), which is responsible for the death of nigrostriatal DA neurons, similar to that seen in PD patients. Initially, MPP+ was used as a herbicide under the trade name cyperquat

[26]. MPTP is metabolised to an intermediate MPP+ by the action of a specific DA catabolising enzyme in glial cells [27]. MPP+ blocks the activity of mitochondrial complex 1 of the electron transport chain [28]. It is one of the highest-appreciated models for PD. It shows almost all pathological features of PD in humans except the Lewy body formation. There are many routes and doses of MPTP administration based on the objectives of the study [29]. MPTP mimics many cardinal features of sporadic PD, including a decrease in DA Levels and tyrosine hydroxylase (TH) cell number. Aberrant autophagy and increased level of inflammatory markers in SNPc and stratum of mice are widely reported in this model [30, 31]. It is considered a gold standard model for studying PD because it is well studied in many animals such as dogs, cats, sheep, rats, guinea pigs, mice, and nonhuman primates [32–34]. Mice are more sensitive than rats, and neuronal cell death observed in monkeys is similar to that in human PD patients [35, 36]. MPTP exerts its toxicity by induced apoptosis, impaired autophagy [37], and mitochondrial dysfunction [3]. It also activates microglia, which also leads to neurodegeneration [38].

2.2.3 Rotenone

Rotenone is a neurotoxin isolated from leaves, roots, and stems of the *Fabaceae* family of plants such as *Derris involuta* and *Derris eliptica*, commonly used as a fish poison [39]. Rotenone was the first chemical model that linked environmental toxicity to a PD study that shows the Lewy body deposition [40]. It is one of the widely used pesticides and inhibits complex 1 of mitochondria. Its nature is lipophilic, so it can easily enter the brain. It also shows most of the anatomical features of sporadic PD, such as aggregate formation inside the cells that resemble Lewy body formation, behavioral dysfunction, increased oxidative stress, and inflammation [41, 42]. Rotenone can be administered in animals by various routes such as intravenous, intraperitoneal, and subcutaneous injections. It can also be injected directly into the brain by stereotactic procedure [43]. Rotenone has an inhibitory effect on oxidative phosphorylation by reducing the interaction between nicotinamide adenine dinucleotide (NADH)-ubiquinone reductase activity and the PSST subunit of the multi-polypeptide enzyme complex 1 inside the mitochondria's electron transport chain [44]. Rotenone can impede mitosis and cellular proliferation through its disruption of microtubule assembly, reduction of GTP hydrolysis rate, and inhibition of in vitro microtubule synthesis from tubulin [45, 46]. Aberrant autophagy, inflammatory, and oxidative parameters include the formation of protein carbonyl and activation of microglia in the SNPc and stratum [31].

2.2.4 Paraquat (1,1′-dimethyl-4-4′-bipyridinium dichloride) and Maneb (Manganese Ethylene Bis-Dithiocarbamate)

Syngenta developed paraquat (PQ) in the early 1960s as a nonselective quaternary ammonium herbicide [47]. PQ toxicity was first reported in 1980 [48]. PQ has the ability to enter the brain but has an inefficient and restricted efficacy. However, it is worth noting that measurable quantities of this herbicide have been observed in the CNS following its systemic administration in mice [49]. PQ is lipophilic in nature. And enter into the brain through L-amino acid transporters, which facilitate the transport of neutral amino acids [50]. It causes oxidative stress and induces Lewy body formation in DA neurons [51, 52]. PQ is a non-specific herbicide having a structural resemblance with the MPTP metabolite, MPP+ [53]. PQ exhibits a pronounced toxicity profile and possesses the ability to be absorbed into the human body through many routes, with variations in the rate of absorption [54]. The maximal absorption of the substance occurs through subcutaneous injection, while the smallest absorption is observed through oral delivery. Furthermore, it has been found to have detrimental effects on many organs such as the lungs, kidneys, and liver [55]. PQ displays the involvement of mitochondria in PD pathogenesis by disturbing the redox cycling, resulting in ROS induction and death of neuronal cells [56, 57]. Exposure of rodents to PQ results in a decrease in behavioral activity, DA depletion in the striatum, TH-containing cell number loss, and increase in the number of microglial cells [56–59]. PD-like characteristics, viz. impairment of UPS and an increase in α-synuclein aggregation, are also observed in PQ-induced PD phenotype [60, 61]. One notable differentiation from other models of PD is that PQ can be utilised independently or in conjugation with other agrochemicals, thereby replicating the exposure circumstances commonly encountered in agriculture with more accuracy.

Maneb and mancozeb are two instances of agrochemicals that are employed in conjugation with PQ to intensify systemic toxicity and induce damage to the neurological systems [47]. However, PQ alone does not produce any change in the nigrostriatal dopaminergic system [53], but in conjugation with Maneb, a fungicide, it potentiates the effect. It also decreases locomotor activity and causes loss of SNPc neurons [62]. Maneb potentiates the neurotoxic effect of paraquat by inhibiting the transportation of neurotransmitters and eventually potentiates motor impairment [63]. These mixtures frequently affect the prodynorphin-kappa opioid (KOP) and nociceptin/orphanin (NOP) receptor systems in the SNPc and caudate putamen in rats, supplying additional proof that agrochemical interaction and chronic exposure maybe involved in the pathophysiological mechanism of sporadic PD [64]. It is also reported that combined treatment of PQ and maneb causes neurotoxic effects via changing the activity of the proteasome by lowering the amounts of the mRNAs for the 1 and Rpt3 subunits of the proteasome. Additionally, their combined treatment modifies the opioid systems by decreasing the expression of the opioid receptor in neuroblastoma cells, highlighting the need for additional research to clarify the mechanism [65]. As maneb and paraquat are frequently used chemicals in agriculture, this model further emphasises the impact of environmental factors in the pathogenesis of PD.

2.2.5 Lipopolysaccharides (LPS)

LPS is an endotoxin that is isolated from gram-negative bacteria and widely used to study neuroinflammation in PD. LPS can be administered either intraperitoneally, by infusion, or by stereotactic injection. Among them, stereotactic injection is widely used. LPS is not a widely used animal model for PD pathogenesis [66]. The LPS model was accidentally discovered when a laboratory researcher was unintentionally exposed to LPS from *Salmonella Minnesota* and developed PD-like symptoms [67]. Later intranigral injection of LPS into rats showed dopaminergic neuron depletion and strengthened the concept of the LPS model of PD [68, 69]. Its efficacy mostly depends on the route of administration [70].

2.2.6 Methamphetamine

It is a synthetic psycho-stimulant that can be prepared by catalytic hydrogenation of the ephedrine [71]. It is also lipophilic in nature and easily enters the brain. It is responsible for DA nerve terminal degeneration in the stratum, nucleus accumbens, and frontal cortex, disturbing their cell bodies in SNPc and VTA. It increases the levels of cytosolic DA, resulting in increased oxidative stress [66]. It degenerates only nerve terminals and does not have any similarity with sporadic PD; therefore, it is not considered an appropriate PD model.

2.2.7 Cypermethrin

Cypermethrin is a chemical pyrethroid which is commonly used in agriculture, globally and domestically. It is lipophilic in nature and can easily enter the brain [72]. It induces alteration in crucial neuroproteins during developmental periods, leading to selective and progressive nigrostriatal dopaminergic neurodegeneration and several behavioral manifestations in adulthood [73]. The exact cause for the loss of nigrostriatal fibers has not yet been deciphered. However, cypermethrin reduces the mitochondrial membrane potential and also inhibits mitochondrial complex 1 activity, which dysregulates the expression of many mitochondrial proteins that are involved in the electron transport chain, carbohydrate metabolism, redox recycling, ion homeostasis, and neuronal survival [1]. It also modulates neuro-inflammation and induces α-synuclein overexpression/aggregation in cypermethrin-exposed animals [1, 74].

2.2.8 Metals

Metals act as a cofactor in the synthesis of DNA, RNA, and proteins, and their deficiencies cause disturbance in normal cellular homeostasis, and their excess presence in the cell could be cytotoxic [75, 76]. An association between metal dyshomeostasis and neurodegeneration was found in various post-mortem and experimental studies showing the deposition of various heavy metals such as Fe, Al, Zn, etc. in DA neurons of the SNPc region [77]. Fe and Cu generate highly reactive hydroxyl radicals through various chemical reactions, which further lead to increased oxidative stress and DNA damage, while cadmium, Pb, and arsenic can bind with sulfhydryl groups of proteins and reduce the level of antioxidants [78–80].

2.3 POSSIBLE TREATMENT FOR PD

Many drugs exhibit protective/curative effects in animal models of PD, but there is no drug available to date that can offer a cure for PD in humans. Drugs available in the market are DA supplements, DA receptor agonists, adenosine receptor antagonists, inhibitors of DA catabolising enzymes, and several anti-inflammatory and neurotrophic agents [81]. Only one drug, levodopa, which is the precursor of DA, is a widely used drug in the treatment of PD phenotype in humans [82, 83]. Pramipexole treatment of PD with levodopa induces dyskinesia because of DA catabolism, which produces free radicals and potentiates neurodegeneration [84]. Levodopa is given in combination with carbidopa, for slowing the release of DA from levodopa in peripheral tissue. DA receptor agonists also have been extensively used to reduce PD phenotype [81, 82]. Pramipexole, a dopamine receptor agonist, acts as the direct antioxidant owing to the presence of hydroxylated benzyl ring structure [83]. MAO-B is responsible for the metabolism of DA, and thus its inhibition can reduce DA oxidation. Selegiline, a MAO-B inhibitor, reduces DA metabolism and is globally used to treat PD [85]. Catechol -o-methyl transferase (COMT) is another dopamine-metabolising enzyme, and inhibitors of COMT, such as entacapone and tolcapone, block DA metabolism and thus provide symptomatic relief from motor fluctuations in PD [86, 87]. Antioxidants reduce the oxidative stress and free radical generation. Several groups of antioxidants are reported to provide relief from the symptoms of PD. Nicotine, caffeine, vitamin C, and resveratrol are potential antioxidants in the various rodent models [88, 89].

REFERENCES

1. Agrawal, S., et al., *Cyclosporine a and MnTMPyP alleviate α-synuclein expression and aggregation in cypermethrin-induced parkinsonism.* Mol Neurobiol, 2015. **52**(3): p. 1619–1628.
2. Tripathi, M.K., M. Kartawy, and H. Amal, *The role of nitric oxide in brain disorders: Autism spectrum disorder and other psychiatric, neurological, and neurodegenerative disorders.* Redox Biol, 2020. **34**: p. 101567.
3. Tripathi, M.K., et al., *Silymarin protects against impaired autophagy associated with 1-methyl-4-phenyl-1,2,3,6-tetrahydropyridine-induced parkinsonism.* J Mol Neurosci, 2020. **70**(2): p. 276–283.
4. Olanow, C.W., and W.G. Tatton, *Etiology and pathogenesis of Parkinson's disease* Annu Rev Neurosci, 1999. **22**: p. 123–44.
5. Fahn, S., and D. Sulzer, *Neurodegeneration and neuroprotection in Parkinson disease* NeuroRx, 2004. **1**(1): p. 139–54.
6. Nicholson, G., A.C. Pereira, and G.M. Hall, *Parkinson's disease and anaesthesia* Br J Anaesth, 2002. **89**(6): p. 904–16.

7. Lang, A.E., and A.M. Lozano, *Parkinson's disease* New England Journal of Medicine, 1998. **339**(16): p. 1130–1143.
8. Riederer, P., and K.W. Lange, *Pathogenesis of Parkinson's disease* Curr Opin Neurol Neurosurg, 1992. **5**(3): p. 295–300.
9. Iacopino, A.M., and S. Christakos, *Specific reduction of calcium-binding protein (28-kilodalton calbindin-d) Gene expression in aging and neurodegenerative diseases.* Proc Natl Acad Sci U S A, 1990. **87**(11): p. 4078–82.
10. Blum, D., et al., *Molecular pathways involved in the neurotoxicity of 6-OHDA, dopamine and MPTP: Contribution to the apoptotic theory in Parkinson's disease.* Progress in Neurobiology, 2001. **65**(2): p. 135–172.
11. Sauer, H., and W.H. Oertel, *Progressive degeneration of nigrostriatal dopamine neurons following intrastriatal terminal lesions with 6-hydroxydopamine: A combined retrograde tracing and immunocytochemical study in the rat* Neuroscience, 1994. **59**(2): p. 401–15.
12. Ungerstedt, U., and G.W. Arbuthnott, *Quantitative recording of rotational behavior in rats after 6-hydroxy-dopamine lesions of the nigrostriatal dopamine system* Brain Res, 1970. **24**(3): p. 485–93.
13. Simola, N., M. Morelli, and A.R. Carta, *The 6-hydroxydopamine model of Parkinson's disease* Neurotox Res, 2007. **11**(3-4): p. 151–67.
14. Przedborski, S., et al., *Dose-dependent lesions of the dopaminergic nigrostriatal pathway induced by intrastriatal injection of 6-hydroxydopamine.* Neuroscience, 1995. **67**(3): p. 631–47.
15. Breese, G.R., and T.D. Traylor, *Depletion of brain noradrenaline and dopamine by 6-hydroxydopamine* British Journal of Pharmacology, 1971. **42**(1): p. 88–99.
16. Luthman, J., et al., *Effects of d-amphetamine and methylphenidate on hyperactivity produced by neonatal 6-hydroxydopamine treatment.* Psychopharmacology, 1989. **99**(4): p. 550–557.
17. Cohen, G., and R.E. Heikkila, *The generation of hydrogen peroxide, superoxide radical, and hydroxyl radical by 6-hydroxydopamine, dialuric acid, and related cytotoxic agents* J Biol Chem, 1974. **249**(8): p. 2447–52.
18. McDowell, K., and M.F. Chesselet, *Animal models of the non-motor features of Parkinson's disease* Neurobiol Dis, 2012. **46**(3): p. 597–606.
19. Mazzio, E.A., R.R. Reams, and K.F. Soliman, *The role of oxidative stress, impaired glycolysis and mitochondrial respiratory redox failure in the cytotoxic effects of 6-hydroxydopamine in vitro* Brain Res, 2004. **1004**(1-2): p. 29–44.
20. Glinka, Y., K.F. Tipton, and M.B. Youdim, *Nature of inhibition of mitochondrial respiratory complex i by 6-hydroxydopamine* J Neurochem, 1996. **66**(5): p. 2004–10.
21. Luthman, J., et al., *Selective lesion of central dopamine or noradrenaline neuron systems in the neonatal rat: motor behavior and monoamine alterations at adult stage.* Behavioural Brain Research, 1989. **33**(3): p. 267–277.
22. Schwarzkopf, S.B., et al., *Effects of haloperidol and SCH 23390 on acoustic startle in animals depleted of dopamine as neonates: Implications for neuropsychiatric syndromes.* Psychopharmacology, 1996. **123**(3): p. 258–266.
23. Takasuna, M., and T. Iwasaki, *Active and passive avoidance learning in rats neonatally treated with intraventricular 6-hydroxydopamine* Behav Brain Res, 1996. **74**(1-2): p. 119–26.
24. Dauer, W., and S. Przedborski, *Parkinson's disease: Mechanisms and models* Neuron, 2003. **39**(6): p. 889–909.
25. Yadav, S., et al., *Rodent models and contemporary molecular techniques: Notable feats yet incomplete explanations of Parkinson's disease pathogenesis.* Mol Neurobiol, 2012. **46**(2): p. 495–512.
26. Di Monte, D.A., *The role of environmental agents in Parkinson's disease* Clinical Neuroscience Research, 2001. **1**(6): p. 419–426.
27. Jackson-Lewis, V., and S. Przedborski, *Protocol for the MPTP mouse model of Parkinson's disease* Nature Protocols, 2007. **2**(1): p. 141–151.
28. Varastet, M., et al., *Chronic MPTP treatment reproduces in baboons the differential vulnerability of mesencephalic dopaminergic neurons observed in Parkinson's disease.* Neuroscience, 1994. **63**(1): p. 47–56.
29. Schmidt, N., and B. Ferger, *Neurochemical findings in the MPTP model of Parkinson's disease* J Neural Transm (Vienna), 2001. **108**(11): p. 1263–82.
30. Mishra, A.K., et al., *Aberrant autophagy and parkinsonism: does correction rescue from disease progression?* Mol Neurobiol, 2015. **51**(3): p. 893–908.
31. Duty, S., and P. Jenner, *Animal models of Parkinson's disease: A source of novel treatments and clues to the cause of the disease* Br J Pharmacol, 2011. **164**(4): p. 1357–91.

32. Przedborski, S., et al., *The parkinsonian toxin 1-methyl-4-phenyl-1,2,3,6-tetrahydropyridine (MPTP): A technical review of its utility and safety.* J Neurochem, 2001. **76**(5): p. 1265–74.
33. Burns, R.S., et al., *The clinical syndrome of striatal dopamine deficiency. Parkinsonism induced by 1-methyl-4-phenyl-1,2,3,6-tetrahydropyridine (MPTP).* N Engl J Med, 1985. **312**(22): p. 1418–21.
34. Zeng, X.S., et al., *The role of thioredoxin-1 in suppression of endoplasmic reticulum stress in Parkinson disease.* Free Radic Biol Med, 2014. **67**: p. 10–8.
35. Giovanni, A., et al., *Studies on species sensitivity to the dopaminergic neurotoxin 1-methyl-4-phenyl-1,2,3,6-tetrahydropyridine. Part 1: Systemic administration.* J Pharmacol Exp Ther, 1994. **270**(3): p. 1000–7.
36. Oleh, H., *Biochemical aspects of Parkinson's disease.* Neurology, 1998. **51**(2 Suppl 2): p. S2.
37. Tripathi, M.K., et al., *Malfunctioning of chaperone-mediated autophagy in Parkinson's disease: Feats, constraints, and flaws of modulators.* Neurotox Res, 2019. **35**(1): p. 260–270. doi: 10.1007/s12640-018-9917-z
38. Huang, D., et al., *Dynamic changes in the nigrostriatal pathway in the MPTP mouse model of Parkinson's disease.* Parkinsons Dis, 2017. **2017**: p. 9349487.
39. Cannon, J.G., et al., *Naturally occurring fish poisons from plants.* Journal of Chemical Education, 2004. **81**(10): p. 1457.
40. Lee, H.J., et al., *Formation and removal of alpha-synuclein aggregates in cells exposed to mitochondrial inhibitors.* J Biol Chem, 2002. **277**(7): p. 5411–7.
41. Sherer, T.B., et al., *Mechanism of toxicity in rotenone models of Parkinson's disease.* J Neurosci, 2003. **23**(34): p. 10756–64.
42. Greenamyre, J.T., et al., *Lessons from the rotenone model of Parkinson's disease.* Trends Pharmacol Sci, 2010. **31**(4): p. 141–2; author reply 142-3.
43. Xiong, N., et al., *Stereotaxical infusion of rotenone: A reliable rodent model for Parkinson's disease.* PLOS ONE, 2009. **4**(11): p. e7878.
44. Schuler, F., and J.E. Casida, *Functional coupling of PSST and ND1 subunits in NADH:Ubiquinone oxidoreductase established by photoaffinity labeling* Biochim Biophys Acta, 2001. **1506**(1): p. 79–87.
45. Srivastava, P., and D. Panda, *Rotenone inhibits mammalian cell proliferation by inhibiting microtubule assembly through tubulin binding* Febs j, 2007. **274**(18): p. 4788–801.
46. Marshall, L.E., and R.H. Himes, *Rotenone inhibition of tubulin self-assembly* Biochim Biophys Acta, 1978. **543**(4): p. 590–4.
47. Bastías-Candia, S., J.M. Zolezzi, and N.C. Inestrosa, *Revisiting the paraquat-induced sporadic Parkinson's disease-like model* Molecular Neurobiology, 2019. **56**(2): p. 1044–1055.
48. Grant, H., P.L. Lantos, and C. Parkinson, *Cerebral damage in paraquat poisoning* Histopathology, 1980. **4**(2): p. 185–95.
49. Corasaniti, M.T., et al., *Paraquat: A useful tool for the in vivo study of mechanisms of neuronal cell death.* Pharmacol Toxicol, 1998. **83**(1): p. 1–7.
50. Shimizu, K., et al., *Carrier-mediated processes in blood–brain barrier penetration and neural uptake of paraquat.* Brain Res, 2001. **906**(1-2): p. 135–42.
51. McCormack, A.L., et al., *Environmental risk factors and Parkinson's disease: Selective degeneration of nigral dopaminergic neurons caused by the herbicide paraquat.* Neurobiol Dis, 2002. **10**(2): p. 119–27.
52. Cicchetti, F., et al., *Systemic exposure to paraquat and maneb models early Parkinson's disease in young adult rats.* Neurobiol Dis, 2005. **20**(2): p. 360–71.
53. Miller, G.W., *Paraquat: The red herring of Parkinson's disease research* Toxicol Sci, 2007. **100**(1): p. 1–2.
54. Naylor, J.L., et al., *Further evidence that the blood/brain barrier impedes paraquat entry into the brain.* Hum Exp Toxicol, 1995. **14**(7): p. 587–94.
55. Raghu, K., et al., *Paraquat poisoning: A case report and review of literature.* J Family Community Med, 2013. **20**(3): p. 198–200.
56. Brooks, A.I., et al., *Paraquat elicited neurobehavioral syndrome caused by dopaminergic neuron loss.* Brain Res, 1999. **823**(1-2): p. 1–10.
57. Gupta, S.P., et al., *Involvement of nitric oxide in maneb- and paraquat-induced Parkinson's disease phenotype in mouse: Is there any link with lipid peroxidation?* Neurochem Res, 2010. **35**(8): p. 1206–13.
58. Kumar, A., et al., *Effect of zinc and paraquat co-exposure on neurodegeneration: Modulation of oxidative stress and expression of metallothioneins, toxicant responsive and transporter genes in rats.* Free Radic Res, 2010. **44**(8): p. 950–65.
59. Singhal, N.K., et al., *Melatonin or silymarin reduces maneb- and paraquat-induced Parkinson's disease phenotype in the mouse.* J Pineal Res, 2011. **50**(2): p. 97–109.

60. Yang, W., and E. Tiffany-Castiglioni, *The bipyridyl herbicide paraquat induces proteasome dysfunction in human neuroblastoma SH-SY5Y cells* J Toxicol Environ Health A, 2007. **70**(21): p. 1849–57.
61. Manning-Bog, A.B., et al., *The herbicide Paraquat causes up-regulation and aggregation of alpha-synuclein in mice: Paraquat and alpha-synuclein.* J Biol Chem, 2002. **277**(3): p. 1641–4.
62. Thrash, B., et al., *Paraquat and maneb induced neurotoxicity.* Proc West Pharmacol Soc, 2007. **50**: p. 31–42.
63. Thiruchelvam, M., et al., *The nigrostriatal dopaminergic system as a preferential target of repeated exposures to combined paraquat and maneb: Implications for Parkinson's disease.* J Neurosci, 2000. **20**(24): p. 9207–14.
64. Bastías-Candia, S., et al., *Combined exposure to agriculture pesticides, paraquat and maneb, induces alterations in the N/OFQ—NOPr and PDYN/KOPr systems in rats: Relevance to sporadic Parkinson's disease.* Environmental Toxicology, 2015. **30**(6): p. 656–663.
65. Caputi, F.F., et al., *Proteasome subunit and opioid receptor gene expression down-regulation induced by paraquat and maneb in human neuroblastoma SH-SY5Y cells.* Environ Toxicol Pharmacol, 2015. **40**(3): p. 895–900.
66. Tieu, K., *A guide to neurotoxic animal models of Parkinson's disease* Cold Spring Harb Perspect Med, 2011. **1**(1): p. a009316.
67. Niehaus, I., and J.H. Lange, *Endotoxin: Is it an environmental factor in the cause of Parkinson's disease?* Occup Environ Med, 2003. **60**(5): p. 378.
68. Castaño, A., et al., *Lipopolysaccharide intranigral injection induces inflammatory reaction and damage in nigrostriatal dopaminergic system.* J Neurochem, 1998. **70**(4): p. 1584–92.
69. Le, W., et al., *Microglial activation And dopaminergic cell injury: An in vitro model relevant to Parkinson's disease.* J Neurosci, 2001. **21**(21): p. 8447–55.
70. Deng, I., et al., *Lipopolysaccharide animal models of Parkinson's disease: Recent progress and relevance to clinical disease.* Brain Behav Immun Health, 2020. **4**: p. 100060.
71. Thrash, B., et al., *Methamphetamine-induced neurotoxicity: The road to Parkinson's disease.* Pharmacol Rep, 2009. **61**(6): p. 966–77.
72. Singh, A.K., et al., *A current review of cypermethrin-induced neurotoxicity and nigrostriatal dopaminergic neurodegeneration.* Curr Neuropharmacol, 2012. **10**(1): p. 64–71.
73. Lee, I., et al., *Developmental neurotoxic effects of two pesticides: Behavior and neuroprotein studies on endosulfan and cypermethrin.* Toxicology, 2015. **335**: p. 1–10.
74. Singh, A., et al., *Ibuprofen abates cypermethrin-induced expression of pro-inflammatory mediators and mitogen-activated protein kinases and averts the nigrostriatal dopaminergic neurodegeneration.* Mol Neurobiol, 2016. **53**(10): p. 6849–6858.
75. Sayre, L.M., G. Perry, and M.A. Smith, *Redox metals and neurodegenerative disease* Curr Opin Chem Biol, 1999. **3**(2): p. 220–5.
76. Sayre, L.M., et al., *The role of metals in neurodegenerative diseases.* Cell Mol Biol (Noisy-Le-Grand), 2000. **46**(4): p. 731–41.
77. Montgomery, E.B. Jr., *Heavy metals and the etiology of Parkinson's disease and other movement disorders* Toxicology, 1995. **97**(1-3): p. 3–9.
78. Jomova, K., and M. Valko, *Advances in metal-induced oxidative stress and human disease* Toxicology, 2011. **283**(2-3): p. 65–87.
79. Jomova, K., et al., *Metals, oxidative stress and neurodegenerative disorders.* Mol Cell Biochem, 2010. **345**(1-2): p. 91–104.
80. Tripathi, M.K., et al., *Arsenic alters nitric oxide signaling similar to autism spectrum disorder and Alzheimer's disease-associated mutations.* Translational Psychiatry, 2022. **12**(1): p. 127.
81. Carvey, P.M., S. Pieri, and Z.D. Ling, *Attenuation of levodopa-induced toxicity in mesencephalic cultures by pramipexole* Journal of Neural Transmission, 1997. **104**(2): p. 209–228.
82. Kitamura, Y., et al., *Protective effects of the antiparkinsonian drugs talipexole and pramipexole against 1-methyl-4-phenylpyridinium-induced apoptotic death in human neuroblastoma SH-SY5Y cells.* Mol Pharmacol, 1998. **54**(6): p. 1046–54.
83. Ogawa, N., et al., *Bromocriptine protects mice against 6-hydroxydopamine and scavenges hydroxyl free radicals in vitro.* Brain Res, 1994. **657**(1-2): p. 207–13.
84. Barzilai, A., E. Melamed, and A. Shirvan, *Is there a rationale for neuroprotection against dopamine toxicity in Parkinson's disease?* Cell Mol Neurobiol, 2001. **21**(3): p. 215–35.
85. Tetrud, J.W., and J.W. Langston, *The effect of deprenyl (selegiline) on the natural history of Parkinson's disease.* Science, 1989. **245**(4917): p. 519–22.

86. Gershanik, O., et al., *Efficacy and safety of levodopa with entacapone in Parkinson's disease patients suboptimally controlled with levodopa alone, in daily clinical practice: An international, multicentre, open-label study*. Prog Neuropsychopharmacol Biol Psychiatry, 2003. **27**(6): p. 963–71.
87. Schrag, A., *Entacapone in the treatment of Parkinson's disease* Lancet Neurol, 2005. **4**(6): p. 366–70.
88. Singh, S., et al., *Nicotine and caffeine-mediated modulation in the expression of toxicant responsive genes and vesicular monoamine transporter-2 in 1-methyl 4-phenyl-1,2,3,6-tetrahydropyridine-induced Parkinson's disease phenotype in mouse*. Brain Res, 2008. **1207**: p. 193–206.
89. Srivastava, G., et al., *Resveratrol potentiates cytochrome P450 2 d22-mediated neuroprotection in maneb- and paraquat-induced parkinsonism in the mouse*. Free Radic Biol Med, 2012. **52**(8): p. 1294–306.

3 Recent Advancement in the Treatment of Parkinson's Disease Cellular Models

Mansi Gandhi
School of Innovation in Biodesign, Translational Health Science and Technology Institute, Faridabad, India
and
Institute of Chemistry, Hebrew University of Jerusalem, Israel

Devhuti Palan
School of Innovation in Biodesign, Translational Health Science and Technology Institute, Faridabad, India

3.1 INTRODUCTION TO CELLULAR MODELS FOR PARKINSON'S DISEASE (PD)

Parkinson's disease is among the foremost devastating and highly ubiquitous neuro disorders. It is the second most prevalent disease associated with bradykinesia, along with tremor, postural instability, and rigidity symptoms. This was initially established in 1817 by the surgeon James Parkinson and was characterised by the pathophysiologist Jean-Martin Charcot. This disorder has even led to a decrease in life expectancy, and there is no casual treatment protocol. Its prevalence is nearly 1% for people of the age between 65 and 69 years and 3% for people above the age of 80 years [1, 2]. Nearly 90,000 people are diagnosed with this disorder every year, and this is estimated to increase by 30% by 2030. This will lead to simultaneously explicit and implicit expenses to society and the financial market due to the increase in the aging population along with its incidence, as predicted to rise in 2030 by 1.2 million [3]. Parkinson's is the sole disease that has been contested with proof of identity of several clinical sub-types, pathogenic genes, and influential surrounding constraints. It is a multi-factorial disorder that has a pivotal role due to coinciding domains of genetic and environmental factors. These include cigarette smoking, caffeine, pesticides, herbicides and heavy metals, genetics, autosomal dominant PD, and autosomal recessive PD. Genetic research can help us in suppressing the PD disorder [4].

Dopamine, a hormone cum neurotransmitter plays a prominent part in the brain by controlling muscle activity. If the degree of dopamine is equivalent to acetylcholine, it leads to a damping effect in the basal ganglia, which will convey signals to the spinal cord, thus controlling muscle activity. But during Parkinson's disease, dopamine is not produced efficiently, thereby creating an imbalance. Furthermore, it leads to muscle tension and tremors in body. The major pathologic hallmark of PD includes recognising intracytoplasmic inclusion bodies as introduced by Frederic Lewy in 1912. He investigated dopamine inadequacy and participation in the parkinsonian animal model [5, 6]. The rate of exploration was accelerated in 1957 when Arvid Carlsson and Oleh Hornykiewicz reported the link between dopamine (present in the mammalian brain) deficiency and PD, which was later supported by proof of concept in 1961. Thus, demonstrating the clinical rescue with intravenous levodopa, followed by the high dosage levodopa (L-3,4-dihydroxyphenylalanine) therapy in 1967 [7].

DOI: 10.1201/9781032620442-3

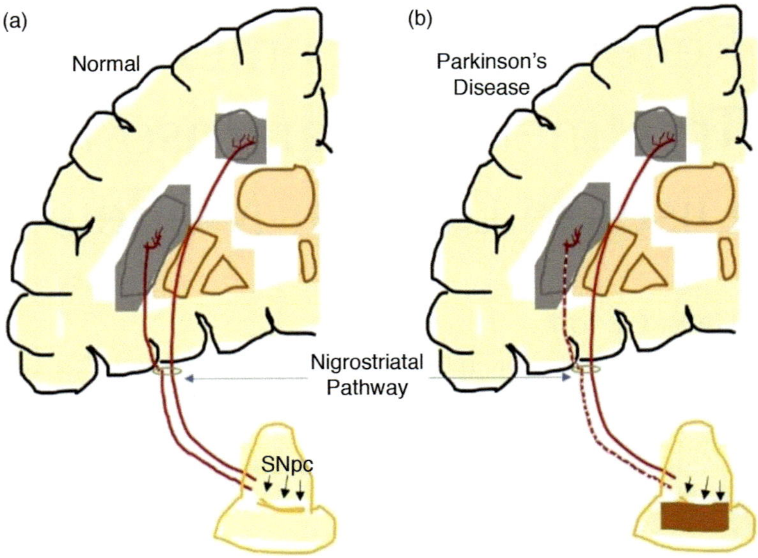

FIGURE 3.1 (a) Normal vs. (b) Parkinson's disease for the nigrostriatal pathway consisting of dopaminergic neurons, which degenerate in the case of the disease.

The line of research culminated with two discoveries in 1982 when William Langston revealed the trigger for medication-induced parkinsonism, 1-methyl-4-phenyl-1,2,3,6-tetrahydropyridine toxicity towards the substantia nigra dopaminergic neurons [8]. The difference in the dopaminergic neuron degeneration is portrayed in **Figure 3.1**, where a comparative view of normal vs. PD is used for the nigrostriatal pathway study. The substantia nigra pars compacta (SNpc) of PD patients causes a nerve-wracking experience, with ~50% to 90% nigra dopaminergic neurons already dead in the disease's early phase with negligible loss thereafter. However, a controversial debate exists about the underlying mechanistic pathway for neuronal death. Post-mortem analysis revealed that patients suffer from ferroptosis, apoptosis, and necrosis. Hence, significant research has been carried out over decades, and the drugs have been tested. Thus, drug trials are conducted extensively for animal models and occupational exposure to potential toxins. Although the discovery of the levodopa drug revolutionised PD therapy, before long, we discovered that after several years, patients tended to develop involuntary movements: dyskinesias (i.e. difficulty controlling movements and greatly diminished quality of life). **Figure 3.2** illustrates the various non-motor symptoms expressed as a mind map for a holistic view.

In extension, Polymeropoulos and co-workers discovered a genetic marker on chromosome 4q21-q23 linked with PD phenotype in Italian kinfolk and three Greek families with autosomal dominant PD, followed by reports of α-synuclein gene that have a genetic etiology [9]. This led to the chain of research for gene hunting and genetic risk loci, sparking the study of various genetic animal models and new therapeutic targets for clinical trials [10].

3.2 EMERGING AVENUES FOR NOVEL THERAPEUTICS FOR PD

Although neurodegenerative Parkinson's extends beyond the nigral dopaminergic neuron, it also encompasses cells in the remaining neural network areas. Thus, PD is a heterogeneous disorder, and no reliable diagnosis is available to date. Developing a new therapeutic strategy involves the prominent method of cellular models. The etiology and pathogenesis of PD were obtained from postpartum tissue (i.e. 1-methyl-4-phenyl-1,2,3,6-tetrahydropyridine (MPTP)-induced dopaminergic neurodegeneration). MPTP is a chemical molecule that stands for 1-methyl-4-phenyl-1,2,3,6-tetrahydropyridine,

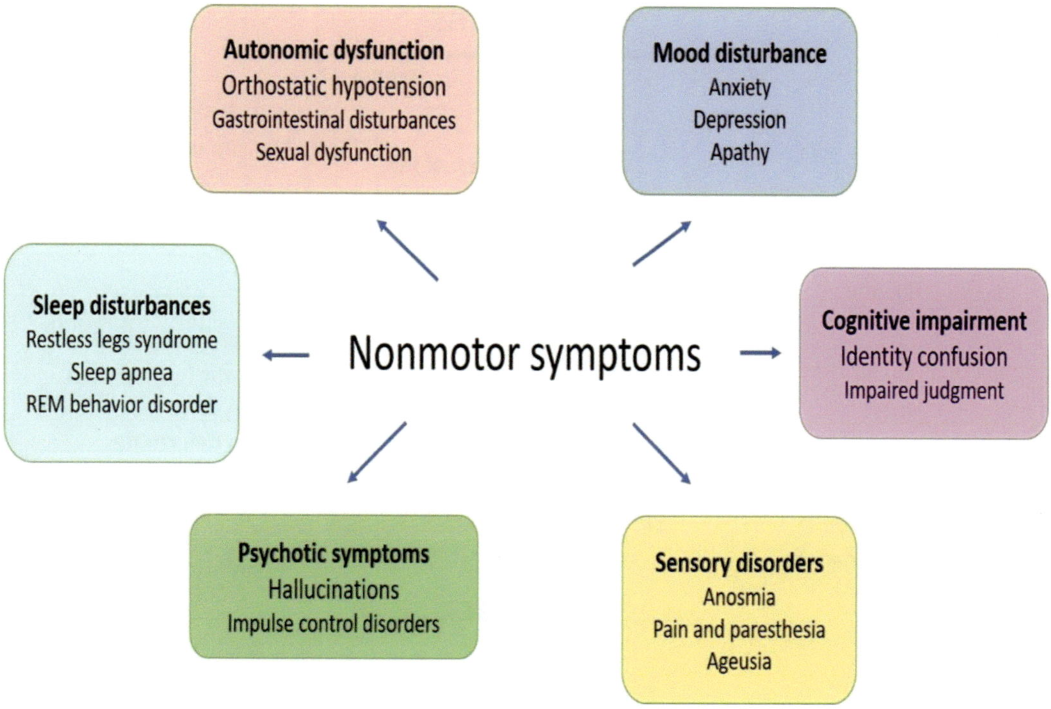

FIGURE 3.2 Timeline of the various non-motor symptoms expressed in the form of a mind map for a holistic view.

as portrayed in **Figure 3.3**. It belongs to the tetrahydropyridine class. It is of relevance as a precursor to the neurotoxin MPP+, which produces life-long Parkinson's disease symptoms by killing dopaminergic neurons in the brain's substantia nigra. It has been used to investigate illness models in a variety of animals.

While MPTP has no psychoactive properties, the chemical may be created inadvertently during the production of MPPP, a synthetic opioid medication having effects similar to morphine and pethidine (meperidine). The Parkinson-inducing effects of MPTP were found after an inadvertent injection of tainted MPPP. The distinct clinical symptoms expressed during the PD are represented in **Figure 3.4**.

Human exposure to MPTP triggers a syndrome resembling the core neurological manifestations and dopamine-related neurodegeneration seen in PD. MPTP toxicity in mice represents the extensively researched arena in animal models with cellular dysfunctions such as oxidative stress, mitochondrial respiration defects, and abnormal protein aggregation. The MPTP monkey model, the gold standard for the preclinical evaluation, produced invaluable functions regarding functional modifications in the basal ganglia circuit subsequent to the striatal dopamine depletion. The use of several animal models is discussed in detail in **Table 3.1**.

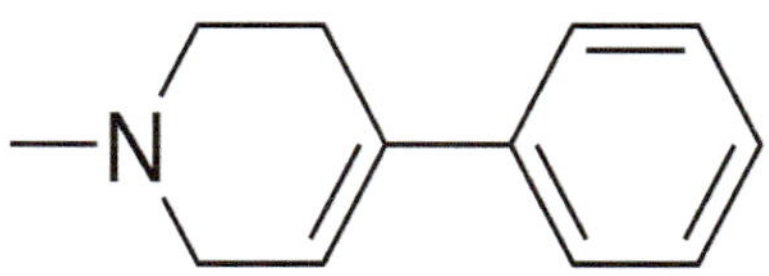

FIGURE 3.3 Structure of MPTP. (Image taken from Wikipedia.)

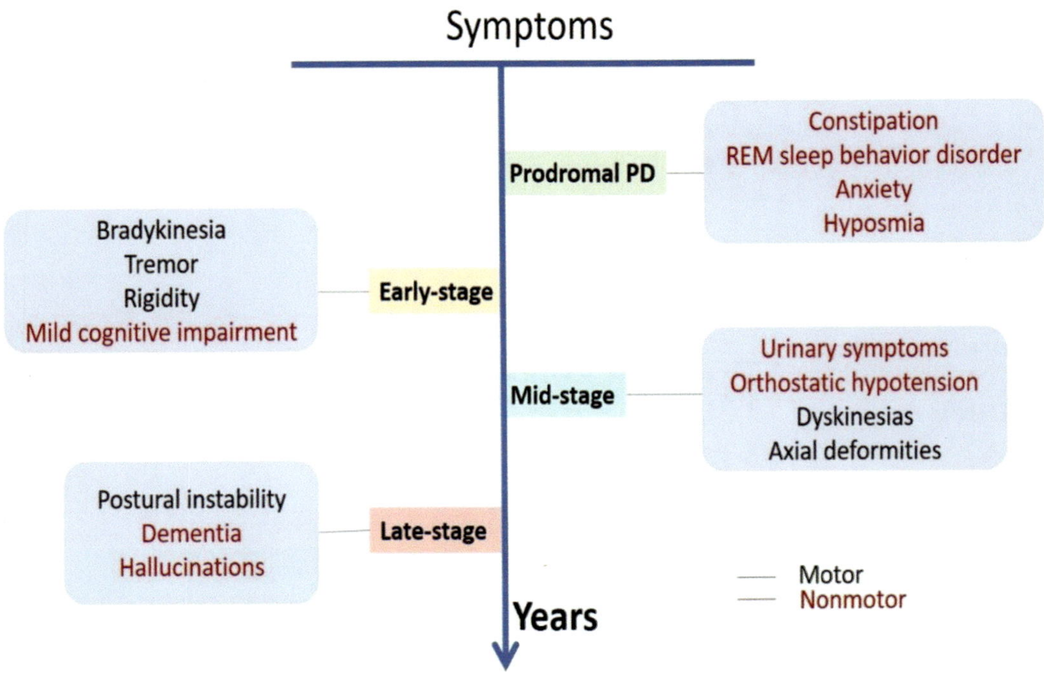

FIGURE 3.4 Timeline of the various clinical signs expressed during Parkinson's.

A handful of model organisms have been confirmed for the research of Parkinson's, among which *Drosophila melanogaster* has well-developed dopaminergic neurons that are the main factor during PD pathogenesis as PD-associated genes exhibit homology in the animals. Thus, the complete understanding of DA neurons from the larvae stages till the adult makes it the most suitable animal model for studying progressive degeneration in Parkinson's. The position of the cell body, dendrite, and number of DA neurons in each cluster distinctly recognise the DA neuron cluster. The brief lifespan makes it a powerful model for the study of epidemiological models and shows a huge prospective for PD genetic and drug screening.

Distinct types of cellular models have been extended for investigating the PD mechanism, drug screening for promoting novel therapeutics development. In addition to this, cell-based models impart strength due to the attributes of easy access along with high throughput screening and helps in displaying features of protein aggregates (α-synuclein) and parallel study of mitochondrial impairment, altered proteolysis, dysfunction mycophagy and apoptosis. Human neuroblast cell lines SH-SY5Y and rat pheochromocytoma cell line PC 12 are the most similar cell lines employed for modelling since they possess the machinery to produce and release catecholamines, which have the ability to develop neuron-like features.

The only clinical criteria were based on the clinical symptoms, while histopathological assessment with identification of the various cellular studies is required, such as α-synuclein containing Lewy bodies or Lewy neurites [11]. Thus, the focal point of therapy primarily relies on symptomatic relief using medications involving drugs directed towards restoration of dopamine hormone levels in the striatum or acting on post-synaptic dopamine receptors. But the involvement of dopamine is not the only neurotransmitter. Further, the investigation for therapies to replenish disease-modifying strategies for development as an effective therapy and improvements in current understanding of molecular and cellular mechanistic pathways involving the pathogenies and progression is a point of emphasis [12].

Based on the genetic factors, it is an idiopathic disorder with ~20% of the cases reporting a family history and about 5% Mendelian inheritance. The genes responsible for the same are PARK.

TABLE 3.1
The Use of Various Animal Models and Their Advantages along with Disadvantages

Comparative Analysis	Cellular Model	*C. elegans*	Drosophila	Teleost Fish	Rodents	Non-Human Primates
Advantages	Easy for genetic and pharmacological invention	Simple nervous system with 8 DA neurons	Complex nervous system with DA neuron clusters	Resemblance of brain structures and functions with mammals, including dopamine system	High conservation and similarity between rodents and humans	Extremely high conservation and similarity between NHPs and human
	Displaying important biological processes and mechanisms	Conservation of basic biological processes and PD-related genes	Conservation of basic biological processes and PD-related genes	-	Ability to mimic and recapitulate the complex pathways and features of PD	Manifesting many hallmarks of PD, such as motor and non-motor symptoms
	Allows high throughput screen	Can exhibit some key PD features and phenotypes	Can exhibit some key PD features and phenotypes	-	-	-
		Allows high throughput screen	Allows high throughput screen	-	-	-
Disadvantages	Cannot develop natural neuronal	Cannot recapitulate the complicated features of human dopamine system and complex PD phenotypes such as non-motor symptoms	Limitations to reproduce the complex features of PD	Relative new PD models need to perform the future evaluation	Restrictions from animal testing committee	Economic and ethical considerations
	Unable to reproduce the PD complexity	Lack of PD-related genes and complex protein families	Lack of some PD-related genes and complex protein families	-	Not suitable for high throughput screening	In-appropriate model for performing preliminary study on PD and high throughput screening

To date, nearly 23 PARK genes have been identified. Furthermore, their mutations signify either the case of autosomal dominant (e.g. SCNA, LRRK2, VPS32) or autosomal recessive inheritance (e.g. PRKN, PINK1, DJ-1). The details are furnished in **Table 3.2**. Most of the genes have not been conclusively confirmed, while many others are considered risk factors. Numerically the critical predisposing genetic risk factors are mutations in GBA1, a gene encoding β-glucocerebrosidase- a lysosomal enzyme responsible for hydrolysis of glucocerebrosidase. The GBA1 is known to cause Gaucher disease, a common lysosomal disorder disease [13–16].

TABLE 3.2
The Various PARK Genes and Their Inheritance Types and Clinical Features

PARK	Gene	OMIM Reference	Inheritance	Description	Clinical Features	Ref
PARK1 PARK4	SNCA	168601	AD	α-synuclein	Conventional PD in patients with early onset symptoms	[10, 17, 18]
PARK8	LRRK2	607060	AD	Leucine-rich repeat kinase 2	Slower progression and less frequent dementia in classical PD	[19]
PARK11	GIGYF2	607688	AD	GRB10 interacting GYF protein 2	Classical PD	[20]
PARK5	UCHL1	613643	AD	Ubiquitin C-terminal hydrolase L1	Classical PD – only one family, no subsequent replications of the findings	[21]
PARK17	VPS35	614203	AD	Vacuolar protein sorting-associated protein 35	Classical PD	[22]
PARK18	EIF4G1	614251	AD	Eukaryotic translation initiation factor 4 gamma 1	Classical PD	[23]
PARK21	DNAJC13	616361	AD	Receptor-mediated endocytosis 8 (RME-8)	Classical PD	[24]
PARK15	FBX07	260300	AR	F-box protein 7	Pallido-pyramidal syndrome – early onset and peculiar symptoms	[25]
PARK2	PRKN	600116	AR	Parkin RBR E3 ubiquitin protein ligase	Slow progression, early onset, and frequent signs of dystonia	[26]
PARK6	PINK1	605909	AR	PTEN-induced putative kinase 1	Early-onset PD, slow progression	[27]
PARK7	DJ-1	606324	AR	Parkinsonism-associated deglycase	Early-onset PD, slow progression	[28]
PARK9	ATP13A2	606693	AR	Cation-transporting ATPase 13A2	Atypical parkinsonism with dementia, supranuclear palsy, and early onset (during adolescence) – Kufor-Rakeb syndrome	[1] [29]
PARK13	HTRA2	610297	AR	HtrA serine peptidase 2	Classical PD	[30]
PARK14	PLA2G6	612593	AR	Calcium-independent phospholipase A2 enzyme	Atypical characteristics and early onset of dystonia-parkinsonism	[31]
PARK19	DNAJC6	615528	AR	HSP40 Auxilin	Early-onset PD, slow progression	[32]
PARK20	SYNJ1	615530	AR	Synaptojanin 1	Dystonic Parkinsonism and Cognitive Decline	[33]
PARK23	VPS13C	616840	AR	Vacuolar protein sorting-associated protein 13C	PD with early start and quick progression	[34]

Abbreviations: AD, autosomal dominant; AR, autosomal recessive; OMIM, Online Mendelian Inheritance in Human database.

Note: PARK3 PARK10, PARK12, PARK16, and PARK22 are not shown in this table because they are thought to be risk factors or because the genes have not yet been discovered.

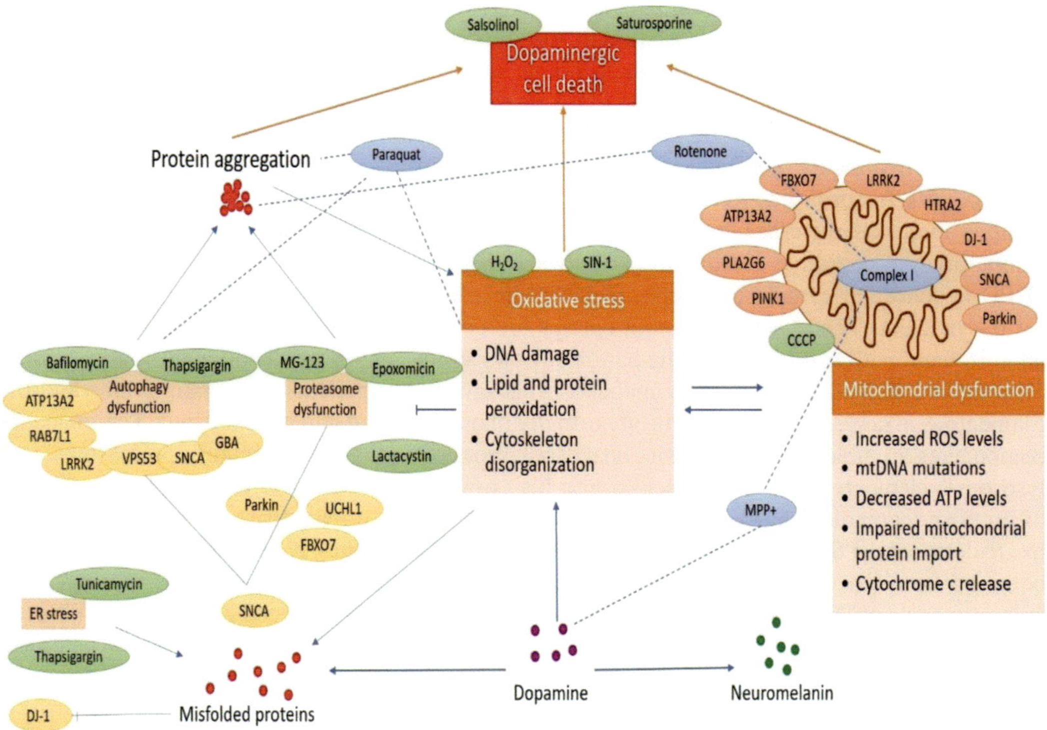

FIGURE 3.5 Pictorial illustration of the molecular mechanisms in Parkinson's disease, which lead to neuronal cell death. Genes associated with PD and the main pathways in the disease have been portrayed. (Copyright, molecular neurodegeneration, 2017 [35].)

Various molecular mechanisms related to PD, namely, abnormal mitochondrial function, oxidative stress, and autophagy, along with protease dysfunction, have been portrayed in **Figure 3.5**. These are usually triggered due to administration of certain drugs as shown in **Figure 3.5**. These drugs knock down the expression of genes corresponding to PD. For example, PINK1 leads to mitochondrial dysfunction genes while the muted forms of α-synuclein impair the protease activity.

Several mechanistic pathways have been applied and tested for pathogenesis, accounting for the involvement of α-synuclein misfolding and aggregation central to the disease phenomenon. Protein misfolding in a particular brain area is a blanket umbrella phrase and is termed 'proteinopathy.' One of the prominent reasons for this might be aging. There is little evidence for a crosstalk between aggregated proteins and PD [36]. During PD, it forms a β-sheet-richamyloid–like structure that is susceptible to aggregation. Indeed, its misfolding turns into a distinct species of α-synuclein.

Another protein of this domain is tau, encoded by the MAPT gene. The 'tau' or the tubulin-associated unit belongs to a lineage of six highly soluble protein isoforms that are derived from the microtubule-associated protein tau gene. They serve a chief role in the integrity of microtubules in axons. They are numerous in neurons of the central nervous system (CNS), with the cerebral cortex having the peak concentration. They are not so common in other tissues. However, they are exhibited in extremely limited quantities in CNS astrocytes and oligodendrocytes.

Pathologies and dementias of the neurological system, such as Alzheimer's and Parkinson's diseases, are linked to tau proteins, which have turned into hyperphosphorylated insoluble masses known as neurofibrillary tangles. Tau proteins were discovered in 1975 as heat-stable proteins required for microtubule assembly and have been studied ever since. In a few events, tau can be abnormally hyperphosphorylated, forming intracytoplasmic inclusions, which are linked to the development of PD. In postpartum patients, a remarkable increment in tau hyper-phosphorylation at Ser262 and Ser396/404 in the striatum of sufferers of PD and its dementia has been revealed in reports [37, 38]. It is noteworthy that the genomic study has led to substantial evidence for a direct relation between MAPT and the risk of PD. Subsequently, longitudinal studies direct that the H1/H1 haplotype of MAPT is a solid predictor of this disease [37].

The new treatments shed light on the PD-related pathways, as discussed by Vingill and his co-workers [39]. According to Porras et al. MPTP causes the well-known NHP model of PD, which exhibits many characteristics of the disease, including DA cell loss and motor and non-motor symptoms like cognitive decline and sleep/wake disorders [40]. An essential distinctive feature of PD is abnormal cytoplasmic deposits within neuronal cell bodies, which are immunoreactive for α-synuclein. The abnormal protein clumps are commonly known as Lewy bodies, which go hand in hand with axonal dystrophic neurites. Lewy bodies are cytoplasmic inclusions consisting of granular, fibrillar core, and filamentousα-synuclein. In PD patients, they gain an amyloid-like filamentous arrangement and become abnormally phosphorylated and accumulated. The molecular components are ubiquitin, tau, parkin, heat shock proteins, cytoskeletal proteins, proteasomal, and lysosomal elements [41, 42]. Franscisco and his co-workers reported effective cellular therapies to improve locomotor ability in DJ-1β mutant flies and reduced oxidative stress-induced lethality in DJ-deficient 1β SH-SY 5Y human cells. One of the sources they have discussed is an alkaloid obtained from leaves of Vinca minor (i.e. vincamine). This compound was able to suppress in animal models of Drosophila and even in human cell models for PD-related phenotypes. This compound reduced the oxidative stress levels in DJ-1-deficient cells and furthermore helped in diminishing the lethality due to a decrease in apoptosis and an increase in mitochondrial viability as well as inhibition of voltage-gated sodium channels [43].

The main staging for PD was introduced by Braak and colleagues in 2003. The study is based on semi-quantitative assessments that have shown that a fraction of PD does not seem to align with the pattern, while efforts to establish a correlation between Braak staging and clinical dysfunction were also unsuccessful. The add-on deprecation of the Braak system is that it is based not on neuronal loss but on the distribution of Lewy-related pathology.

3.3 LIMITATIONS OF CELLULAR MODELS TOWARDS DRUG DISCOVERY

Currently, there are no available blood or laboratory tests for diagnosing non-genetic instances of Parkinson's disease. Typically, physicians rely on a patient's medical history and a comprehensive neurological examination to make a diagnosis. Additionally, an improvement in symptoms following the initiation of medication can serve as an additional sign of Parkinson's.

Numerous conditions can produce symptoms akin to those of Parkinson's disease. Individuals displaying Parkinson's-like symptoms originating from other sources, such as multiple system atrophy and dementia with Lewy bodies, are occasionally classified as having parkinsonism. Though these disorders can initially be misdiagnosed as Parkinson's, specific medical tests and the response to drug therapy can aid in a more accurate determination of the underlying cause. Many other illnesses share similar characteristics but necessitate distinct treatments, underscoring the significance of securing an accurate diagnosis promptly. Despite the absence of a cure for Parkinson's disease, medications, surgical interventions, and various therapies frequently offer relief from certain symptoms.

The known gold standard for treatment of Parkinson's disease is levodopa in combination with a peripheral dopa-decarboxylase inhibitor. However, it is quite a prolonged treatment that has led to the numerous side-effects namely, dyskinesis and motor fluctuations. Kinase are quite easy druggable target systems and nearly 70 FDA-approved small inhibitor drugs called the 'kinase targeting molecules' are reported. Dopamine agonists like ergot derivatives such as bromocriptine, lisuride, pergolide, and cabergoline are few examples, although, the non-ergot compounds such as rotigotine, pramipexole, and ropirinole are frequently used. While the monoamine-oxidase-B

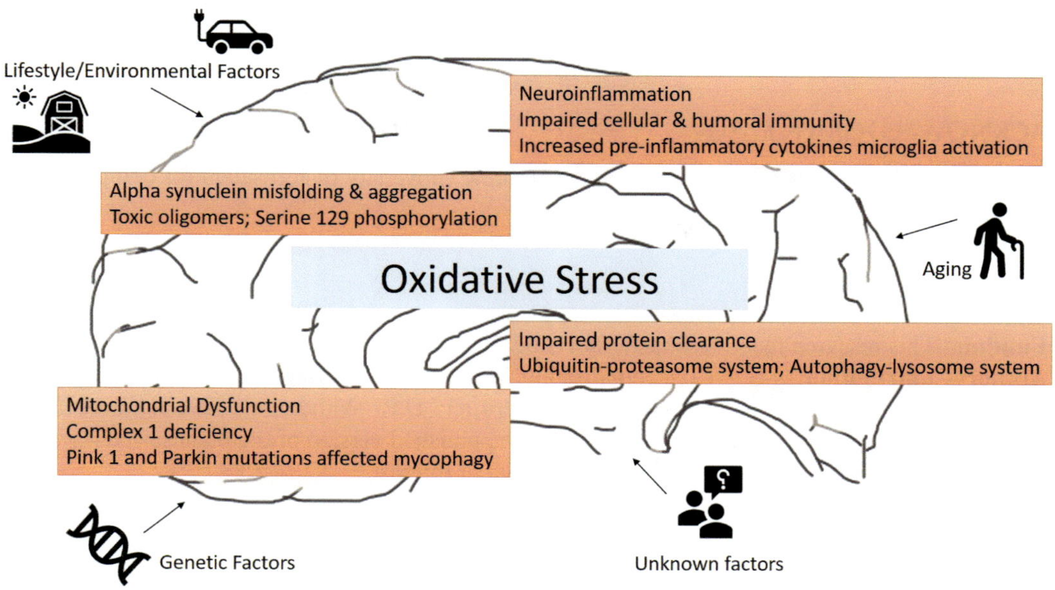

FIGURE 3.6 Pictorial illustration of PD pathogenesis. The cellular mechanisms responsible for oxidative stress with amalgamation of lifestyle/environmental factors, genetic heredities, aging, and a few unknown attributes leading to neurodegeneration. (Copyright, *Journal of Neurology, Neurosurgery & Psychiatry*, 2020 [12].)

inhibitor resagaline is quite effective on effective of early stages. Other therapeutic compounds include catechol-O-methyl transferase inhibitor entacapone and NMDA receptor antagonist amantadine. NDMA blockage alleviates the symptoms of PD. The use of subsequent drugs can contribute to drug toxicity, and their efficacy is quite critical in the process of drug discovery for neurodegeneration. **Figure 3.6** illustrates the PD pathogenesis due to various factors wherein the cellular models are quite important from the point of futuristic approach.

However, the major drawback of using cellular models is that the culture cells do not develop the natural neuronal network and lack the interaction of different cell types and cellular microenvironments. Hence, they are unable to reproduce the complexity of PD disease [44].

3.4 CONCLUSION AND FUTURE PERSPECTIVE

PD is the second most prevalent neurodegenerative disorder for which the major constraints of etiologic, pathological, and molecular mechanisms still remain to be unveiled with time and research. This disease is more prevalent in males than females. A minute portion of PD patients possess a single-gene origin for their illness or abnormality. Environmental influences contribute to PD risk, such as lifestyle habits, pesticides, mental stress, climate change, and other attributes, which are poorly studied. Medications are potentially amenable and therapeutically relevant classes for targeted pharmacological areas. Still, there is quite a need for new targets required for its treatment as the available systems only help in symptomatic treatments but are unresponsive to the underlying neurodegenerative process. Despite this, the scientific community has shown the ability to transform the kinase inhibitor into repurposed medications. Many kinases are overexpressed in the human genome and often result as a cardinal biomarker of PD. The main obstacle in their therapy development is due to an incomplete grasp of the crucial molecular occurrences that lead to degeneration. In addition, the voltage-gated sodium channels can be a potential target for the search for new compounds to treat PD, as experimentally verified in one of the research articles discussed in this chapter. Therefore, understanding these crucial cellular processes can allow recognition of the new therapeutic objective and, thus, make headway in disease-modifying protocols in the near future.

ACKNOWLEDGMENT

Mansi Gandhi would like to acknowledge the Hebrew University of Jerusalem for completing her postdoc research.

DECLARATIONS

Funding: This research received no external funding.

Authors Contribution: M.G.—Conceptualisation, Methodology, Writing, Project Administration, Validation, Investigation, Supervision and Writing; D.P.—Validation, Investigation, and Writing. All authors have read and agreed to the published version of the manuscript.

Conflict of Interest: The authors declare no conflict of interest.

REFERENCES

1. A. Ramirez, A. Heimbach, J. Gründemann, B. Stiller, D. Hampshire, L.P. Cid, I. Goebel, A.F. Mubaidin, A.L. Wriekat, J. Roeper, A. Al-Din, A.M. Hillmer, M. Karsak, B. Liss, C.G. Woods, M.I. Behrens, C. Kubisch, Hereditary parkinsonism with dementia is caused by mutations in ATP13A2, encoding a lysosomal type 5 p-type ATPase, Nat Genet. 38 (2006) 1184–1191. https://doi.org/10.1038/ng1884.
2. D.P. Perl, C.W. Olanow, D. Calne, Alzheimer's disease and Parkinson's disease: Distinct entities or extremes of a spectrum of neurodegeneration? in: Ann Neurol, Lippincott Williams and Wilkins, 1998. https://doi.org/10.1002/ana.410440705.
3. Parkinson's Foundation, 2022. https://www.parkinson.org/about-us/news/incidence-2022 (accessed October 6, 2023).
4. Parkinson's disease_ etiology, neuropathology, and pathogenesis - Parkinson's Disease - NCBI Bookshelf, (n.d.).
5. S.N. Gomperts, Lewy body dementias: Dementia with lewy bodies and Parkinson disease dementia, 2016. www.ContinuumJournal.com.
6. E. Engelhardt, M. da M. Gomes, Lewy and his inclusion bodies: Discovery and rejection, Dementia e Neuropsychologia. 11 (2017) 198–201. https://doi.org/10.1590/1980-57642016dn11-020012.
7. S. Fahn, The 200-year journey of Parkinson disease: Reflecting on the past and looking towards the future, Parkinsonism RelatDisord. 46 (2018) S1–S5. https://doi.org/10.1016/j.parkreldis.2017.07.020.
8. J.W. Langston, The MPTP story, J Parkinsons Dis. 7 (2017) S11–S19. https://doi.org/10.3233/JPD-179006.
9. A.P. Kiely, Y.T. Asi, E. Kara, P. Limousin, H. Ling, P. Lewis, C. Proukakis, N. Quinn, A.J. Lees, J. Hardy, T. Revesz, H. Houlden, J.L. Holton, A-synucleinopathy associated with G51D SNCA mutation: A link between Parkinson's disease and multiple system atrophy? ActaNeuropathol. 125 (2013) 753–769. https://doi.org/10.1007/s00401-013-1096-7.
10. M.H. Polymeropoulos, C. Lavedan, E. Leroy, S.E. Ide, A. Dehejia, A. Dutra, B. Pike, H. Root, J. Rubenstein, R. Boyer, E.S. Stenroos, S. Chandrasekharappa, A. Athanassiadou, T. Papapetropoulos, W.G. Johnson, A.M. Lazzarini, R.C. Duvoisin, G. Di Iorio, L.I. Golbe, R.L. Nussbaum, Mutation in the α-synuclein gene identified in families with Parkinson's disease, Science(1979). 276 (1997) 2045–2047. https://doi.org/10.1126/science.276.5321.2045.
11. C. Váradi, Clinical features of Parkinson's disease: The evolution of critical symptoms, Biology (Basel. 9 (2020). https://doi.org/10.3390/biology9050103.
12. J. Jankovic, E.K. Tan, Parkinson's disease: Etiopathogenesis and treatment, J NeurolNeurosurg Psychiatry. 91 (2020) 795–808. https://doi.org/10.1136/jnnp-2019-322338.
13. H. Deng, P. Wang, J. Jankovic, The genetics of Parkinson disease, Ageing Res Rev. 42 (2018) 72–85. https://doi.org/10.1016/j.arr.2017.12.007.
14. C. Schulte, T. Gasser, Genetic basis of Parkinson's disease: Inheritance, penetrance, and expression, Application of Clinical Genetics. 4 (2011) 67–80. https://doi.org/10.2147/TACG.S11639.
15. W.C. Nichols, N. Pankratz, D.K. Marek, B.M.W. Pauciulo, V.E. Elsaesser, C.A. Halter, A. Rudolph, J. Wojcieszek, R.F. Pfeiffer, T. Foroud, Mutations in GBA are associated with familial Parkinson disease susceptibility and age at onset, 2009. www.neurology.org.

16. E. Sidransky, G. Lopez, The link between the GBA gene and parkinsonism, 2012. www.thelancet.com/neurology.
17. A. Puschmann, O.A. Ross, C. Vilariño-Güell, S.J. Lincoln, J.M. Kachergus, S.A. Cobb, S.G. Lindquist, J.E. Nielsen, Z.K. Wszolek, M. Farrer, H. Widner, D. van Westen, D. Hägerström, K. Markopoulou, B.A. Chase, K. Nilsson, J. Reimer, C Nilsson, A Swedish family with de novo α-synuclein A53T mutation: Evidence for early cortical dysfunction, Parkinsonism RelatDisord. 15 (2009) 627–632. https://doi.org/10.1016/j.parkreldis.2009.06.007.
18. A.B. Singleton, M. Farrer, J. Johnson, A Singleton, S. Hague, J. Kachergus, M. Hulihan, T. Peuralinna, A. Dutra, R. Nussbaum, S. Lincoln, A. Crawley, M. Hanson, D. Maraganore, C. Adler, M.R. Cookson, M. Muenter, M. Baptista, D. Miller, J. Blancato, J. Hardy, K. Gwinn-Hardy, α-Synuclein locus triplication causes Parkinson's disease, Science(1979). 302 (2003) 841. https://doi.org/10.1126/science.1090278.
19. A. Zimprich, S. Biskup, P. Leitner, P. Lichtner, M. Farrer, S. Lincoln, J. Kachergus, M. Hulihan, R.J. Uitti, D.B. Calne, A.J. Stoessl, Mutations in LRRK2 cause autosomal-dominant parkinsonism with pleomorphic pathology families, we have found six disease-segregating mutations (five missense and one putative splice site mutation) in a gene encoding a large, multifunctional Pro, 44 (2004) 601–607. http://www.neuron.org/cgi/content/full/44/4/.
20. C. Lautier, S. Goldwurm, A. Dürr, B. Giovannone, W.G. Tsiaras, G. Pezzoli, A. Brice, R.J. Smith, Mutations in the GIGYF2 (TNRC15) Gene at the PARK11 locus in familial Parkinson disease, Am J Hum Genet. 82 (2008) 822–833. https://doi.org/10.1016/j.ajhg.2008.01.015.
21. K. Nishikawa, H. Li, R. Kawamura, H. Osaka, Y.L. Wang, Y. Hara, T. Hirokawa, Y. Manago, T. Amano, M. Noda, S. Aoki, K. Wada, Alterations of structure and hydrolase activity of parkinsonism-associated human ubiquitin carboxyl-terminal hydrolase L1 variants, BiochemBiophys Res Commun. 304 (2003) 176–183. https://doi.org/10.1016/S0006-291X(03)00555-2.
22. C. Wider, L. Skipper, A. Solida, L. Brown, M. Farrer, D. Dickson, Z.K. Wszolek, F.J.G. Vingerhoets, Autosomal dominant dopa-responsive parkinsonism in a multigenerational Swiss family, Parkinsonism RelatDisord. 14 (2008) 465–470. https://doi.org/10.1016/j.parkreldis.2007.11.013.
23. M.C. Chartier-Harlin, J.C. Dachsel, C. Vilariño-Güell, S.J. Lincoln, F. Leprêtre, M.M. Hulihan, J. Kachergus, A.J. Milnerwood, L. Tapia, M.S. Song, E. Le Rhun, E. Mutez, L. Larvor, A. Duflot, C. Vanbesien-Mailliot, A. Kreisler, O.A. Ross, K. Nishioka, A.I. Soto-Ortolaza, S.A. Cobb, H.L. Melrose, B. Behrouz, B.H. Keeling, J.A. Bacon, E. Hentati, L. Williams, A. Yanagiya, N. Sonenberg, P.J. Lockhart, A.C. Zubair, R.J. Uitti, J.O. Aasly, A. Krygowska-Wajs, G. Opala, Z.K. Wszolek, R. Frigerio, D.M. Maraganore, D. Gosal, T. Lynch, M. Hutchinson, A.R. Bentivoglio, E.M. Valente, W.C. Nichols, N. Pankratz, T. Foroud, R.A. Gibson, F. Hentati, D.W. Dickson, A. Destée, M.J. Farrer, Translation initiator EIF4G1 mutations in familial Parkinson disease, Am J Hum Genet. 89 (2011) 398–406. https://doi.org/10.1016/j.ajhg.2011.08.009.
24. C. Vilariño-Güell, A. Rajput, A.J. Milnerwood, B. Shah, C. Szu-Tu, J. Trinh, I. Yu, M. Encarnacion, L.N. Munsie, L. Tapia, E.K. Gustavsson, P. Chou, I. Tatarnikov, D.M. Evans, F.T. Pishotta, M. Volta, D. Beccano-Kelly, C. Thompson, M.K. Lin, H.E. Sherman, H.J. Han, B.L. Guenther, W.W. Wasserman, V. Bernard, C.J. Ross, S. Appel-Cresswell, A.J. Stoessl, C.A. Robinson, D.W. Dickson, O.A. Ross, Z.K. Wszolek, J.O. Aasly, R.M. Wu, F. Hentati, R.A. Gibson, P.S. McPherson, M. Girard, M. Rajput, A.H. Rajput, M.J. Farrer, DNAJC13 mutations in Parkinson disease, Hum Mol Genet. 23 (2014) 1794–1801. https://doi.org/10.1093/hmg/ddt570.
25. A. Di Fonzo, M.C.J. Dekker, P. Montagna, A. Baruzzi, E.H. Yonova, B.L. CorreiaGuedes, A. Szczerbinska, B.T. Zhao, L.O.M. Dubbel-Hulsman, C.H. Wouters, E. De Graaff, W.J.G. Oyen, E.J. Simons, G.J. Breedveld, B.A. Oostra, M.W. Horstink, V. Bonifati, FBXO7 mutations cause autosomal recessive, early-onset parkinsonian-pyramidal syndrome, 2009. www.neurology.org.
26. T. Kitada, S. Asakawa, N. Hattori, H. Matsumine, Y. Yamamura, S. Minoshima, M. Yokochi, Y. Mizuno, N. Shimizu, Mutations in the parkin gene cause autosomal recessive juvenile parkinsonism, 392 (1998) 605–608.
27. E.M. Valente, S. Salvi, T. Ialongo, R. Marongiu, A.E. Elia, V. Caputo, L. Romito, A. Albanese, B. Dallapiccola, A.R. Bentivoglio, PINK1 mutations are associated with sporadic early-onset parkinsonism, Ann Neurol. 56 (2004) 336–341. https://doi.org/10.1002/ana.20256.
28. V. Bonifati, P. Rizzu, M.J. van Baren, O. Schaap, G.J. Breedveld, E. Krieger, M.C. J Dekker, F. Squitieri, P. Ibanez, M. Joosse, J.W. van Dongen, N. Vanacore, J.C. van Swieten, A. Brice, G. Meco, C.M. van Duijn, B.A. Oostra, P. Heutink, Mutations in the DJ-1 Gene Associated with Autosomal Recessive Early-Onset Parkinsonism, 1995. https://www.science.org.
29. D.R. Williams, A. Hadeed, A.S. Najim al-Din, A.L. Wreikat, A.J. Lees, KuforRakeb disease: Autosomal recessive, levodopa-responsive parkinsonism with pyramidal degeneration, supranuclear gaze palsy, and dementia, Movement Disorders. 20 (2005) 1264–1271. https://doi.org/10.1002/mds.20511.

30. K.M. Strauss, L.M. Martins, H. Plun-Favreau, F.P. Marx, S. Kautzmann, D. Berg, T. Gasser, Z. Wszolek, T. Müller, A. Bornemann, H. Wolburg, J. Downward, O. Riess, J.B. Schulz, R. Krüger, Loss of function mutations in the gene encoding Omi/HtrA2 in Parkinson's disease, Hum Mol Genet. 14 (2005) 2099–2111. https://doi.org/10.1093/hmg/ddi215.
31. C. Paisan-Ruiz, K.P. Bhatia, A. Li, D. Hernandez, M. Davis, N.W. Wood, J. Hardy, H. Houlden, A. Singleton, S.A. Schneider, Characterization of PLA2G6 as a locus for dystonia-parkinsonism, Ann Neurol. 65 (2009) 19–23. https://doi.org/10.1002/ana.21415.
32. S. Olgiati, M. Quadri, M. Fang, J.P.M.A. Rood, J.A. Saute, H.F. Chien, C.G. Bouwkamp, J. Graafland, M. Minneboo, G.J. Breedveld, J. Zhang, F.W. Verheijen, A.J.W. Boon, A.J.A. Kievit, L.B. Jardim, W. Mandemakers, E.R. Barbosa, C.R.M. Rieder, K.L. Leenders, J. Wang, V. Bonifati, DNAJC6 mutations associated with early-onset Parkinson's disease, Ann Neurol. 79 (2016) 244–256. https://doi.org/10.1002/ana.24553.
33. M. Quadri, M. Fang, M. Picillo, S. Olgiati, G.J. Breedveld, J. Graafland, B. Wu, F. Xu, R. Erro, M. Amboni, S. Pappatà, M. Quarantelli, G. Annesi, A. Quattrone, H.F. Chien, E.R. Barbosa, B.A. Oostra, P. Barone, J. Wang, V. Bonifati, Mutation in the SYNJ1 gene associated with autosomal recessive, early-onset parkinsonism, Hum Mutat. 34 (2013) 1208–1215. https://doi.org/10.1002/humu.22373.
34. S. Lesage, V. Drouet, E. Majounie, V. Deramecourt, M. Jacoupy, A. Nicolas, F. Cormier-Dequaire, S.M. Hassoun, C. Pujol, S. Ciura, Z. Erpapazoglou, T. Usenko, C.A. Maurage, M. Sahbatou, S. Liebau, J. Ding, B. Bilgic, M. Emre, N. Erginel-Unaltuna, G. Guven, F. Tison, C. Tranchant, M. Vidailhet, J.C. Corvol, P Krack, A.L. Leutenegger, M.A. Nalls, D.G. Hernandez, P. Heutink, J.R. Gibbs, J. Hardy, N.W. Wood, T Gasser, A. Durr, J.F. Deleuze, M. Tazir, A. Destée, E. Lohmann, E. Kabashi, A. Singleton, O. Corti, A. Brice, F. Tison, M. Vidailhet, J.C. Corvol, Y. Agid, M. Anheim, A.M. Bonnet, M. Borg, E. Broussolle, P. Damier, A Destée, A. Dürr, F. Durif, P. Krack, S. Klebe, E. Lohmann, M. Martinez, P. Pollak, O. Rascol, C. Tranchant, M. Vérin, F. Viallet, S. Lesage, E. Majounie, F. Tison, M. Vidailhet, J.C. Corvol, A. Nalls, D.G. Hernandez, J.R Gibbs, A. Dürr, S. Arepalli, R.A. Barker, Y. Ben-Shlomo, D. Berg, F. Bettella, K. Bhatia, R.M.A. de Bie, A. Biffi, B.R. Bloem, Z. Bochdanovits, M. Bonin, J.M. Bras, K. Brockmann, J. Brooks, D.J. Burn, G. Charlesworth, H. Chen, P.F. Chinnery, S. Chong, C.E. Clarke, M.R. Cookson, C. Counsell, P. Damier, J.F. Dartigues, P. Deloukas, G. Deuschl, D.T. Dexter, K.D. van Dijk, A. Dillman, J. Dong, F. Durif, S. Edkins, V. Escott-Price, J.R. Evans, T. Foltynie, J. Gao, M. Gardner, A. Goate, E. Gray, R. Guerreiro, C. Harris, J.J. van Hilten, A. Hofman, A. Hollenbeck, P. Holmans, J. Holton, M. Hu, X. Huang, H. Huber, G. Hudson, S.E. Hunt, J. Huttenlocher, T. Illig, P. V. Jónsson, L.L. Kilarski, I.E. Jansen, J.C. Lambert, C. Langford, A. Lees, P. Lichtner, P. Limousin, G. Lopez, D. Lorenz, S. Lubbe, C. Lungu, M. Martinez, W. Mätzler, A. McNeill, C. Moorby, M. Moore, K.E. Morrison, E. Mudanohwo, S.S. O'sullivan, M.J. Owen, J. Pearson, J.S. Perlmutter, H. Pétursson, V. Plagnol, P. Pollak, B. Post, S. Potter, B. Ravina, T. Revesz, O. Riess, F. Rivadeneira, P. Rizzu, M. Ryten, M. Saad, J. Simón-Sánchez, S. Sawcer, A. Schapira, H. Scheffer, C. Schulte, M. Sharma, K. Shaw, U.M. Sheerin, I. Shoulson, J. Shulman, E. Sidransky, A. Spencer, H. Stefánsson, K. Stefánsson, J.D. Stockton, A. Strange, K. Talbot, C.M. Tanner, A. Tashakkori-Ghanbaria, D. Trabzuni, B.J. Traynor, A.G. Uitterlinden, D. Velseboer, R. Walker, B. van de Warrenburg, M. Wickremaratchi, C.H. Williams-Gray, S. Winder-Rhodes, I. Wurster, N. Williams, H.R. Morris, P. Heutink, N.W. Wood, T Gasser, A.B Singleton, A Brice, Loss of VPS13C function in autosomal-recessive parkinsonism causes mitochondrial dysfunction and increases PINK1/Parkin-dependent mitophagy, Am J Hum Genet. 98 (2016) 500–513. https://doi.org/10.1016/j.ajhg.2016.01.014.
35. H. Xicoy, B. Wieringa, G.J.M. Martens, The SH-SY5Y cell line in Parkinson's disease research: A systematic review, MolNeurodegener. 12 (2017) 1–11. https://doi.org/10.1186/s13024-017-0149-0.
36. A. Elobeid, S. Libard, M. Leino, S.N. Popova, I. Alafuzoff, Altered proteins in the aging brain, J NeuropatholExp Neurol. 75 (2016) 316–325. https://doi.org/10.1093/jnen/nlw002.
37. A. Goris, C.H. Williams-Gray, G.R. Clark, T. Foltynie, S.J.G. Lewis, J. Brown, M. Ban, M.G. Spillantini, A. Compston, D.J. Burn, P.F. Chinnery, R.A. Barker, S.J. Sawcer, Tau and α-synuclein in susceptibility to, and dementia in, Parkinson's Disease, Ann Neurol. 62 (2007) 145–153. https://doi.org/10.1002/ana.21192.
38. F.J. Hart de Ruyter, T.H.J. Morrema, J. den Haan, G. Gase, J.W.R. Twisk, J.F. de Boer, P. Scheltens, F.H. Bouwman, F.D. Verbraak, A.J.M. Rozemuller, J.J.M. Hoozemans, α-synuclein pathology in postmortem retina and optic nerve is specific for α-synucleinopathies, NPJ Parkinsons Dis. 9 (2023). https://doi.org/10.1038/s41531-023-00570-5.
39. S. Vingill, D. Brockelt, C. Lancelin, L. Tatenhorst, G. Dontcheva, C. Preisinger, N. Schwedhelm-Domeyer, S. Joseph, M. Mitkovski, S. Goebbels, K. Nave, J.B. Schulz, T. Marquardt, P. Lingor, J. Stegmüller, Loss of FBXO 7 (PARK 15) results in reduced proteasome activity and models a parkinsonism-like phenotype in mice, EMBO J. 35 (2016) 2008–2025. https://doi.org/10.15252/embj.201593585.

40. G. Porras, Q. Li, E. Bezard, Modeling Parkinson's disease in primates: The MPTP model, Cold Spring HarbPerspect Med. 2 (2012). https://doi.org/10.1101/cshperspect.a009308.
41. M. Gómez-Benito, N. Granado, P. García-Sanz, A. Michel, M. Dumoulin, R. Moratalla, Modeling Parkinson's disease with the alpha-synuclein protein, Front Pharmacol. 11 (2020). https://doi.org/10.3389/fphar.2020.00356.
42. Y.R. Butler, Y. Liu, R. Kumbhar, P. Zhao, K. Gadhave, N. Wang, Y. Li, X. Mao, W. Wang, α-synuclein fibril-specific nanobody reduces prion-like α-synuclein spreading in mice, Nat Commun. 13 (2022). https://doi.org/10.1038/s41467-022-31787-2.
43. F.J. Sanz, C. Solana-Manrique, N. Paricio, disease-modifying effects of vincamine supplementation in drosophila and human cell models of Parkinson's disease based on DJ-1 deficiency, ACS ChemNeurosci. 14 (2023) 2294–2301. https://doi.org/10.1021/acschemneuro.3c00026.
44. J.D. Lambris, Advances in Experimental Medicine and Biology, n.d. http://www.springer.com/series/5584.

4 Relevance of Genetic Model and Linked Therapeutics in Parkinson's Disease

Kakarla Ramakrishna, Samuel Junior Osei Asokwah, Grace Mwihaki Njuguna, Hayiru Mohammed, Naresh Dumala, Matte Kasi Viswanadh, Guntupalli Chakaravarthi, Buchi N. Nalluri, and Vinjavarapu Lakshmi Anusha
Department of Pharmacy, Koneru Lakshmaiah Education Foundation, Vaddeswaram, Andhra Pradesh, India

4.1 INTRODUCTION

Parkinson's disease (PD) is a devastating neurodegenerative condition that manifests as tremors, postural instability, and dyskinesia because of the death of dopamine neurons of the substantia nigra (SN) [1]. Oxidative stress, mitochondrial dysfunction endoplasmic reticulum (EPR) stress, and genetic mutations aggravate the dopaminergic neuron loss leading to PD [2]. Apart from the loss of dopamine neurons, the development of Lewy bodies composed of α-synuclein (α-syn/SNCA) protein is the key contributor to the progress of PD [3]. In healthy neuronal tissues, α-syn levels are tightly regulated in multiple ways, including autophagy induction by synaptic vesicles of axons and by lysosomes. There are multiple anti-PD drugs available in clinics, including levodopa, carbidopa, entacapone, tolacapone, selegenil, and amantadine, etc. However, these drugs are used to manage the symptoms associated with PD [4]. In fact, these medications do not completely relieve symptoms or stop the disease's course, nor do they replace the destroyed dopaminergic neurons in the substantia nigra [5]. Therefore, there is an urgent requirement for disease-modifying therapies.

Multiple factors make treating PD difficult. Drug development is more demanding since, for instance, the pathogenesis of PD is only partially known. Since PD is a chronic condition, symptoms can change over time and even from person to person [6]. Consequently, it is challenging to create a better medication that treats all individuals and various illness stages [7]. Moreover, existing anti-PD drugs have significant side effects, including dyskinesia, hallucinations, impulse control disorders, end-of-dose wearing-off, and on-off phenomena [8]. Besides, the blood-brain barrier (BBB) may not permit the anti-PD drugs to enter the brain. Therefore, these challenges must be overcome to ensure better treatment for PD patients [9]. Conventional anti-PD therapy in multiple ways attempt to mitigate the PD progression using diet, antioxidants, mitochondrial protective mechanisms, and neurotrophins along with physical, occupational, and speech therapies. These may improve the behaviors in PD patients but do not impede the PD progression [10].

According to existing literature, the combination of conventional therapy and coadministration of neuroprotective agents might halt the PD progress as well as symptoms [5]. Neuroprotection is the strategy of preventing or slowing neuronal cell death and is widely applied in CNS disorders such as Alzheimer's disease [11, 12], PD, multiple sclerosis, amyotrophic lateral sclerosis, traumatic brain injury, stroke [13, 14], and spinal cord injury [3, 8]. These are also applied for short-time management of neurotoxin consumption, which eventually slows down the effects of neurotoxin on the body [15]. Numerous neuroprotective strategies have been used thus far, such as growth factors, antioxidants, ion channel blockers, anti-inflammatory drugs, neurotransmitter modulators,

DOI: 10.1201/9781032620442-4

and gene therapy [8, 16, 17]. While preclinical studies have yielded outstanding results for these neuroprotective techniques, most of these have not held up well in clinical trials. The failure of neuroprotective treatments can be attributed to a variety of factors. Preclinical research allows for the prediction of disease progression and symptom onset, which in turn facilitates the start of treatment. On the other hand, patients typically don't have symptoms until after they have experienced moderate to severe neuronal loss. Therefore, achieving maximum neuroprotection is challenging. There may have already been moderate to severe brain damage before a diagnosis is made [3]. As a result, developing neuroprotective treatments for neurodegenerative illnesses is extremely difficult [18]. Moreover, the majority of neuroprotective drugs exhibit broad-spectrum actions, resulting in the selection of several pathways and targets as well as a multitude of adverse effects [19]; hence, it is challenging to decide a single target for a chosen neuroprotective drug [20]. Further, neuroprotection is also impacted by the dose and varies according to age; sex; weight [21]; environmental, genetic, and lifestyle factors; and BBB [22, 23]. Therefore, all these factors need to be considered before considering neuroprotective approaches in human PD treatment.

4.2 CONVENTIONAL PD MODELS

Parkinson's disease is mimicked in preclinical animal models using neurotoxins such as 6-hydroxydopamine (6-OHDA), rotenone (mitochondrial toxin), and MPTP [24]. In preclinical studies, acute exposure to these neurotoxins produced a significant neuronal cell death at nigrostriatal dopaminergic neurons, resulting in motor dysfunction [25]. These neurotoxins cause the loss of dopaminergic neurons in multiple ways by triggering inflammation, mitochondrial functional impairment, oxidative stress, EPR stress, and lysosomal dysfunction, and eventually cause dopamine neurons death in SN region [26].

MPTP is routinely employed to induce PD in experimental models. Due to its lipophilic nature, MPTP can easily cross the BBB and is taken up by glial cells. Monoamine oxidase-B (MAO-B) transforms MPTP to MPDP+ and MPDP+ converts into MPP+ via spontaneous oxidation. MPP+ selectively transports to dopaminergic cells through the vesicular monoamine transporters (VMAT) [27]. Then, MPP+ blocks mitochondrial complex I enzyme, which further enhances oxidative stress and neuroinflammation, leading to dopaminergic neurodegeneration. In fact, most of the reports indicate that MPTP was employed to trigger PD in mice [27]. However, due to resistance produced by MPTP moderate doses, MPTP is not suitable to induce PD in rats. However, high doses of MPTP enhanced the mortality rate in rat PD model [28]. Therefore, MPTP use to induce PD is limited to mice. MPTP acute intraperitoneal injection causes SNPC dopaminergic neuron loss in hours to one week. Moreover, MPTP-induced motor deficits recover a few days after MPTP injection, suggesting that the MPTP model would not be suitable to study motor deficits in PD. Further, repeated subacute intraperitoneal injections of MPTP cause progressive accumulation of α-synuclein, leading to nigrostriatal neurodegeneration, which mimics PD with the hallmarks of α-synuclein deposition and neuroinflammation and the extent of lesions is closely associated with motor deficits [27]. 6-hydroxydopamine (6-OHDA) is the analog of dopamine and norepinephrine, also widely used to induce PD in experimental models. Unlike MPTP, 6-OHDA cannot cross the BBB; therefore, it should be administered through intracerebral injection [29]. Similar to MPTP, 6-OHDA also cause mitochondrial complex I inhibition, which eventually generates high levels of reactive oxygen species (ROS), including superoxide radicals, hydroxyl (OH) radicals, and hydrogen peroxide. 6-OHDA injection into the SNPC kills 90% of dopaminergic neurons in a couple of days [30], thereby producing PD in less span. As 6-OHDA produce PD in short span, this model may not be suitable to study the underlying mechanisms of progressive degeneration in PD. Rotenone is a mitochondrial complex I inhibitor that promotes α-synuclein accumulation, dopaminergic neuron degeneration, and motor deficits. It has been reported that 3 mg/kg intraperitoneal injection for six to ten days induces dopaminergic neuron loss (45%) in SNPC, depletion of striatal dopamine, bradykinesia, rigidity and postural instability, and sleep disturbances [31].

4.3 LIMITATIONS OF CONVENTIONAL PRECLINICAL PD MODELS

Conventional preclinical PD models have been widely practiced across the globe. However, there are several disadvantages of these models that limit their use. The following are the disadvantages of using conventional PD models [32]; First, the chemicals that mimic PD in rodents (rats and mice) do not completely replicate the PD pathology that is encountered in humans due to biological differences when compared to humans [32]. Therefore, the translation of preclinical data to clinical settings is challenging. Second, the existing PD models have limited predictive validity; therefore, drugs that exhibit promising protection against dopaminergic neurodegeneration in animal models may not have similar protection in PD patients [33]. Third, PD is accompanied by genetic and environmental factors in humans. However, preclinical models such as, rodents and other species are genetically homogenous. Therefore, conventional PD models do not offer heterogeneity in their genetics, which makes it difficult to study the genetic contribution of PD that occur in humans [26]. Fourth, preclinical models aim for specific outcomes in PD pathology, particularly dopaminergic loss. However, conventional PD models do not specify the crosstalk between cellular molecular events involved in PD pathogenesis [33]. Fifth, the current preclinical PD models exhibit PD symptoms acutely and chronically. The acute PD models are failing because they do not study for chronic and progressive nature of the PD pathogenesis. Therefore, these conventional models do not provide the complete underlying mechanisms of PD [34]. Sixth, the success of induction of PD in animal models by neurotoxins can be studied based on the behavioral outcomes such as motor deficits cognitive and mood disturbances. However, the conventional models exhibit only motor deficits [27]. Therefore, there is a need for a preclinical model, where all these symptoms can be produced and studied. Seventh, individual PD patients exhibit a variety of symptoms, but the conventional preclinical models exhibit homogeneous symptoms, which affects the personalised approach to PD treatment [35]. Hence, the above-mentioned disadvantages of conventional preclinical models must be overcome.

4.4 GENETICS IN PD

Genetic factors are the lead contributors to PD development and progression. Genetic mutations contribute to familiar cases of PD (30%) and sporadic forms of PD (3–5%). Based on the available literature, the following gene have been involved (at least 20 genes) in PD development or progression, including SNCA, LRRK2, UCHL1, PARK7, DJ-1, PINK1, POLG, HTRA2, ATP13A2, FBX07, GIGYF2, GBA, PLA2G6, EIF4G1, VPS35, DNAJC6, SYNJ1, DNAJC13, TMEM230, VPS13C, and VPS13C [36]. Genome-wide association studies (GWAS) nominated the SNCA, LRRK2, GBA, and VPS13C genes that are commonly involved in PD development and progression. Out of these 20 genes, some genes are dominant (SNCA, UCGC1, LRRK2, POLG, HTRA2, GIGYF2, GBA, EIF4G1, VPS35, DNAJC13, TNEN230, and LRP10), and the remaining genes are recessive [26, 37].

4.4.1 Alpha-synuclein

α-synuclein (SNCA) is a small protein that consists of 140 amino acids that are mapped to chromosome 4922.1 [38]. The SNCA family of proteins consists of three domains, including an amino-terminal associated with the membrane, a nonamyloidogenic core domain, and an unorganised carboxy-terminal domain responsible for fibrillisation and aggregation inhibitory regions [39]. Four different SNCA proteins/transcripts have been discovered, including SNCA 98, SNCA 112, SNCA 126, and SNCA 140. Among them, SNCA 140 is the major transcript involved in the PD pathogenesis [40]. SNCA protein is predominantly expressed in CNS neurons, cytosol, and presynaptic terminals (synaptic vesicle membranes) and detected in the CSF and peripheral plasma. SNCA's physiological function is not fully studied. However, this is thought to be involved in vesicle trafficking, preservation of the reserve synaptic vesicles of primary hippocampal neurons, and release of neurotransmitters, and regulation of DA homeostasis [41]. SNCA protein undergoes pre- or post-translation modifications, resulting in the formation of Lewis bodies (LB) in the brain, which eventually causes PD. Currently, five pathogenic mutations, genomic duplication, and triplication with an overall 57 SNCA-related pedigrees have been identified which are causing the PD, myoclonus, cognitive, and psychiatric symptoms. The development of the autosomal dominant type of familial PD is caused by both N-terminal point mutations (A53E, A53T, A53V, A30P, E46K, H50Q, and G51D) and genomic SNCA locus triplication. As of now, these genes are thought to be the lead contributors of PD pathogenesis [40].

Given the importance of SNCA, several animal models with knockout and overexpression of human SNCA via transgenic approaches have been developed to understand the PD pathogenesis. Several preclinical models have been developed and studied by modifying the SNCA in *C. elegance*, *Drosophila*, rat, mice and monkeys to understand the role of SNCA in PD pathogenesis [42, 43]. Most of the PD cases detected are sporadic (with no known cause of genetic abnormalities) cases associated with SNCA gene mutations. Using the above genetic models of SNCA, it has been confirmed that SNCA mutations, overexpression, post-translational modifications leading to misfolding or aggregation of SNCA thereby causing dopaminergic neuron loss through the activation of ubiquitous protease system/or autophagy-lysosomal mediated degradation which are the key reasons for SNCA-mediated PD [44]. Indeed, SNCA cause mitochondrial dysfunction, oxidative stress, neuroinflammation, and endoplasmic reticulum stress, leading to impaired dopamine metabolism [45, 46].

Several compounds are being investigated in clinical trials that target SNCA release, modifications, and mutations. SNCA is released by endocytosis, spread by microglia and its toll-like receptors (TLR), and gap junctions which facilitate the internalisation of SNCA. Further, SNCA is localised in endosomal compartments where they are degraded by autophagosome and lysosomal pathways. Then, these SNCA fibrils move to cytosolic compartments where they misfold and form SNCA aggregates induced by lipids [47]. Given importance of SNCA, several drugs and new chemical entities have been tested and shown promising SNCA inhibition properties in preclinical and clinical studies. For example, endocytosis inhibitors (Dynasore and Sertraline), autophagy and lysosomal inducers (Rapamycin, Metformin, Amrboxol AR, Nilotininb, Trehalose, KYP-2407), misfolding inhibitors (SynuClean D, Apigenin, Anle1386, EGCG, Clr01, NPT200-11), lipid-induced aggregation inhibitors (Trodusquenine, NPT200-11, and ENT-01), gap junction blockers (Gap2409, Gap2605, Gap3211, and CBX), microglia activation inhibitors (Candesartan cilexetil, Lenalidomide, and Hypoestoxide), and toll-like receptors (in microglia) inhibitors (CU-CPT22 and NPT520-34) are proven to reduce the SNCA protein aggregation, thereby mitigating the PD [47].

Collectively, these events disrupt neuron-to-neuron communication, causing neuronal cell death (neurodegeneration). Therefore, several attempts were made to halt the SNCA-mediated PD at earlier stages. Indeed, from the clinical investigations, SNCA mediated the synaptic dysfunction. It is noteworthy to mention that these synaptic network dysfunctions cannot be overcome in the advanced stages of PD either in the preclinical or clinical studies. Moreover, SNCA-mediated alterations are identified in the initial stages of the normal history of PD in multiple areas of the brain, eventually producing motor, cognitive, and other psychiatric symptoms associated with PD. However, further genetic manipulation of SNCA validity could be investigated in clinics.

4.4.2 LRRK2

Leucine rich-repeat kinase 2 (LRRK-2) contains a GTPase domain, kinase domain, and multiple protein-protein domains physiologically. LRRK2 exhibits multiple functions, including kinase activity (addition of phosphate group to the other proteins) [48]. These kinase activities of LRRK2 are involved in cell survival, cytoskeletal dynamics, and vesicle trafficking. The GTPase activity helps to maintain its kinase activity. LRRK2 maintains endolysosomal-mediated vesicle trafficking and dynamics, which eventually maintains cellular homeostasis [49]. LRRK2 also regulates the mitochondrial dynamics and functions, assists microglial-mediated immune function in the brain, regulates autophagy, and helps in the recycling of cellular components [50]. The leucine-rich repeat (IRR) interaction domain serves as protein binding and scaffold. The above-mentioned physiological roles of LRRK2 are due to GTPase and kinase activity. The mutations of LRRK2 are linked to PD pathogenesis. For example, these mutations in LRRK2 impair the vesicle trafficking (particularly dopamine), mitochondrial dysfunction, and microglial mediated neuroinflammation, and enhance the autophagy, leading to neurodegeneration, thus contributing to the pathogenesis of PD [51].

The vast literature suggests that LRRK2 mutations severely impair protein translation, neurite outgrowth, and vesicle trafficking, autophagy, and cytoskeletal dynamics. The LRRK2 role has been studied in several preclinical models such as *Drosophila* (transgenic and knockout), zebrafish, *C. elegance*, mouse (knockout and transgenic), rat (knockout and transgenic), and viral models (herpes simplex virus, and adenovirus) [49, 51]. From this vast literature, it has been identified that three regions of LRRK2 are mutated and are responsible for PD pathogenesis [51]. These include the GTPase domain with R1441C/G (R1441G, R1441C, R1441H, and N1437H) mutation, c-terminal of Ras (COR) with Y1699C mutation, and kinase domain with G2019S and I2020T mutation. Among these five mutations, the G2019S mutation accounts for 5–6% of PD cases that are autosomal dominant and causes 1% of late-onset PD in sporadic cases [51]. All these PD patients presented with Lewy bodies. However, the remaining four mutations vary on Lewy body formation indicating that the severity of PD is distinct to each mutation.

Currently, several LRRK inhibitors are being explored for the management and treatment of PD. Based on the available literature, the LRRK inhibitors are categorised into zero-, first-, second-, and third-generation [52]. The zero-generation chemicals such as H-1152, GW5074, sunitinib, and staurosporine show poor potency with no BBB penetration properties. The first-generation chemicals such as LRRK2-IN-1, CZC-54252, CZC-25146, and TTT-3002 show improved potency and BBB penetration. The second-generation chemicals such as HG-10-102-1, JH-II-127, GSK2578215A, GNE-7915, GNE-0877, GNE-9605, and PF-06447475 show high potency and BBB penetration, but the half-life is short. The third-generation chemicals such as PF-06685360 and MLi-2 show high potency and high BBB penetration and are currently in the stage of dose optimisation [52]. Modeling of LRRK2 in PD models provided unparalleled insights into the role of LRRK2-mediated neurodegeneration in PD and suggested that LRRK2 alter protein translation, neurite outgrowth, autophagy, cytoskeletal dynamics, and vesicle trafficking [51]. Indeed, all the above-mentioned genetic models failed to fulfil all key events of PD. For example, the crosstalk between LRRK2 and SNCA has not been fully addressed. Rodent DA neurons are resistant to LRRK2 toxicity, and the rodents may also have compensatory mechanisms to prevent DA neuron loss. Further, in human PD, LRRK2 mutations are partially effective implying that other factors may contribute to DA neuron degeneration. The viral-mediated LRRK2 rodent models are associated with robust loss of DA neurons [52]. Hence, further, research is needed to address the above issues to fully understand the LRRK2 in the pathogenesis of PD.

4.4.3 PINK 1 and Parkin

Phosphatase and tensin homolog (PTEN)- induced putative kinase 1 (PINK 1) is a mitochondrial Ser/Thr protein kinase. It is cleaved and freely released into the cytosol for degradation. Parkin RBRE3 ubiquitin-protein ligase (PARKIN) and PINK 1 genes are involved in the familial cases of PD. PINK 1, also known as PARK 6 and BRPK, is localised in chromosome 1, which is composed of eight exons encoding 581 amino acids [53]. PINK 1 consists of five domains named N-terminal mitochondrial targeting sequences (MTS), N-terminal regulatory domain (NT), N-lobe and C-lobe transmembrane domain (TM), and C-terminal domain. PARKIN, also called PARK 2, PDJ, AR-JP, and LPRS2, is located on chromosome 6 with 14 exons encoding 456 amino acids. Both PINK 1 and PARKIN are controlled by the mitochondria [54]. Mitochondrial dysfunction and mutations of PINK 1 and PARKIN are linked to PD. Healthy mitochondria have relatively high mitochondrial membrane potential (MMP), which promote the degradation of PINK 1. On the contrary, unhealthy mitochondria have low MMP leading to PINK 1 accumulation that can recruit PARKIN from the cytoplasm, which further cause mitophagy [55]. The mitophagy mediated by PINK 1 and PARKIN is mainly governed by phosphorylation, ubiquitination, and post-transcriptional modifications, which eventually impair the energy balance and impair the protein quality [56, 57].

Several reports suggest that mutation in PINK 1 and PARKIN is linked to mitochondrial damage in PD. Briefly, both knockout mice models of PINK 1 and PARKIN did not show any abnormalities in the number of DA neurons, striatal dopamine content, and DA receptors and not exhibiting oxidative and nitrosative stress [57]. Nitric oxidative stress-mediated S-nitrosylation of PINK 1 and PARKIN leads to dysfunctional mitochondria. Mouse and rat PARKIN knockout models exhibit no detectable neurochemical abnormalities and neurodegeneration when compared to wild types. However, these findings are obtained from the single knockout of PARKIN or PINK 1 [53]. Stephenson et al. created a double knockout of PARKIN, and PINK 1 has not been observed as pathological abnormalities of PD [58]. Therefore, PINK 1 and PARKIN are considered reliable genetic models to study the pathophysiology of PD. However, PINK 1 and PARKIN genetic models have some advantages and disadvantages. The advantages include that this model can be used to study the mitochondrial-mediated neurodegeneration and understand E3 ubiquitin ligase-mediated degradation mechanisms in PD. The disadvantages include that both genetic models involve complex pathways, which hinders the development of therapies for PD.

4.4.4 Glucosylceramidase Beta A

The glucosylceramide (GlcCer) hydrolysis into glucose and ceramide, and hydrolysis of D-glucosyl-N-acylsphingosine to D-glucose and N-acylsphingosine, are catalysed by glucocerebrosidase (GCase) [59]. The presence of Gcase promote the weakening of α-syn binding to lipid membranes in neurons. However, in the pathological conditions of PD, Gcase activity has been reduced and further accelerates the risk of PD development and progression.

Glucosylceramidase beta A (GBA) gene encodes Gcase, which is a lysosomal enzyme. GBA mutants are involved in 5%–15% of PD patients. Numerous reports suggest that GBA is involved in sporadic PD [60]. At least 300 GBA mutations have been identified. GCase is primarily detected as the pathological hallmark of Gaucher disease, where mutations of GCase impair the catabolism of glucosylceramide and glycosylsphingosine [61]. Later, it was discovered that GCase in PD has been studied in several preclinical models. These findings conclude that mutations in GCase lead to its misfolding, which is accumulated in the endoplasmic reticulum, and eventually triggers the unfolded protein response (UPR). The ER-mediated UPR promotes mitochondrial imbalance, α-synaccumulation, and neuroinflammation and eventually causes dopaminergic neurons degeneration [62].

The GBA gene is situated on chromosome 1 and comprises 11 exons and composed of 497 residues with 59–69 KDa molecular weight. This molecular weight difference is due to posttranslational modifications [63, 64]. There are three major mutations reported on GBA that significantly contribute to PD pathogenesis, including N370S, L444P, and E326K [65]. Based on the available clinical and preclinical data, N370S is associated with lower PD penetrance with mild clinical phenotype [66]. E326K is reported to have a worse clinical phenotype [67]. Among these three mutations, N370S and L44P cause the loss of GCase activity-mediated triggering of UPR and produce α-syn, whereas E326K reduces the GCase activity to a much lesser extent, which further produces neurodegeneration [64]. Therefore, given the importance of the role of GBA in PD pathology, targeting GBA mutations opened a new avenue for PD therapy. Drugs like Ambroxol, Isofagomine, Miglustat, Venglustat (sensual molecular chaperones), Miglustat, Venglustat, G2667161 (reduces glycosphingolipid accumulation), AAV mediated delivery of recombinant GCase (replaces GCase activity and proteins), BIA, and 28-6156/LTI-291 (GCase activator) are currently under the clinical investigations to combat Gcase-mediated PD [64].

Understanding the relation between GBA mutations and PD is a crucial bridge between the pathogenesis and treatment of PD. Based on the available genetic models of GBA, it is evident that

GBA mutations cause dopaminergic neurodegeneration in multiple ways, including α-syn accumulation, mitochondrial dysfunction, ER stress-mediated neuronal cell death, neuroinflammation, and impaired brain lipid metabolism. Further, an understanding of the cause of GBA mutations needs to be investigated to find better treatment approaches for PD.

4.4.5 DJ-1

DJ-1 protein is encoded by PARK7 gene, which is an autosomal recessive gene that has been reported to be involved in the pathology of PD. In healthy conditions, DJ-1 reduce the oxidative stress, thereby, offers neuroprotection [68]. It has been reported that mutations in PARK7 which encodes the DJ-1 promote mitochondrial dysfunction, oxidative stress, neuroinflammation, cell death, dopamine loss, and impaired autophagy [69]. Mutations at L166P, E64D, D14A, M26I, and C106 of DJ-1 reported have impaired DJ-1 activity. Further, it has been reported that DJ-1 mutations accounts for at least 1% of inherited recessive early onset of PD. Several knockout animal models suggest that DJ-1 deletion enhanced the above-mentioned PD pathological events. Whereas overexpression or upregulation of DJ-1 or removal of mutation mitigated these pathological events in PD [70]. Though, DJ-1 exerts potent neuroprotection through the mitigation of oxidative stress and other pathological events, the complete neuroprotective mechanism is not clear. Further, the antioxidant effects are relatively slow [71]. Cumulatively, DJ-1 acts as an antioxidant in PD, thereby reducing its progression. Further studies need to investigate the underlying mechanism of DJ-1 neuroprotective effects. The detailed mechanisms of genetic mutations in the development of PD are represented in **Figure 4.1**.

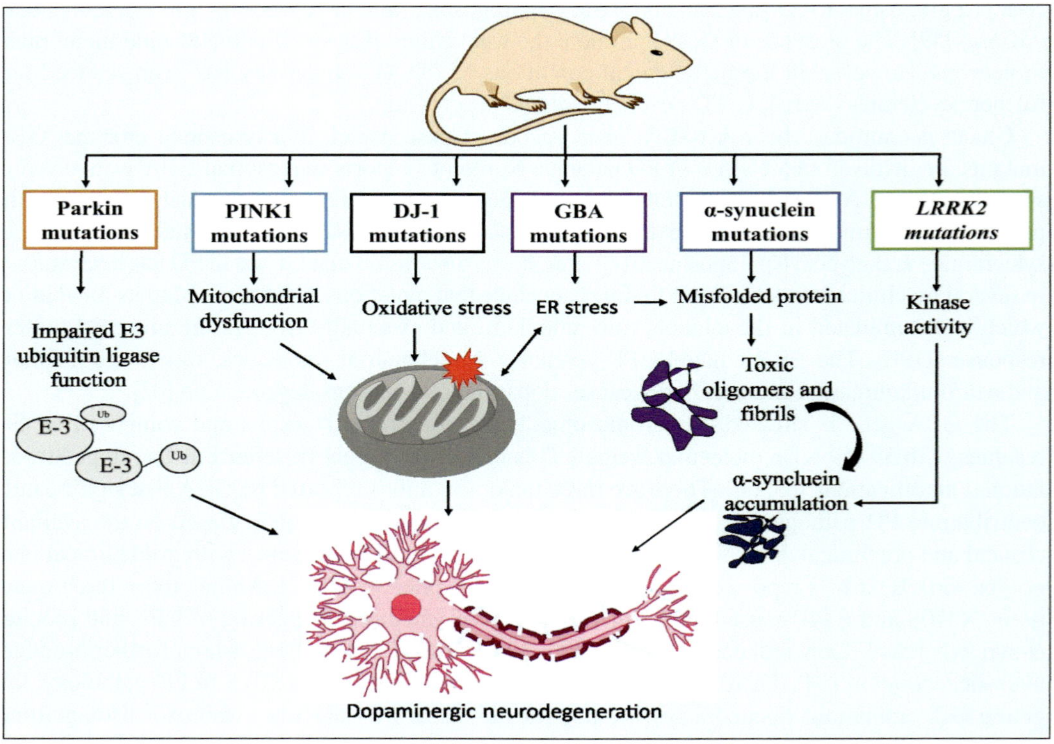

FIGURE 4.1 PD is a progressive neurodegenerative disease, where several genetic mutations contribute to dopaminergic neurodegeneration. For example, parkin mutations impair the E3 ubiquitin ligase function; PINK1 mutations cause mitochondrial dysfunction; DJ-1 mutations cause oxidative stress; GBA mutation triggers the ER stress; SNCA mutation generates the Lewy bodies; and LRRK2 mutations promote the SNCA accumulation, which eventually causes dopaminergic neurodegeneration leading to the development of PD.

4.5 CONCLUSION AND FUTURE DIRECTIONS

PD mainly occurs because of autosomal dominant gene mutations such as SNCA, LRRK2, and GBA and autosomal recessive mutations of parkin, PINK1, and DJ-1. Several animal and cellular models have been reported to provide useful insight into how mutations lead to neurodegeneration in PD. Currently, several potential drugs and new chemical entities show promising results in mitigating these mutations and pathophysiological aspects associated with the above-mentioned genes in preclinical and clinical studies. Though these genetic models overcome the limitations of conventional PD models, and these have few limitations. Although non-human primates and rodents are valuable models for studying Parkinson's disease, conclusions should be interpreted with caution due to the limitations of these models. Further, these animal models are physiologically, anatomically, and behaviorally similar to humans, but their usage is limited due to ethical and financial considerations. PD is commonly modeled in rats and mice, but no chemical or genetic model fully replicates the pathophysiology observed in PD patients. PD is believed to be influenced by both genetic vulnerability and environmental variables; thus, models that combine exposure to toxins and genetic susceptibility may be the most promising. Moreover, SNCA, LRRK2, GBA, Parkin, PINK1, and DJ-1 mutations and their role in the pathogenesis have been better understood. However, the remaining autosomal dominant and recessive gene mutations and their roles in the development of PD are not fully understood; therefore, future studies need to be conducted to evaluate their roles in PD pathogenesis.

REFERENCES

1. Mishra, A., and S. Krishnamurthy, *Neurorestorative effects of sub-chronic administration of ambroxol in rodent model of Parkinson's disease* Naunyn-Schmiedeberg's Archives of Pharmacology, 2020. **393**: p. 429–444.
2. Simon, D.K., C.M. Tanner, and P. Brundin, *Parkinson disease epidemiology, pathology, genetics, and pathophysiology* Clinics in Geriatric Medicine, 2020. **36**(1): p. 1–12.
3. Ramakrishna, K., et al., *WNT-β catenin signaling as a potential therapeutic target for neurodegenerative diseases: Current status and future perspective.* Diseases, 2023. **11**(3): p. 89.
4. Singh, N., V. Pillay, and Y.E. Choonara, *Advances in the treatment of Parkinson's disease* Progress in Neurobiology, 2007. **81**(1): p. 29–44.
5. Armstrong, M.J., and M.S. Okun, *Diagnosis and treatment of parkinson disease: A review* Jama, 2020. **323**(6): p. 548–560.
6. Kakarla, R., et al., *Current understanding and future directions of cruciferous vegetables and their phytochemicals to combat neurological diseases.* Phytotherapy Research, 2024. **38** (3), 1381–1399.
7. Tolosa, E., et al., *Challenges in the diagnosis of Parkinson's disease.* The Lancet Neurology, 2021. **20**(5): p. 385–397.
8. Rai, S.N., and P. Singh, *Advancement in the modelling and therapeutics of Parkinson's disease* Journal of Chemical Neuroanatomy, 2020. **104**: p. 101752.
9. Krishnan, U.M., *Biomaterials in the treatment of Parkinson's disease* Neurochemistry International, 2021. **145**: p. 105003.
10. Church, F.C., *Treatment options for motor and non-motor symptoms of Parkinson's disease* Biomolecules, 2021. **11**(4): p. 612.
11. Gutti, G., et al., *Discovery of novel series of 2-substituted benzo [d] oxazol-5-amine derivatives as multi-target directed ligands for the treatment of Alzheimer's disease.* European Journal of Medicinal Chemistry, 2019. **182**: p. 111613.
12. Gutti, G., et al., *Discovery of triazole-bridged aryl adamantane analogs as an intriguing class of multifunctional agents for treatment of Alzheimer's disease.* European Journal of Medicinal Chemistry, 2023. **259**: p. 115670.
13. Ramakrishna, K., K. Srinivasan, and S.S. Sharma, *Chronic treatment of 4-phenylbutyric acid ameliorates cognitive impairment after focal cerebral ischemia/reperfusion injury in rats* Indian Journal of Physiology and Pharmacology, 2021. **64**(3): p. 188–194.
14. Ramakrishna, K., S.K. Singh, and S. Krishnamurthy, *Diindolylmethane ameliorates ischemic stroke-induced brain injury by peripheral and Central mechanisms* Current Neurovascular Research, 2022. **19**(5): p. 462–475.

15. Ahmad Bhat, S., et al., *Synopsis on managment strategies for neurodegenerative disorders: Challenges from bench to bedside in successful drug discovery and development.* Current Topics in Medicinal Chemistry, 2017. **17**(12): p. 1371–1378.
16. Rehman, M.U., et al., *Neuroprotective strategies for neurological disorders by natural products: An update.* Current Neuropharmacology, 2019. **17**(3): p. 247–267.
17. Bhanukiran, K., et al., *Vasicinone, a pyrroloquinazoline alkaloid from Adhatoda vasica nees enhances memory and cognition by inhibiting cholinesterases in Alzheimer's disease.* Phytomedicine Plus, 2023. **3**(2): p. 100439.
18. Katsuno, M., et al., *Preclinical progression of neurodegenerative diseases.* Nagoya Journal of Medical Science, 2018. **80**(3): p. 289.
19. Youdim, M.B., and J.J. Buccafusco, *Multi-functional drugs for various CNS targets in the treatment of neurodegenerative disorders* Trends in Pharmacological Sciences, 2005. **26**(1): p. 27–35.
20. Levin, L.A., et al., *Neuroprotection in neurodegenerations of the brain and eye: Lessons from the past and directions for the future.* Frontiers in Neurology, 2022. **13**: p. 964197.
21. Dhir, N., et al., *Pre-clinical to clinical translational failures and current status of clinical trials in stroke therapy: A brief review.* Current Neuropharmacology, 2020. **18**(7): p. 596–612.
22. DeGraba, T.J., and L.C. Pettigrew, *Why do neuroprotective drugs work in animals but not humans?* Neurologic Clinics, 2000. **18**(2): p. 475–493.
23. Leclerc, M., S. Dudonné, and F. Calon, *Can natural products exert neuroprotection without crossing the Blood–Brain barrier?*International Journal of Molecular Sciences, 2021. **22**(7): p. 3356.
24. Tripathi, P., et al., *Generation of wild-type rat glucocerebrosidase homology modeling: Identification of putative interactions site and mechanism for chaperone using combined in-silico and in-vitro studies.* Bioorganic Chemistry, 2022. **126**: p. 105871.
25. Blandini, F., and M.T. Armentero, *Animal models of Parkinson's disease* The FEBS Journal, 2012. **279**(7): p. 1156–1166.
26. Vázquez-Vélez, G.E., and H.Y. Zoghbi, *Parkinson's disease genetics and pathophysiology* Annual Review of Neuroscience, 2021. **44**: p. 87–108.
27. Mustapha, M., and C.N.M. Taib, *MPTP-induced mouse model of Parkinson's disease: A promising direction for therapeutic strategies.* Bosnian Journal of Basic Medical Sciences, 2021. **21**(4): p. 422.
28. Zhang, J., et al., *Alpha-lipoic acid improved motor function in MPTP-induced parkinsonian mice by reducing neuroinflammation in the nigral and spinal cord.* Neuroscience Letters, 2022. **781**: p. 136669.
29. Mishra, A., and S. Krishnamurthy, *Rebamipide mitigates impairments in mitochondrial function and bioenergetics with α-synuclein pathology in 6-OHDA-induced hemiparkinson's model in rats* Neurotoxicity Research, 2019. **35**(3): p. 542–562.
30. Mishra, A., et al., *Dopamine receptor activation mitigates mitochondrial dysfunction and oxidative stress to enhance dopaminergic neurogenesis in 6-OHDA lesioned rats: A role of wnt signalling.* Neurochemistry International, 2019. **129**: p. 104463.
31. Innos, J., and M.A. Hickey, *Using rotenone to model Parkinson's disease in mice: A review of the role of pharmacokinetics* Chemical Research in Toxicology, 2021. **34**(5): p. 1223–1239.
32. Prasad, E.M., and S.-Y. Hung, *Behavioral tests in neurotoxin-induced animal models of Parkinson's disease* Antioxidants, 2020. **9**(10): p. 1007.
33. Buhidma, Y., et al., *Potential of animal models for advancing the understanding and treatment of pain in Parkinson's disease.* NPJ Parkinson's Disease, 2020. **6**(1): p. 1.
34. Belloli, S., et al., *Translation imaging in Parkinson's disease: Focus on neuroinflammation.* Frontiers in Aging Neuroscience, 2020. **12**: p. 152.
35. Jennings, D., et al., *Preclinical and clinical evaluation of the LRRK2 inhibitor DNL201 for Parkinson's disease.* Science Translational Medicine, 2022. **14**(648): p. eabj2658.
36. Shalaby, K.E., and O.M. El-Agnaf, *Gene-Based therapeutics for Parkinson's disease* Biomedicines, 2022. **10**(8): p. 1790.
37. Safari, F., et al., *CRISPR system: A high-throughput toolbox for research and treatment of Parkinson's disease.* Cellular and Molecular Neurobiology, 2020. **40**: p. 477–493.
38. Pihlstrøm, L., and M. Toft, *Genetic variability in SNCA and Parkinson's disease* Neurogenetics, 2011. **12**: p. 283–293.
39. Ozansoy, M., and A.N. Başak, *The central theme of Parkinson's disease: α-synuclein* Molecular Neurobiology, 2013. **47**: p. 460–465.
40. Deng, H., and L. Yuan, *Genetic variants and animal models in SNCA and Parkinson disease* Ageing Research Reviews, 2014. **15**: p. 161–176.

41. Beyer, K., et al., *Identification and characterization of a new alpha-synuclein isoform and its role in lewy body diseases*. Neurogenetics, 2008. **9**: p. 15–23.
42. Betarbet, R., T.B. Sherer, and J.T. Greenamyre, *Animal models of Parkinson's disease* Bioessays, 2002. **24**(4): p. 308–318.
43. Guatteo, E., et al., *Pathophysiological features of nigral dopaminergic neurons in animal models of Parkinson's disease*. International Journal of Molecular Sciences, 2022. **23**(9): p. 4508.
44. Cherian, A., K. Divya, and A. Vijayaraghavan, *Parkinson's disease–genetic cause* Current Opinion in Neurology, 2023. **36**(4): p. 292–301.
45. Siddiqui, I.J., N. Pervaiz, and A.A. Abbasi, *The parkinson disease Gene SNCA: Evolutionary and structural insights with pathological implication* Scientific Reports, 2016. **6**(1): p. 24475.
46. Dorszewska, J., et al., *Oxidative stress factors in Parkinson's disease*. Neural Regeneration Research, 2021. **16**(7): p. 1383.
47. Menon, S., et al., *Alpha-synuclein targeting therapeutics for Parkinson's disease and related synucleinopathies*. Front Neurol, 2022. **13**: p. 852003.
48. Cookson, M.R., *The role of leucine-rich repeat kinase 2 (LRRK2) in Parkinson's disease* Nature Reviews Neuroscience, 2010. **11**(12): p. 791–797.
49. Mata, I.F., et al., *LRRK2 in Parkinson's disease: Protein domains and functional insights*. Trends in Neurosciences, 2006. **29**(5): p. 286–293.
50. Madureira, M., N. Connor-Robson, and R. Wade-Martins, *LRRK2: Autophagy and lysosomal activity*. Frontiers in Neuroscience, 2020. **14**: p. 498.
51. Xiong, Y., T.M. Dawson, and V.L Dawson, *Models of LRRK2-associated Parkinson's disease*. Leucine-Rich Repeat Kinase, 2017. **LRRK2**(LRRK2): p. 163–191.
52. Zhao, Y., and N. Dzamko, *Recent developments in LRRK2-targeted therapy for Parkinson's disease* Drugs, 2019. **79**(10): p. 1037–1051.
53. Quinn, P.M., et al., *PINK1/PARKIN signalling in neurodegeneration and neuroinflammation*. Acta Neuropathologica Communications, 2020. **8**(1): p. 1–20.
54. Seirafi, M., G. Kozlov, and K. Gehring, *Parkin structure and function* The FEBS Journal, 2015. **282**(11): p. 2076–2088.
55. Pickrell, A.M., and R.J. Youle, *The roles of PINK1, parkin, and mitochondrial fidelity in Parkinson's disease* Neuron, 2015. **85**(2): p. 257–273.
56. Poole, A.C., et al., *The PINK1/Parkin pathway regulates mitochondrial morphology*. Proceedings of the National Academy of Sciences, 2008. **105**(5): p. 1638–1643.
57. Whitworth, A.J., and L.J. Pallanck, *The PINK1/Parkin pathway: A mitochondrial quality control system?*Journal of Bioenergetics and Biomembranes, 2009. **41**: p. 499–503.
58. Stephenson, S.E., J.M. Taylor, and P.J Lockhart, *Parkinson's disease and parkin: Insights from Park2 knockout mice* Mechanisms in Parkinson's Disease-Models and Treatments, 2012: p. 251–276.
59. Sidransky, E., and G. Lopez, *The link between the GBA gene and parkinsonism* The Lancet Neurology, 2012. **11**(11): p. 986–998.
60. Sidransky, E., et al., *Multicenter analysis of glucocerebrosidase mutations in Parkinson's disease*. New England Journal of Medicine, 2009. **361**(17): p. 1651–1661.
61. Hruska, K.S., et al., *Gaucher disease: Mutation and polymorphism spectrum in the glucocerebrosidase Gene (GBA)*. Human Mutation, 2008. **29**(5): p. 567–583.
62. Tripathi, P., et al., *Identification of novel glucocerebrosidase chaperone for potential treatment of Parkinson's disease: An approach using in silico virtual screening, molecular docking and molecular dynamics, and in vitro studies*. International Journal of Biological Macromolecules, 2023. **228**: p. 453–466.
63. Reczek, D., et al., *LIMP-2 is a receptor for lysosomal mannose-6-phosphate-independent targeting of β-glucocerebrosidase*. Cell, 2007. **131**(4): p. 770–783.
64. Smith, L., and A. Schapira, *GBA variants and parkinson disease: Mechanisms and treatments Cells*, 2022. ***11*** *(8)*: p. 1261.
65. Behl, T., et al., *Cross-talks among GBA mutations, glucocerebrosidase, and α-synuclein in GBA-associated Parkinson's disease and their targeted therapeutic approaches: A comprehensive review*. Translational Neurodegeneration, 2021. **10**: p. 1–13.
66. García-Sanz, P., et al., *N370S-GBA1 mutation causes lysosomal cholesterol accumulation in Parkinson's disease*. Movement Disorders, 2017. **32**(10): p. 1409–1422.
67. Park, J., et al., *The E326K Mutation and Gaucher disease: Mutation or polymorphism?* Clinical Genetics, 2002. **61**(1): p. 32–34.

68. Repici, M., et al., *DJ-1 in Parkinson's disease: Clinical insights and therapeutic perspectives*, Journal of Clinical Medicine, 2019. **8** (9): p. 1377.
69. Skou, L.D., et al., *Pathogenesis of DJ-1/PARK7-mediated Parkinson's disease.* Cells, 2024. **13** (4): p. 296.
70. Batelli, S., et al., *The Parkinson's disease-related protein DJ-1 protects dopaminergic neurons in vivo and cultured cells from alpha-synuclein and 6-hydroxydopamine toxicity.* Neurodegenerative Diseases, 2015. **15**: 13–23.
71. Gibson R., et al., *DJ-1 and Parkinson's disease.* Brain Disorders, 2021. **3**:100020.

5 Importance of Newer 3D Organoids Model-Based Therapeutic Approach in Parkinson's Disease

Kakarla Ramakrishna, Hayiru Mohammed, Grace Mwihaki Njuguna, Samuel Junior Osei Asokwah, Vineeta Singh, Naresh Dumala, Matte Kasi Viswanadh, Vinjavarapu Lakshmi Anusha, Guntupalli Chakaravarthi, and Buchi N. Nalluri

Department of Pharmacy, Koneru Lakshmaiah Education Foundation, Vaddeswaram, Andhra Pradesh, India

5.1 INTRODUCTION

Parkinson's disease (PD) is a progressive neurodegeneration disease that occurs owing to loss of dopamine neurons of the substantia nigra (SN) [1]. Due to the neurodegeneration of dopamine neurons, dopamine (DA) availability in the brain is reduced, resulting in motor dysfunction. The commonly seen motor deficits in PD patients include muscular rigidity, bradykinesia, and uncontrolled tremors [2]. These severe motor dysfunctions impair the quality of life in PD patients. Currently, drugs such as levodopa, carbidopa, tolcapone, entacapone, amantadine, and selegiline are widely used in clinics for the management of symptoms associated with PD [3]. However, the above-mentioned drugs have some disadvantages. First, these drugs failed to restore the dopaminergic neuron loss. Second, these drugs failed to impede the disease progression and also did not provide complete symptomatic relief. Hence, the quest for disease-modifying agents to combat PD is still on [4].

Pathophysiological events such as oxidative stress, endoplasmic reticulum (ER) stress, defective mitochondria, excitotoxicity, and inflammation contribute to PD [5]. Apart from these pathological abnormalities, genetic mutations are also cause for familial PD (30%) and sporadic forms of PD (3%–5%) [6, 7]. Based on the literature, PD occurs because of at least 20 genetic mutations. Autosomal dominant genes, such as SNCA, LRRK2, and GBA, and autosomal recessive genes, such as PINK1 and PARKIN, and DJ mutations are the main contributors to the development and progression of PD [8]. Given the significance of genetic contribution to PD pathogenesis, several approaches, such as gene mutation inhibitors, have been developed [9].

Organoids are derived from stem cells or tissue-specific progenitor cells after growing in in vitro conditions [10]. Brain-derived specific progenitor cells or stem cell organoids are useful in mimicking the complexity of the brain. The intrinsic property of stem cells or progenitor cells to become self-organising structures into specific organs allows the organoids to be used for diagnosis and treatment of neurodegenerative diseases, including PD. Therefore, with the help of organoids, personalised medicine for individual patients could be developed [11].

DOI: 10.1201/9781032620442-5

5.2 ORGANOIDS

An organoid is a self-forming 3D tissue, usually harvested from stem cells that mimic the structural and functional aspects of an organ. Cells composed of organoids are derived from induced pluripotent stem cells (iPSCs) or tissue-derived cells, including stem/progenitor cells, normal cells, differentiated cells, and cancer cells. Ensuring patient uniqueness is facilitated by traditional two-dimensional culture, animal models, and organoid culture, which precisely replicate *in vitro* tissue-like structures and functions found *in vivo* [10]. Organoid cultures are widely accessible for understanding thorough biological research compared to other preclinical models. As a result, organoid cultures have been exploited for many applications, for example cell therapy, drug discovery, and personalised diagnostics. Organoids can be classified into tissue-derived stem cells and pluripotent stem cells, in which tissue-derived organoids are typically derived from adult tissues, while pluripotent organoids are derived from embryonic stem cells (ESCs) or iPSCs [12, 13]. 3D cultures have been developed using two different technologies: One is scaffold free and the other one is scaffold based. Scaffold-based 3D organoid self-assembling spherical clusters do not contain any biological material, whereas scaffold-free organoid models also are spherical structures generated by their extracellular matrix (ECM), which replicates the cellular niche's natural development as it occurs in the body. Scaffold-based organoid models are widely used because their physical-chemical properties enable cell adhesion, proliferation, differentiation, and survival [14]. Moreover, the culturing of scaffold-based organoids has some additional advantages, such as enhanced nutrient and oxygen permeability, electrical conductivity, and ease of handling. Additionally, these organoids have high rates of reproducibility, develop new blood vessels (angiogenesis), and do not exhibit any immunological reactions [15].

5.3 2D AND 3D ORGANOID CULTURE AND PARKINSON'S DISEASE

2D cell culture and 3D organoid culture are used to study human biology and disease. 2D cell cultures are grown on a flat surface, while 3D organoid cultures are grown in a three-dimensional matrix. 2D cell cultures are simpler to grow and maintain than 3D organoids. They are also more accessible to experimental techniques, such as microscopy and drug screening. 2D cell cultures are simpler, less representative of human organs, less resistant, easier to grow and maintain, and more accessible [16]. 2D cultures have several disadvantages: there is low accuracy in mimicking the structure and function of human organs; modeling of culturing of patient-derived organoids is impossible; and limitations in cell morphology, survival, proliferation, and differentiation of developed 2D cultures [17]. 3D organoids are more complex and more closely resemble human organs in structure and function. They are also more resistant to stress and damage. However, 3D organoid cultures are more difficult to grow and maintain than 2D cell cultures [18]. Whereas 3D organoid cultures are less accessible and expensive to culture and maintain, modeling organoids derived from human iPSCs (hiPSCs) for the understanding of PD is growing [19–21]. After the development of organoids, formation of dopaminergic neurons, oligodendrocytes, and astrocytes shall be assessed for understanding the formation of neural networks (synaptic connections) [22, 23].

5.3.1 SNCA Mutant Organoids

α-Synuclein (SNCA) elevation is one of the primary pathological hallmarks of PD and is paramount in the formation of Lewy bodies in PD patients [24]. N-terminal point mutations at A53E, A53T, A53V, A30P, E46K, H50Q, G5ID, and genomic triplication of SNCA mutations in the SNCA locus contribute to autosomal dominant familial PD cases [25]. iPSC-derived dopamine neurons of PD patients confirmed that multiple pathways are interrupted in PD patients, for example, disruption in synaptic connectivity and signaling genes [26], reduced SNCA tetramer to monomer ratio, and elevated mitochondrial dysfunction and nitrative stress, oxidative stress, oxidised dopamine levels, and decreased lysosomal Gcase enzyme activity and lysosomal hydrolase trafficking [27]. Besides, SNCA triplication mutations have been identified with elevated permeability and opening of mitochondrial membrane transition pores, accelerating the ER stress, and impairing ER-mitochondrial communication [28, 29]. Accumulation of this evidence shows that SNCA triplication and duplication cause mitochondrial dysfunction, oxidative stress, and ER stress and eventually cause dopaminergic neuron degeneration.

5.3.2 GBA-Mutant Organoids

GBA gene encode the enzyme named glucocerebrosidase (GCase), which converts the glucosyl ceramidase to glucose and ceramide and the hydrolysis of D-glucosyl-N-acylsphingosine into D-glucose and N-acylsphingosine [30]. GBA mutations such as insertion, deletion, point, and frameshift mutations are linked to PD [31]. N370S and L444P/WT iPSC- DA neurons indicate intensified accumulation of SNCA and its aggregation [32]. Further, mutant GBA-iPSC-resulting neurons show deposition of lipids such as glucosylsphingosine, glucosylceramide, and GCase substrates [33]. Furthermore, GBA-mutant iPSC-DA neurons exhibit impaired ER, and mitochondrial function. Particularly, the rate of mitochondrial consumption of oxygen, complex I activity, and mitochondrial morphology are altered [34]. Moreover, GBA-mutant iPSC DA neurons exhibit ER stress, which is evidenced by imbalanced calcium homeostasis and unfolded protein response (UPR) [32]. Besides, GBA mutation-derived organoids show impaired HDAC4 and MAO-B mRNA and protein levels [35]. Accumulating these findings, organoids derived from GBA-mutant iPSC-DA neurons show a variety of pathological events [36]. Further studies need to be performed to develop potential therapies based on the GBA mutations.

5.3.3 DJ-1–Mutant Organoids

DJ-1 encoded by the PARK7 gene is highly expressed in reaction astrocytes [37]. DJ-1 consists of 180 amino acids. DJ-1 regulates autophagy, apoptosis, inflammation, and oxidative stress [38]. At least 20 DJ-1 mutations have been identified to be involved in the early onset of PD and loss of DJ-1 function due to mutations results in autosomal recessive PD [39]. DJ-1 mainly regulates mitochondrial-mediated oxidative stress [40]. DJ-1–mutant iPSC-DA neurons show increased mitochondrial oxidation and oxidised DA levels. Furthermore, DJ-1–mutant iPSC-DA neurons exhibit increased SNCA accumulation [41]. These outcomes imply that organoids derived from DJ-1–mutant iPSC-DA neurons provide a better insight of DJ-1 mutations in the pathophysiology of PD. Future studies need to find the potential therapeutic agent to prevent or treat the DJ-1 mutations.

5.3.4 LRRK2–Mutant Organoids

Leucine-rich repeat kinase 2 (LRRK2) is involved in the pathology of PD. Structurally LRRK2 contains three domains, named Ras-GTpase domain, protein binding domain, and a kinase domain [42]. Genetic mutations of LRRK2 were reported to cause autosomal dominant forms of familial PD mutation at the GTpase domain at R1441C, R1441H, R1441G, and N1437H, kinase domain at I2020T and G2019S, and protein-binding domain at Y1699C are reported to be causes of PD [43, 44]. Culturing of LRRK2 iPSC-DA neurons with G2019S and R1441G shows dysfunctional NF-kB signaling [45]. Culturing of LRRK2 G2019S and I2020T mutated iPSC-DA neurons exhibited increased autophagy genes, autophagosomes, apoptosis, mitophagy, mitochondria DNA damage, and SNCA aggregation and also reduced neurite outgrowth [36]. Culturing of LRRK2 R1441G/C without iPSC-DA neurons shows disrupted synaptic vesicle density and endocytosis [46]. Therefore, study of the LRRK2 organoids suggests that culturing the LRRK2 organoid could help in understanding the pathogenesis of PD and its therapeutic opportunities.

5.3.5 PINK 1 and PARKIN–Mutant Organoids

Phosphatase and tension homolog (PTEN)–induced putative kinase 1, which is commonly known as PINK1, facilitates parkin phosphorylation in the mitochondria [47]. The mutation at PINK1 and

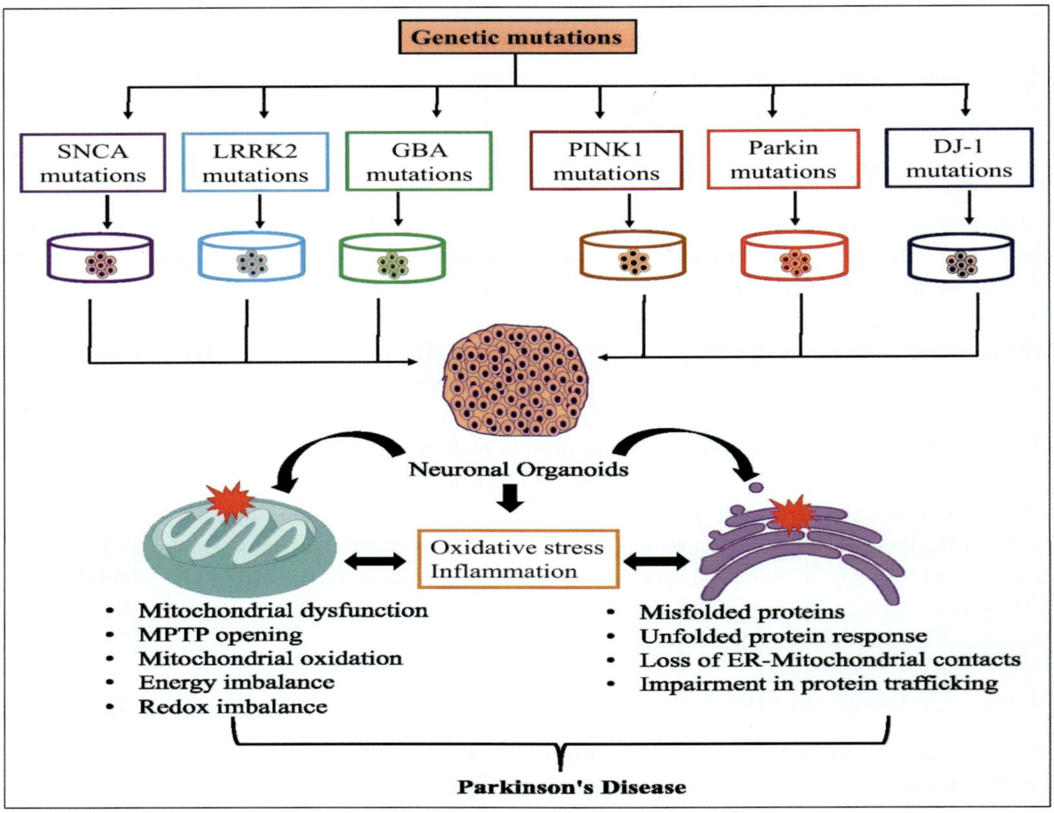

FIGURE 5.1 Parkinson's disease is mainly triggered by mutations in SNCA, LRRK2, PINK1, PARKIN, GBA, and DJ-1 proteins. iPSC-DA neurons organoids from the above-mentioned proteins exhibit mitochondrial dysfunction, oxidative stress, inflammation, and ER stress. These pathological hallmarks contribute to PD pathogenesis.

parkin impairs the mitochondrial function leading to autosomal recessive and early onset of PD [48]. PINK1 expresses Q456X and VI70G mutations, which are non-sense and uni-sense mutations associated with PD pathogenesis [49]. iPSC-DA neurons of PINK1 mutations of Q456X and VI70G exhibit improved parkin recruitment to mitochondria and increased the PGC-1α mitochondrial biogenesis marker [49]. Further iPSC-DA neurons show that s-nitrosylated PINK1 decreases parkin translocation to mitochondria, suggesting that the culturing of PINK1 and parkin could offer therapeutic opportunities [50]. The evidence also suggested that parkin-mutant iPSC-DA neurons also exhibit mitochondrial dysfunction, impairments in ER-mitochondrial communication, increased inflammation, and defects in calcium transfer, and promote SNCA accumulation [51]. Surprisingly, iPSC-DA neurons obtained from parkin V324A mutant show elevated forms of oxidised dopamine with decreased GCase activity [52]. Hence, these findings suggest that culturing of PINK1 and parkin-mutant iPSC-DA neurons is significantly useful in understanding pathogenesis and development of potential modulators of PINK1 and parkin mutations. The detailed genetic mutations and organoids developed from mutant genes involved in the PD pathogenesis are illustrated in **Figure 5.1**.

5.4 ORGANOID-MEDIATED DRUG DEVELOPMENT FOR PD

Organoids derived from the mutant dopaminergic neurons are developed to model the disease, diagnosis, and treatment of PD. A detailed list of mutant organoids developed by various researchers has been provided in **Table 5.1**. Briefly, mutant SNCA, LRRK2, GBA, PINK1, Parkin, and

TABLE 5.1
A Detailed List of Mutant Organoids Developed by Various Researchers

S. No.	Mutant Organoid	Observation
1	Mutant SNCA triplication	Increased oxidised dopamine, SNCA aggregation, mitochondrial dysfunction, ER stress, and oxidative stress and decreased lysosomal hydrolase trafficking and GCase enzyme activity.
2	Mutant LRRK2	Decreased synaptic vesicle endocytosis, neurite length, increased oxidised dopamine, upregulation of several autophagic markers, inflammation, mitochondrial dysfunction, α-synuclein, and oxidative stress response genes.
3	Mutant PINK1	Decreased DA uptake, oxidative stress, abnormal mitochondrial parkin recruitment, and increased expression of PGC-1α.
4	Mutant PARKIN	Increased oxidative stress, mitochondrial elongation and enlargement, and apoptosis.
5	Mutant GBA	Increased oxidised DA, reduced Gcase activity, and SNCA deposition.
6	Mutant DJ-1	Increased mitochondrial oxidation, oxidised DA, decreased basal mitochondrial respiration, Gcase activity, and SNCA deposition.

DJ-1 reported having increased oxidised dopamine, mitochondrial dysfunction, oxidative stress, ER stress, and SNCA accumulation [53]. Based on understanding of pathophysiology, using 3D organoids help to develop or identify potential drugs or chemicals. For example, Glucocerebrosidase (GCase) modulators (NCGC758, NCGC607, quinazoline derivatives, and S-181), antioxidants (mito-TEMPO NAC), and SNCA oligomerisation modulators (NPT100–18A, NPT100–14A, and ELN48228) are currently under investigation [36].

5.5 CONCLUSION AND FUTURE DIRECTIONS

Replicating important PD pathogenic features, iPSC-generated neurons and organoids taken from familial and idiopathic patient is an essential paradigm for studying and discovering new neural pathways implicated in PD. Even though most studies to date have employed 2D neuronal cultures, the utilisation of 3D organoids that approximate the midbrain will be critical for evaluating diseases and neural complexity that are not reproduced in 2D models. These technologies also provide new models for reassessing the neurotoxic and neuroprotective substances that are currently being studied. This is important since new methods for drug screening have been made possible by these technologies. Therefore, the use of patient-derived cell cultures are crucial in identifying the disease pathology that may be used as a therapeutic avenue for complicated disorders like PD. Neuroprotective effects have also been successfully identified through drug screening against iPSC-DA neurons, and this approach may also present awareness on the efficacy of drugs against human diseases.

Lastly, the use of organoids in drug development provides a more realistic 3D model for studying the modulation of different cell types by medicines. In order to better understand neuronal pathways, identify relevant therapeutic targets, and provide essential models for testing novel therapies, future research on PD utilising iPSC-DA neuron and organoid models would be highly advantageous. In the future, establishment of biobanks that contain iPSCs from specific patient cell lines and disease-related mutations would be required. Standardised differentiation procedures and indicators for the generation of midbrain DA neurons will also be crucial. More complex brain modeling will be required for studying the integration of glial cells and other neuronal subtypes as organoid systems. Further research into the pathophysiology of PD will benefit from using many clones and patient lines, perhaps allowing for comparisons of different PD genes and mutations. Finally, further studies employing iPSC neurons from individuals with idiopathic PD are necessary in promoting the development of novel therapies and understanding the pathological pathways in sporadic PD.

REFERENCES

1. Bloem, B.R., M.S. Okun, and C. Klein, *Parkinson's disease* The Lancet, 2021. **397**(10291): p. 2284–2303.
2. Sveinbjornsdottir, S., *The clinical symptoms of Parkinson's disease* Journal of Neurochemistry, 2016. **139**: p. 318–324.
3. Ramakrishna, K., et al., *WNT-β catenin signaling as a potential therapeutic target for neurodegenerative diseases: Current status and future perspective.* Diseases, 2023. **11**(3): p. 89.
4. Boucherie, D.M., et al., *Parkinson's disease drug development since 1999: A story of repurposing and relative success.* Journal of Parkinson's Disease, 2021. **11**(2): p. 421–429.
5. Kakarla, R., et al., *Current understanding and future directions of cruciferous vegetables and their phytochemicals to combat neurological diseases.* Phytotherapy Research, 2024. **38** (3), 1381–1399.
6. Naren, P., et al., *Pathological and therapeutic advances in Parkinson's disease: Mitochondria in the interplay.* Journal of Alzheimer's Disease, 2023. **94**(s1): p. S399–S428.
7. Zhao, Y., et al., *Genetic analysis of transcription factors in dopaminergic neuronal development in Parkinson's disease.* Chinese Medical Journal, 2023, 137(4): 450–456,: p. 10.1097.
8. Shalaby, K.E., and O.M.A. El-Agnaf, *Gene-based therapeutics for Parkinson's disease.* Biomedicines, 2022. **10**(8).
9. Jusop, A.S., et al., *Development of brain organoid technology derived from iPSC for the neurodegenerative disease modelling: A glance through.* Frontiers in Molecular Neuroscience, 2023. **16**.
10. Zhao, Z., et al., *Organoids.* Nature Reviews Methods Primers, 2022. **2**(1): p. 94.
11. de Souza, R.T.V., et al., *Brain organoids as models of Alzheimer's and Parkinson's diseases: a narrative review on perspectives for regenerative and personalized medicine.* 2023. https://doi.org/10.11606/issn.2176-7262.rmrp.2023.203797.
12. Kim, J., and J.A. Knoblich, *Human organoids: Model systems for human biology and medicine* Nature Reviews Molecular Cell Biology, 2020. **21**(10): p. 571–584.
13. Sun, A.X., H.H. Ng, and E.K. Tan, *Translational potential of human brain organoids* Annals of Clinical and Translational Neurology, 2018. **5**(2): p. 226–235.
14. Knight, E., and S. Przyborski, *Advances in 3D cell culture technologies enabling tissue-like structures to be created in vitro* Journal of Anatomy, 2015. **227**(6): p. 746–756.
15. Hopkins, A.M., et al., *3D in vitro modeling of the central nervous system.* Progress in Neurobiology, 2015. **125**: p. 1–25.
16. Kim, S.-h., and M.-Y. Chang, *Application of human brain organoids—Opportunities and challenges in modeling human brain development and neurodevelopmental diseases* International Journal of Molecular Sciences, 2023. **24**(15): p. 12528.
17. Biju, T.S., V.V. Priya, and A.P. Francis, *Role of three-dimensional cell culture in therapeutics And diagnostics: An updated review* Drug Delivery and Translational Research, 2023. **13**: 2239–2253.
18. Centeno, E.G., H. Cimarosti, and A. Bithell, *2D versus 3D human induced pluripotent stem cell-derived cultures for neurodegenerative disease modelling* Molecular Neurodegeneration, 2018. **13**: p. 1–15.
19. Martínez-Morales, P.L., and I. Liste, *Stem cells as in vitro model of Parkinson's disease.* Stem Cells International, 2012. **2012**.
20. Cooper, O., et al., *Differentiation of human ES and Parkinson's disease iPS cells into ventral midbrain dopaminergic neurons requires a high activity form of SHH, FGF8a and specific regionalization by retinoic acid.* Molecular and Cellular Neuroscience, 2010. **45**(3): p. 258–266.
21. Son, M.Y., et al., *Distinctive genomic signature of neural and intestinal organoids from familial Parkinson's disease patient-derived induced pluripotent stem cells.* Neuropathology and Applied Neurobiology, 2017. **43**(7): p. 584–603.
22. Monzel, A.S., et al., *Derivation of human midbrain-specific organoids from neuroepithelial stem cells.* Stem Cell Reports, 2017. **8**(5): p. 1144–1154.
23. Recasens, A., et al., *Lewy body extracts from Parkinson's disease brains trigger α-synuclein pathology and neurodegeneration in mice and monkeys.* Annals of Neurology, 2014. **75**(3): p. 351–362.
24. Kasten, M., and C. Klein, *The many faces of alpha-synuclein mutations* Movement Disorders, 2013. **28**(6): p. 697–701.
25. Polymeropoulos, M.H., et al., *Mutation in the α-synuclein gene identified in families with Parkinson's disease.* Science, 1997. **276**(5321): p. 2045–2047.
26. Kouroupi, G., et al., *Defective synaptic connectivity and axonal neuropathology in a human iPSC-based model of familial Parkinson's disease.* Proceedings of the National Academy of Sciences, 2017. **114**(18): p. E3679–E3688.
27. Vidović, M., and M.G. Rikalovic, *Alpha-synuclein aggregation pathway in Parkinson's disease: Current status and novel therapeutic approaches* Cells, 2022. **11**(11): p. 1732.

28. Devine, M.J., et al., *Parkinson's disease induced pluripotent stem cells with triplication of the α-synuclein locus.* Nature Communications, 2011. **2**(1): p. 440.
29. Fuchs, J., et al., *Phenotypic variation in a large Swedish pedigree due to SNCA duplication and triplication.* Neurology, 2007. **68**(12): p. 916–922.
30. Smith, L., and A. Schapira, *GBA variants and Parkinson disease: Mechanisms and treatments Cells,* 2022. ***11*** *(8)*: p. 1261.
31. Murphy, K.E., et al., *Reduced glucocerebrosidase is associated with increased α-synuclein in sporadic Parkinson's disease.* Brain, 2014. **137**(3): p. 834–848.
32. Schöndorf, D.C., et al., *iPSC-derived neurons from GBA1-associated Parkinson's disease patients show autophagic defects and impaired calcium homeostasis.* Nature Communications, 2014. **5**(1): p. 4028.
33. Aflaki, E., et al., *A new glucocerebrosidase chaperone reduces α-synuclein and glycolipid levels in iPSC-derived dopaminergic neurons from patients with Gaucher disease and parkinsonism.* Journal of Neuroscience, 2016. **36**(28): p. 7441–7452.
34. Kim, S., *Misregulation of Mitochondria-Lysosome Contact Sites in GBA-linked Parkinson's Disease.* 2021, Northwestern University.
35. Lang, C., et al., *Single-cell sequencing of iPSC-dopamine neurons reconstructs disease progression and identifies HDAC4 as a regulator of Parkinson cell phenotypes.* Cell Stem Cell, 2019. **24**(1): p. 93–106. e6.
36. Marotta, N., S. Kim, and D. Krainc, *Organoid and pluripotent stem cells in Parkinson's disease modeling: An expert view on their value to drug discovery* Expert Opinion on Drug Discovery, 2020. **15**(4): p. 427–441.
37. Antipova, D., and R. Bandopadhyay, *Expression of DJ-1 in neurodegenerative disorders* DJ-1/PARK7 Protein: Parkinson's Disease, Cancer and Oxidative Stress-Induced Diseases, 2017. **1037**: p. 25–43.
38. Huang, M., and S. Chen, *DJ-1 in neurodegenerative diseases: Pathogenesis and clinical application* Progress in Neurobiology, 2021. **204**: p. 102114.
39. Malgieri, G., and D. Eliezer, *Structural effects of Parkinson's disease linked DJ-1 mutations* Protein Science, 2008. **17**(5): p. 855–868.
40. Wang, X., et al., *Parkinson's disease-associated DJ-1 mutations impair mitochondrial dynamics and cause mitochondrial dysfunction.* Journal of Neurochemistry, 2012. **121**(5): p. 830–839.
41. Bonifati, V., et al., *Mutations in the DJ-1 gene associated with autosomal recessive early-onset parkinsonism.* Science, 2003. **299**(5604): p. 256–259.
42. Mata, I.F., et al., *LRRK2 in Parkinson's disease: Protein domains and functional insights.* Trends in Neurosciences, 2006. **29**(5): p. 286–293.
43. Xiong, Y., T.M. Dawson, and V.L Dawson, *Models of LRRK2-associated Parkinson's disease.* Leucine-Rich Repeat Kinase, 2017. **LRRK2**(LRRK2): p. 163–191.
44. Li, J.Q., L. Tan, and J.T. Yu, *The role of the LRRK2 gene in parkinsonism* Mol Neurodegener, 2014. **9**: p. 47.
45. López de Maturana, R., et al., *Mutations in LRRK2 impair NF-κB pathway in iPSC-derived neurons.* J Neuroinflammation, 2016. **13**(1): p. 295.
46. Goveas, L., et al., *Mind the gap: LRRK2 phenotypes in the clinic vs. in patient cells.* Cells, 2021. **10**(5).
47. Pickrell, A.M., and R.J. Youle, *The roles of PINK1, parkin, and mitochondrial fidelity in Parkinson's disease* Neuron, 2015. **85**(2): p. 257–273.
48. Whitworth, A.J., and L.J. Pallanck, *The PINK1/parkin pathway: A mitochondrial quality control system?* Journal of Bioenergetics and Biomembranes, 2009. **41**: p. 499–503.
49. Siuda, J., et al., *Early-onset Parkinson's disease due to PINK1 p.Q456X mutation–clinical and functional study.* Parkinsonism Relat Disord, 2014. **20**(11): p. 1274–8.
50. Oh, C.K., et al., *S-nitrosylation of PINK1 attenuates PINK1/parkin-dependent mitophagy in hiPSC-based Parkinson's disease models.* Cell Rep, 2017. **21**(8): p. 2171–2182.
51. Quinn, P.M.J., et al., *PINK1/PARKIN signalling in neurodegeneration and neuroinflammation.* Acta Neuropathologica Communications, 2020. **8**(1): p. 189.
52. Chung, S.Y., et al., *Parkin and PINK1 patient iPSC-derived midbrain dopamine neurons exhibit mitochondrial dysfunction and α-synuclein accumulation.* Stem Cell Reports, 2016. **7**(4): p. 664–677.
53. Klein, C., and A. Westenberger, *Genetics of Parkinson's disease* Cold Spring Harb Perspect Med, 2012. **2**(1): p. a008888.

6 Chemical-Induced Alzheimer's Disease Models with an Emphasis on Therapeutic Significance and Limitation of Natural Products

Ashutosh Gupta, Aakansha Pathak, and Neeraj Kumar
ICMR-National Institute of Child Health and Development Research, Sajdarjung Hospital Campus, New Delhi, India

6.1 INTRODUCTION

Alzheimer's disease (AD) is a disorder related to neurological dysfunction that primarily affects speech, cognitive ability, memory, and movement [1]. It is recognised by degradation of cholinergic neurons present in the hippocampus, cerebral cortex, basal ganglia, and amygdala of the brain, which suppresses formation and excretion of the neurotransmitter acetylcholine (Ach) and amyloid beta (Aβ) deposition, leading to dementia (60%–80%) [2]. Dementia is the key indicator of AD in older persons. Besides, AD disorder compromises the whole lifestyle and ultimately decreases living prospects [3]. Approximately 50 million people are suspected to be detected with AD, and it is predicted that by 2050 this rate might surpass 152 million throughout the world [4]. At present, AD is the fifth foremost cause of mortality in individuals aged 65 and more and the sixth leading cause of death across individuals of all ages. Usually, AD cases are more commonly reported in developing countries, with nearly 68% of total AD patients reported in South Asian countries [5]. Regarding healthcare costs, around $100 billion is spent on the care of AD patients every year. A study reported that AD exerted an immense economic and social burden on individuals, because of delayed diagnosis and lack of primary care [6]. According to a report, less than 50% of AD patients are identified in timely and get primary treatment. Generic medicines are unsuccessful, and individuals emerging countries cannot even afford modern AD treatment [7]. On the other hand, commercially available drugs are very costly and are not easily accessible, while synthetic drugs have some side effects that lead to several cellular abnormalities [8]. The unremitting rise in the occurrence, related disability, and threatened health status with high mortality rate in developed and developing nations prompted a huge interest in the research to explore targeted compounds or drugs for managing AD in respect of effectiveness, availability, and safety. Reports published in past decades revealed that the occurrence of AD is less common in patients consuming natural products of a medicinal nature in their diet [9].

Natural sources like microbes, plants, marine organisms, and animal-derived compounds exhibited plentiful bioactive compounds along with complex structures and novel medicinal features [10, 11]. They are highly therapeutic, with fewer side effects than synthetic drugs [12]. The intake of natural products or bioactive compounds that enhance brain functioning will aid in the overall AD management [13, 14]. Bioactive compounds and their derivatives as one of the chief sources of new drugs have been comprehensively examined to develop more efficient drugs for AD treatment [15]. In fact, galantamine, a natural compound, or rivastigmine, a semi-synthetic compound itself has a cholinesterase inhibitory property [16, 17]. Natural products or combinations have more beneficial effects compared

 DOI: 10.1201/9781032620442-6

to single natural compounds alone. At the same time, combinations of natural products have numerous target sites for AD treatment [18]. Growing evidence has recommended that natural compounds or their products might play a significant role in providing cognitive aids to AD patients [19]. In recent years, several medicinal herbs and their extracts were studied in animal models as well as in AD patients to understand their mode of action [9, 20]. This chapter highlights some Alzheimerogenic chemicals and therapeutic action of natural product for the purpose of enriching existing knowledge.

6.2 AD PATHOGENESIS

AD is a complex neurodegenerative problem and one of the leading causes of dementia throughout the world (**Figure 6.1**) [21]. It is identified by hyperphosphorylated τ-protein and excessive formation of Aβ-proteins, which causes damage of synaptic connections and neurons in the cerebral cortex and hippocampus [22]. The environmental factors and genetic vulnerability are also accountable for the delayed onset of sporadic AD [23]. Various research have tried to explain the mechanism of AD, which helps in the development of disease-managing drug in later stage. Generally, abnormal tau (τ) protein and β-amyloid protein, or in some cases both of them, are primarily responsible for AD development. The excessive formation and storage of Aβ causes lipid peroxidation, apoptosis, alterations of cell functions, inflammation, and synthesis of neurofibrillary tangles (NFTs) [24, 25]. The hyperphosphorylated τ-protein causes tangles and deposition in the hippocampus region and ultimately leads to cell death [26]. Moreover, various biochemical modifications within cells, like inflammation, metabolic disturbances, oxidative stress, changes in Ca^{2+} homeostasis, as well as the storage of misfolded/unfolded proteins, have also been supposed to trigger neuronal cell degradation in AD patients [27].

6.3 COMMONLY USED CHEMICALS TO INDUCE AD-LIKE SYMPTOMS

6.3.1 ALUMINUM

Aluminum (Al) is ubiquitously distributed in the environment and associated with numerous cellular disorders like anemia, osteomalacia, and neurological problems [28]. Al is most frequently

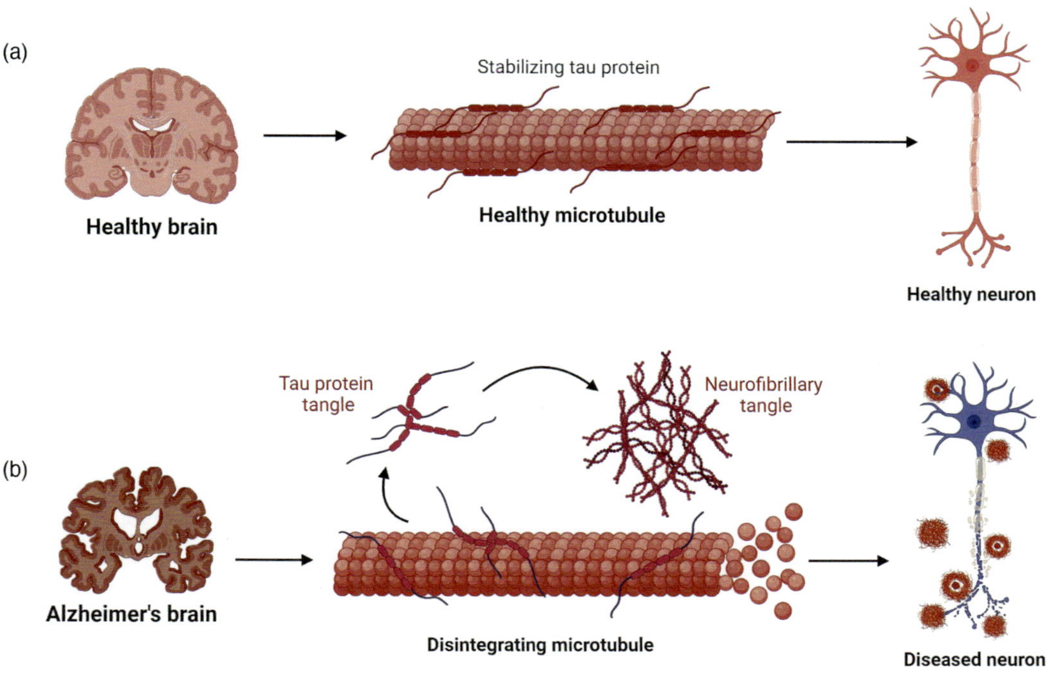

FIGURE 6.1 Pathophysiology of (a) healthy brain (b) brain of Alzheimer's disease.

used in various areas such as food processing, medicines, cosmetics, medical preparations, etc. In humans, Al makes association with plasma transferrin and citrate molecules that can facilitate the movement of Al to the brain. The higher concentration of Al in the brain develops AD and causes adverse toxic effects, including encephalopathy. Al can also behave as a cross-linker of Aβ-protein and causes oligomerisation, consequently inducing neurotoxicity. The underlying mechanisms of Al(III) neurotoxicity were shown to involve oxidative and endoplasmic reticulum stress, mitochondrial dysfunction, inflammation, cell death, interaction with Aβ and α-synuclein, cytoskeletal abnormalities as well as alteration of synaptic plasticity and signal transduction through interference with neurotransmitter systems. It has been observed that prolonged feeding with Al-containing intravenous solutions to infants showed impaired neurological development [29]. A study suggested that rising exposure to Al was related to a decline in the Mental Development Index (MDI), with an adjusted loss of one point per day of intravenous feeding for infants receiving the standard solutions [30]. Furthermore, oral administration of Al (300 mg/kg) triggered accumulation of Aβ and NFTs, oxidative stress, and cholinergic deficiency in rats' brains [31]. Diffuse gliosis and histopathological alterations along with pericellular edema in the cerebral region have been observed, in addition to neuronophagia and loss of neurons in Al-abused rat groups. These findings demonstrated an enhanced activity and expression of malondialdehyde (MDA) and acetylcholine-esterase (AChE) with remarkable reduction in the activity and expression of glutathione reductase (GSH), glutathione-s-transferase (GST), and glutathione peroxides (GPx) (**Figure 6.2**).

6.3.2 Colchicine

Colchicine is a natural alkaloid, commonly isolated from the plants of Liliaceae family and primarily helpful in gouty arthritis treatment. During the course of evolution, colchicine was established as a treatment for Mediterranean fever attacks; after decades, colchicine was successfully used for amyloidosis treatment [32]. It is a microtubule-disrupting compound that leads to massive damage to mossy fibers, granule cells of hippocampal and septo-hippocampal pathways. Colchicine binding to tubulin protein triggers neurofibrillary deterioration that results in impairment of learning and memory [33]. A study proposed that colchicine-mediated memory impairment may be an effect of

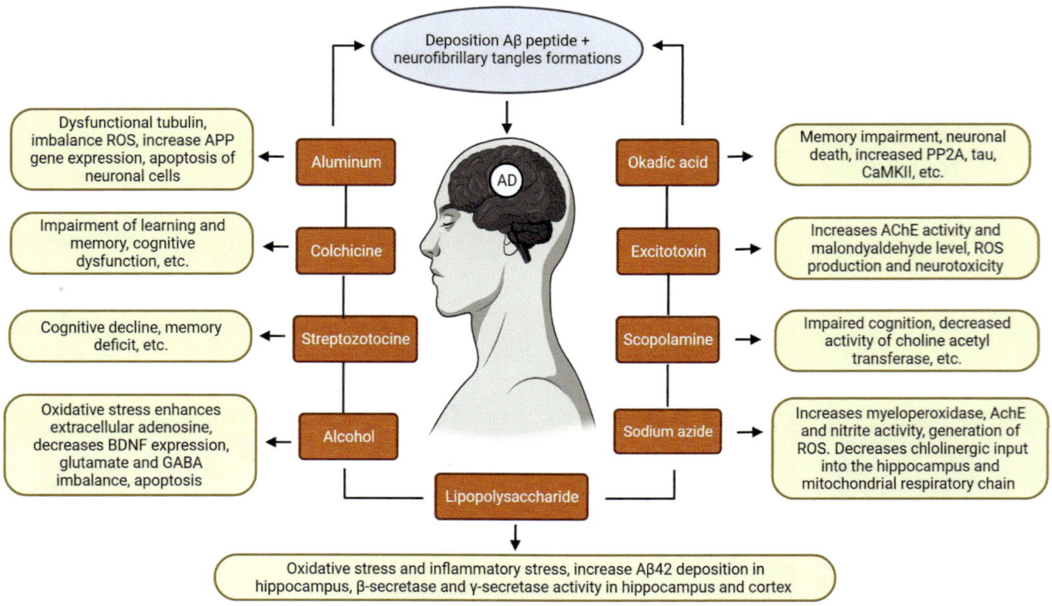

FIGURE 6.2 Alzheimerogenic chemicals and their mode of action to induce-AD.

a reduced serotonin, norepinephrine, and dopamine within the hippocampus, caudate nucleus, and entire cerebral cortex [34]. In addition, colchicine causes overproduction of protein carbonyl following lipid peroxidation. Some studies also suggested that colchicine increases the expression levels of cyclooxygenase-1 (COX-1) and COX-2 and reactive oxygen species (ROS) formation [35, 36]. Colchicine prompts the ratio of glutamate/γ-Aminobutyric acid (GABA) in the cortex and stimulates MDA receptors that are responsible for rapid increase in a Ca^{2+} influx, causing enzyme activation that depends on Ca^{2+} (e.g. COX, phospholipases A2, proteases, xanthine oxidase, and protein kinases) [37]. A study suggested that intracerebroventricular (ICV) injection of colchicine (15 μg in 5μL) leads to cognitive memory reduction in rats [38]. A remarkable deficit of memory was noticed two weeks after induction of cognitive impairment using colchicine [33]. In addition, ICV injection of colchicine (3 g) in mice promoted spatial memory damage [4]. These findings demonstrated that colchicine developed symptoms of sporadic Alzheimer's type of dementia in animal models, related to those reported in human subjects.

6.3.3 Alcohol

Excessive alcohol consumption is linked with several issues, such as motor dysfunction, language impairment and social skills, attention deficit, hyperactivity, and learning failures [39]. Prolonged excessive alcohol intake leads to higher ROS production that triggers neuroimmune response and cellular apoptosis and necrosis via lipid peroxidation, mitochondrial, protein or DNA damage. Long term alcohol administration also upregulates glutamate receptors, glucocorticoids and reduces reuptake of glutamate in the central nervous system, resulting in glutamate excitotoxicity, and eventually mitochondrial injury and cell death [40]. In addition, ethanol could impair cholinergic neurons and the hippocampal region with a consequential action on the sensory-motor system along with disturbance of learning and memory [41]. A study reported that excessive ethanol intake resulted in high nitric oxide (NO) production, which in the long-term damaged memory and learning ability, while a higher concentration of ethanol elevated GABAergic transmission and altered the glutamatergic system in the brain that is associated with memory damage [41]. Another study reported that ethanol also triggers the adenosine level that causes memory impairment and stimulates active and spontaneous motion [42]. Furthermore, mice treated with ethanol demonstrated a noteworthy increase in total swimming distance and escape latency, with a sharp decline in cross time during the Morris water maze (MWM) test. Hence, excessive alcohol consumption can expressively disturb the cognition and learning of spatial capability in mice [43]. Moreover, memory damage associated with persistent alcohol administration can also be examined by anti-inflammatory efforts and the control of glutamate and GABA equilibrium.

6.3.4 Lipopolysaccharide

Lipopolysaccharide (LPS), is derived from the outer membrane of gram-negative bacteria and is responsible for inflammatory response. LPS has been established as a potent endotoxin with resistance to denaturation by enzymes, hence causing persistent inflammatory stimuli and synthesising pro-inflammatory cytokines [44]. These pro-inflammatory cytokines trigger both neuroendocrine and neuroimmune systems, which leads to nearly identical responses formed by behavioral stress [45, 46]. ICV injections of LPS in rats elevated interleukine (IL)-1β levels in the diencephalon and brainstem two hours after injection. While six hours after injection, the IL-1β level increased in all regions of the brain except the cerebellum [47]. In addition, upregulated IL-1β mRNA expression in the nucleus basalis and hippocampus was also seen, along with the presence of mRNA for the tumor necrosis factor-α (TNF-α) gene in the nucleus basalis [48]. Similarly, injection of LPS with Aβ-fibrils at three cortical locations (frontal, sensorimotor, and parietal cortices) of marmoset monkeys triggered the process of amyloidosis, while LPS-injected monkeys showed an early AD immune blood cell expression profile on CD95 and CD45RA [49]. In various other studies, inflammatory proteins in the blood, namely C reactive protein (CRP) and IL-6, have appeared to upsurge many years before the onset of dementia [50, 51]. The appearance of Aβ1-40/42 and LPS in amyloid plaques in gray and white matter of AD-affected brains has also been studied [52]. A study demonstrated that LPS is ample in the hippocampus and neocortex of AD affected individuals and that there is a strong adherence of LPS to the nuclear periphery in AD brain cell nuclei [53]. The presence of LPS was also confirmed in lysates of the superior temporal lobe neocortex and hippocampus of the brains [54]. Available literature recommended that the levels of IL-6 and TNF-α, IL-1β expression within the hippocampus are increased in contrast to a normal brain following three days of LPS administration [55].

6.3.5 Okadaic Acid

Okadaic acid (OA) is a natural marine toxin synthesised by various phytoplanktonic species, mainly from the dinoflagellates group [56]. The ICV administration of OA has been suggested to trigger memory weakening in rats, which makes OA suitable to examine anti-dementia drugs [57]. Previous studies reported that OA has preventive and moderate antagonists of serine/threonine phosphatases-1 and -2A (PP1, PP2A) [58, 59]. It is linked with short-term or long-term memory disorders in animal models [60] and activates tau hyperphosphorylation and neuronal cell death, *in vivo* [61] and *in vitro* [62]. OA prevents the progression of synaptic plasticity and decline of basal synaptic transmission [63]. Recent data observed in hippocampal neurons in culture have demonstrated that OA may enhance the ionic influx by the ionotropic non-N-methyl-D-aspartate (NMDA) type of excitatory amino acid receptors, possibly by preventing PP1 and PP2A activity [64]. A study revealed that OA triggers the ROS formation in the hippocampus and reduces the activity as well as the potency of mitochondria, finally resulting in mitochondrial dysfunction in the rat brain. Moreover, OA has been identified as a phosphatases inhibitor and causes hyperphosphorylation of proteins, which consequently develop neuronal stress and finally lead to neural injury [57].

Previously published reports revealed that OA-associated decline in rats' memory is coupled with impaired pro-inflammatory cytokines (IL-1 and TNF-α), iNOS expression, and total nitrite level in the hippocampus and cortex region of the brain [65]. Moreover, bilateral infusion of OA in the hippocampus region developed spatial cognitive deficiency as a result of reduced GSH, elevated Glial fibrillary acidic protein (GFAP) expression, and boosted mitogen-activated protein kinase 38 (p38 MAPK) and carbonylation of proteins [66]. Various kinases, comprising cyclin-dependent kinase 5 (CDK5), MAPK, and glycogen synthase kinase-3 beta (GSK-3β), have been recognised to be associated with the τ-protein phosphorylation at several positions as seen in AD hyper-phosphorylated tau [67, 68]. In addition, dephosphorylation of tau protein is catalysed via PP2A and other phosphatases (PP1, 2B, and 5) [69, 70]. Studies suggested that OA-mediated selective inhibition of PP2A causes Alzheimer's-like aggregation and hyperphosphorylation of tau *in vivo* [71] and *in vitro* [72, 73]. Alterations in memory because of intra-hippocampal injection of OA has been found to be linked with recognisable neuropathological alterations, including the formation of helical filament-like phosphorylated tau, Aβ peptide-like structures, and neuronal damage in the hippocampus [66]. Currently, drugs that act by inhibiting tau phosphorylation are not known. Therefore, it is proposed that OA could be a suitable alternative for exploring therapeutic approaches for AD pathogenesis [74].

6.3.6 Excitotoxin

Excitotoxins are endogenously synthesised amino acids that, under physiological conditions act as excitatory neurotransmitters but when produced in large amounts become toxic to the neuronal cells. Ibotenic acid is a potent neurotoxin that provokes signs and pathophysiology analogous to AD [75, 76]. It is an agonist of glutamate receptors and precisely acts at both the trans-1-Amino-1,3-dicarboxycyclopentane (ACPD) and NMDA receptor sites in multiple systems in the CNS. It is a suitable toxic agent to study drug efficiency in evading AD pathology. A previous study reported that intra-hippocampal injection of ibotenic acid (5μg/μL) leads to memory deficiency and consequences in elevated AChE activity and MDA levels, thus initiating neuronal toxicity [77]. Various other cholinotoxins and neurotoxins that exhibit AD-like signs are NMDA antagonist dizocilpine, anti-nerve growth factor (NGF), kainic acid, quinolinic acid, and α-amino-3-hydroxy-5-methyl4-isoxazolepropionic acid (AMPA). According to a study, intra-hippocampal injection of kainic acid (0.4μg/2μL) causes oxidative damage (as confirmed by the high hippocampal nitrite level, lipid peroxidation, and reduced superoxide dismutase (SOD)). This study proposed that kainic acid promotes memory loss and oxidative damage in rats [78]. In a diseased condition, stimulation of glutamate receptors, particularly NMDA receptors, can lead to decline in energy, neuronal loss, ion homeostasis disruption, and cell death (excitotoxicity) by an unusually high glutamate level [79, 80].

6.3.7 Scopolamine

Scopolamine, also known as hyoscine, is a natural or synthetically derived tropane alkaloid. It is an anticholinergic drug used for the treatment of motion sickness, vomiting, and postoperative nausea [81]. Scopolamine induces AD by enhancing inflammation, oxidative stress, and AChE levels, which are precisely related to memory and cognitive impairment [82]. Furthermore, studies have suggested that scopolamine blocks the adhesion site of acetylcholine (ACh) muscarinic receptors non-selectively in the cerebral cortex and causes insufficient release of ACh, which damages hippocampal neurons and dose-dependently promotes memory and learning disability in mice [3, 83], although the major effects of scopolamine occur due to the inhibition of some receptors (M1 and M5) because of their discrete diffusion in the brain [84]. In some cases, recommended scopolamine doses led to confusion, hallucinations, agitation, delusions, incoherent speech, and paranoid behaviors [85]. Recent report suggested that in memory deficit there is an active participation of NMDA receptor of the dorsal hippocampus because of scopolamine [86]. DNA methyl transferase-1 and histone deacetylase-2 are essential for synaptic plasticity triggered by scopolamine-mediated memory decline [87]. Scopolamine injection on the hippocampal region blocks long-term potentiation and hinders spatial encoding [88, 89]. Moreover, scopolamine application into the medial septum alters ACh release and obstructs learning in the hippocampus [90]. Its injection in the CA3 region of the hippocampus selectively promotes encoding damage while memory recovery was not affected during the Hebb-William maze test [91]. Similarly, scopolamine leads to the excitation of GSK-3β, inadequate spine maturation, and the arborisation of dendrites connected with modification in CREC, Homerl, and AMPA [92].

6.3.8 Streptozotocin

Streptozotocin (STZ) is a glucosamine nitrosourea-derived compound isolated from the *Streptomyces achromogenes* strain. ICV administration of STZ in rodents exerts a critical and permanent action on the metabolism, biochemistry, functions, and cytoarchitecture of the brain [93]. STZ also resembles certain features of nitrosourea, an anticancer agent that has hypoglycemic action apart from involvement in memory impairment [94]. Initiation of neurodegeneration and hyperphosphorylation of tau is triggered by STZ, causing overproduction of ROS and reactive nitrogen species (RNS) [95]. Outcomes have suggested that STZ alters the activity of glycolytic enzymes in the brain, which ultimately leads to decline in phosphocreatine concentrations and ATP levels. This lowered the acetyl CoA synthesis and devastated energy system can prevent cholinergic conductance [96]. STZ-abused rats demonstrated to have increased AChE activity in their brains and reduced Ach concentration. On the other hand, STZ activates Aβ peptide-like aggregates by altering the GSK α/β [97]. In addition, there is significant evidence that the expression of genes related with the development of brain-derived neurotrophic factors, glial-derived neurotropic factor, and integrin-α-M becomes elevated by STZ administration, while gene expression of NGE-IB and metallothionein

1/2 is down-regulated and ultimately causes alteration in cell survival and apoptosis process [98]. ICV injection of STZ causes a remarkable fall in learning skills and memory and also develop oxidative stress in rats [99]. Similarly, ICV injection of STZ (3 mg/kg) mimicked the sporadic dementia of AD in rats. Memory deficiency was assessed by the MWM test during the last phase of the experiment in the rats [100]. Furthermore, decline in cholinergic markers, compromised glucose metabolism, and oxidative stress were also observed in the rats' brains.

6.3.9 Sodium Azide

Sodium azide is a chemical of rapidly growing commercial importance with a high acute toxicity and an unknown mechanism of action [101]. Sodium azide prevents cytochrome oxidase enzymes and induces mitochondrial dysfunction [102]. Cytochrome oxidase performs an essential function in the mitochondrial electron transport system; its alteration obstructs mitochondrial complex-IV and reduces ATP levels, which leads to formation of ROS and metabolic alterations, finally developing some symptoms like AD [103]. Neurodegeneration in the CA3 region of the hippocampus is the key characteristic used to endorse the AD induction. Previously published reports revealed that administration of sodium azide weakens memory and learning, along with rises in AChE levels in the brain [102]. Sodium azide causes oxidative injury that eventually leads to neuronal degradation. The progressive loss of neurons and necrosis were seen in the hippocampal and cortical regions of treated rats' brains, which again confirmed the neurotoxicity of sodium azide. Further findings demonstrated that intraperitoneal injection of sodium azide for 15 days in rats develops dementia-like features, as confirmed by implanted osmotic minipumps, and supports the active role of sodium azide in inducing neurodegenerative complications [104].

6.4 THERAPEUTIC SIGNIFICANCE OF NATURAL PRODUCTS

Therapeutic herbs have been well documented to reduce the advancement of numerous cellular disorders, including AD [10, 105]. Extensive research has been done to explore the therapeutic properties of medicinal plants and their extracts or compounds on AD [15). Available literature has demonstrated that bioactive compounds exert anti-neuroinflammatory action by prevention of microglia activation, suppression of the release of pro-inflammatory cytokines from activated microglia, or by prevention of NF-κB and p38 MAPK activation. The known therapeutic attributes of medicinal herbs are the outcome of anti-neuroinflammatory or anti-oxidative actions, which act through inhibiting the accumulation of Aβ and tau peptides and enhancing cholinergic signaling. Some common medicinal plants that have significant neuroprotective properties and will be used for the treatment of AD are discussed under in this section (**Table 6.1**).

6.4.1 *Centella asiatica*

Centella asiatica (gotu kola) is a tropical medicinal plant belonging to the family Apiaceae. Ancient medicinal systems have used *C. asiatica* to restore the neuronal cells, enhancing intelligence, longevity, and memory [106]. Isolated bioactive compounds such as asiatic acid and asiaticoside have shown the capability to reduce H_2O_2-mediated cell cytotoxicity, reduce the level of free radicals, and prevent β-amyloid cell injuries *in vitro*. Aqueous extract of *C. asiatica* whole aerial part (100–300 mg/kg) lowered the pathogenesis of β-amyloid and suppressed the oxidative stress in rats' brain [107]. In Aβ1-40–induced neurotoxicity, *C. asiatica* ethanol extracts (25–100 μg/mL) protect neuronal cells (rat PC12 pheochromocytoma cells and human IMR32 neuroblastoma cells), suppresses ROS formation, and regulates the antioxidant defense system by augmenting the activities of glutathione peroxidase (GPx), glutathione reductase (GSH), SOD, catalase, and levels of glutathione and glutathione disulfide (GSSG) [108, 109]. These findings promote the therapeutic role of *C. asiatica* for the prevention and management of AD.

TABLE 6.1
Anti-Alzheimer's Activity of Medicinal Plants

Natural Product	Plant Part Used	Experimental Mode/Dose	Mechanism of Action	References
Centella asiatica	Whole aerial part	PC12 cell line and mice	Suppress oxidative stress and β-amyloid pathology; protect neurons against the neurotoxicity, diminish ROS production, and increases antioxidant-related enzyme activity.	107–109
Moringa oleifera	Leaf	Rats	Inhibit Aβ accumulation, improve memory and learning ability.	110
Ginkgo biloba	Leaf	PC12 cell line, rats and clinical trial	Scavenge ROS-regulated mitochondrial alterations, activated JNK and ERK pathways, and neuronal apoptosis inhibition.	112–114
Juglans regia	Defatted protein	Rats	Reduce pro-inflammatory cytokines and NF-κB expression, decrease AChE level, and restore antioxidant enzymes activity.	115–116
Allium sativum	Aged garlic extract	Rats	Reduce microglia activation and IL-1 level, minimise the inflammatory response, and reduce psychological stress.	119–121
Punica granatum	Juice and extracts	SH-SY5Y cell line and mice	Mitigate oxidative stress and brain inflammation, decrease the accumulation of soluble Aβ42, and suppress amyloid deposition in the hippocampus.	123–126
Bacopa monnieri	Stem and leaf	Cortical cell culture, rats and clinical	Act as an anticholine esterase, antidementia, inhibition of Aβ accumulation.	128–132
Panax ginseng	Root	*In vitro*, mice and clinical trials	Reduce Aβ formation and aggregation, inhibit AChE, restore synaptophysin and ChAT activity.	133–138
Uncaria rhynchophylla	Whole aerial part	Rats	Scavenge free radicals and prevent lipid peroxidation, reduce microglial activation, nNOS, iNOS, and apoptosis, inhibit Aβ fibril formation and dissemble preformed Aβ fibrils	140–143

6.4.2 *Moringa oleifera*

Moringa oleifera (MO) is a prehistoric medicinal plant belonging to the family Moringaceae. Its leaves have been reported as a good reservoir of antioxidants, along with β-carotene, vitamin C, and E. In a study, it has been found that ICV infusion of colchicine exhibited cognitive damage and memory impairment in rats, with unfavorable changes in antioxidant enzymes. Remarkably, administration of MO leaf extract (250 mg/kg) for 14 days showed enhanced learning ability and memory. A noteworthy rise in antioxidant enzymes like catalase and SOD were observed along with a remarkable decline in the level of lipid peroxidase. Findings suggested that the improved antioxidant action may be a reason for the improvement in cognitive behavior. However, clinical trials are required to reveal the mechanism of action of MO [110].

6.4.3 *Ginkgo biloba*

Ginkgo biloba L., known as a "living fossil,"has been utilized for over 2000 years as a therapeutically valuable plant for human health. Its leaves have been used for boosting memory and age-associated problems. The therapeutic action of *G. biloba* leaf extracts against β-amyloid induced neurotoxicity has been demonstrated to promote cytotoxicity and quenching of ROS. Leaf extract improve mitochondrial function, activate c-Jun N-terminal kinase (JNK) and extracellular signal-regulated kinase (ERK) pathways, and prevent neuronal apoptosis. Bilobalide derived from leaf has significant therapeutic effects on neurons and Schwann cells [111]. A study suggested that, bilobalide down-regulates p53, Bax, and caspase-3 proteins expression and inhibits ROS-mediated apoptosis in PC12 cells [112]. Moreover, bilobalide induced synaptogenesis and neurogenesis by enhancing transcription factors levels, namely, phosphorylated neurotrophin, brain-derived neurotrophic factor (BDNF), and cAMP response element-binding protein (CREB) in neuronal cells [113]. In addition, the clinical examination of dietary items that contain *G. biloba* extract was found to be useful in managing cognitive failure in the progression of aging and in the initial phases of AD [114].

6.4.4 *Juglans regia*

Juglans regia (walnut) is a medicinal plant copious in valuable fatty acids, polyphenols, alpha-tocopherol, and vitamins. Epidemiological evidence had demonstrated that a walnut-rich diet lowered the prevalence of neurodegenerative disorders. The preventive action of walnut extract against the formation of Aβ fibril was examined by electron microscopy and thioflavin T (ThT) fluorescent spectroscopy. After two or three days of incubation, methanolic extraction of walnuts exhibits 90% inhibition of Aβ fibril formation along with defibrillisation of preformed Aβ fibrils. A study proposed that the phytoconstituent obtained from walnuts may be accountable for the amyloidogenic activity [115]. Walnut extract (2-4 μg GAE) incubated with Aβ1-42 for 24 hours reduced the Aβ-mediated oxidative stress and toxicity in PC12 cell lines. Further analysis demonstrated that the extract showed protection against Aβ-mediated cell death. The mechanism of *J. regia* action might be related with the mitigation of ROS overproduction and may diminish membrane impairment and DNA injury. Hence, a walnut-rich diet may be helpful for inhibition and delaying the onset of AD [116].

6.4.5 *Allium sativum*

Allium sativum (garlic) has a long history of use in the medicine and food industries. In a Tg2576 mouse model, aged garlic extract enhanced memory in the hippocampal region [117]. Besides, garlic extract quenched the free radicals; improved antioxidant enzyme levels, such as SOD, Catalase, GPx, and glutathione levels; prevented lipid peroxidation; and suppressed the production of prostaglandins [118]. In addition, the extract also inhibited 3-hydroxy-3-methylglutaryl-CoA (HMG-CoA) reductase and lowered cholesterol formation. The extract protected neurons from β-amyloid-related neurotoxicity and apoptosis, regulated cognitive imbalance, ischemia or reperfusion-mediated neuronal damage, as well as improved learning and memory retention [119]. Separately, it has been observed that aged garlic extracts were beneficial in improving short-term memory recognition and lowered the Aβ-mediated neuro-inflammation as well as attenuate neuro-inflammatory responses in microglial cells of rats [120, 121].

6.4.6 *Punica granatum*

Punica granatum L. (pomegranate) is a polyphenolic-rich fruit of the family Lythraceae that has been widely used in traditional medicine [122]. It has been reported that ellagic acid, a bioactive compound, can significantly regulate Aβ-triggered neurotoxicity in SH-SY5Y cells. Further investigation demonstrated that more fibrils formation in Aβ42 samples co-incubated with ellagic acid in earlier phases of aggregation., [123]. Ellagic acid can inhibit neurotoxicity by triggering the formation of Aβ-fibril and noteworthy oligomer loss [124]. Ellagic acid and punicalagin both are found in pomegranate husk, which inhibits the activity β-secretase and exhibit anti-dementia activity [125]. During a study, a remarkable reduction in Aβ42 levels and extracellular amyloid plaques was noticed in the hippocampus region when pomegranate juice was given to transgenic mice (APPsw/Tg2576). Furthermore, enhanced spatial learning and cognitive activities were also confirmed by several behavioral investigations, such as water maze, CUED water maze, and spatial water maze tests [126]. Because of the inconsistencies in the molecular mechanism behind the reported neuroprotective action, additional advanced investigations are required to reveal the therapeutic significance of pomegranate juice in AD treatment.

6.4.7 *Bacopa monnieri*

Bacopa monnieri is frequently utilised as a cardiotonic agent, nerve tonic, and diuretic drug and also for the management of rheumatism, epilepsy, insomnia, and asthma [127]. Extensive investigation on nootropic action and neuropharmacological attribute of *B. monnieri* extracts have been performed throughout the world [106]. *B. monnieri* enhances the protein kinase activity in the hippocampus region of rats' brains, which provides a nootropic action [128]. In rats, *B. monnieri* extract suppressed the degeneration of cholinergic neurons and exhibited cognition-enhancing action [129]. Separately, *B. monnieri* (5 and 10 mg/kg) elevated ACh levels and prevented the activity of AChE. In addition, *B. monnieri* promoted a dose-mediated rise in GPx, SOD, and catalase activities in the brain (frontal cortical, striatal, and hippocampal regions) after 14 and 21 days [130]. *B. monniera* extracts (100–600 μg/mL) also provide protection against β-amyloids-triggered neuronal cells injuries. Extract-treated neuronal cells displayed significant reduction in ROS production that confirms the mitigation of intracellular oxidative stress by *B. monnieri* [131]. A clinical trial on AD patients revealed that *B. monnieri* extract containing a polyherbal formulation proficiently enhanced cognitive activities and reduced inflammation and oxidative stress [132].

6.4.8 *Panax ginseng*

Panax ginseng has been commonly used in eastern countries as traditional medicine and exhibits numerous pharmacological activities. A study reported that *P. ginseng* extract could prevent Aβ-induced neurotoxicity *in vitro* and dose-dependently attenuated the accumulation of Aβ in mice brains [133]. In a mice AD model (ICR and transgenic), fermented root extracts (400 and 800 mg/kg) suppressed the Aβ formation in brain and enhanced memory function [134]. The extracts of white, red, and black ginseng (200 mg/kg) protected from memory dysfunction and inhibited AChE activity in ICR mice caused by hippocampal Aβ oligomer injection [135]. Oral administration of extracts restored the synaptophysin and ChAT activities in mice injected with Aβ oligomer [133]. In a clinical trial, long-term administration of red ginseng extract (4.5–9 g/day) with AD drugs, progressively enhanced cognitive functions with negligible adverse action on AD patients [136]. Similarly, more than 12 weeks' administration of red ginseng extract (4.5 g/day) was believed to restore frontal cortical functioning in elderly AD patients [137]. Extensive studies have confirmed the efficiency of different type of ginseng extracts on the management of AD symptoms. It has been observed that certain bioactive compounds present in *P. ginseng* may inhibit cognitive abnormality by lowering Aβ formation and aggregation. Ginsenosides and gintonin are the major bioactive compounds that provide protection against Aβ-triggered ROS production and neurotoxicity, exhibit anti-inflammatory function, promoted hippocampal neurogenesis, enhance the formation of sAPPα instead of Aβ, and upregulate cholinergic systems and cognitive functions [138]. These findings supposed that *P. ginseng* can be used for the AD management and treatment.

6.4.9 *Uncaria rhynchophylla*

Uncaria rhynchophylla (Miq.) Jacks (Rubinaceae) is a medicinal herb commonly used for convulsions, hypertension, epilepsy, eclampsia, and other cerebral diseases [139]. In Sprague Dawley rats, *U. rhynchophylla* extract (1g/kg) showed free radical scavenging and lipid peroxidation inhibition activity that confirmed anticonvulsant activity [140]. It has also been observed that *U. rhynchophylla* crude aqueous extract (0.5 and 1.0g/kg) provides protection against kainic acid-mediated neuronal damage in Sprague Dawley rats by suppressing microglial activation, nNOS, iNOS, and apoptosis [141]. Furthermore, it inhibits Aβ fibril aggregation and also disassembles preformed Aβ fibrils in Aβ1-40 and Aβ1-42-induced AD models [142]. Active phytoconstitutents reported in *U. rhynchophylla* are primarily alkaloids like hirsutine, rhynchophylline, isorhynchophylline, etc. [11]. Rhynchophylline and isorhynchophylline are broadly investigated and have been widely accepted as neuroprotective compounds [143].

6.4.10 *Paeonia suffruticosa*

P. suffruticosa is a therapeutic herb well known for its antipyretic and anti-inflammatory activities.1,2,3,4,6-Penta-O-galloyl-β-D-glucopyranose (PGG) is the main polyphenol present in *P. suffruticosa*. The neuroprotective action of PGG against Aβ1-42–induced neurotoxicity suggested that PGG prevents the oligomerisation of Aβ1-42. MTT assay confirmed that Aβ1-42 peptide-treated SK-N-SH cells reinstated viability of cell after incubation with PGG. Furthermore, *P. suffruticosa* extract treatment remarkably reduced Aβ peptide concentration and improved memory in Tg2576 mice. Besides, lower concentration of PGG promotes destabilisation and inhibits the formation of Aβ fibril [144]. Hence, the *P. suffruticosa* plant and its bioactive compounds might act as a natural medicine for AD treatment.

6.5 LIMITATIONS

Natural products are valuable resources for discovering new drugs for the treatment of AD. Some major challenges and limitations related to plants or their products that might be accountable for clinical effectiveness are physicochemical instability, rapid metabolism, low bioavailability, limited water solubility, and distribution to the CNS [145–148]. The existence of the blood-brain barrier (BBB) needs sufficient lipophilicity from the natural products to be capable to cross the CNS, which may form further obstacles for some neuroprotective natural products, comprising polyphenols and polysaccharides, and regulate their clinical effectiveness. It has also proved very challenging to translate the ongoing preclinical neuroprotective findings of natural products to clinical applications. To enhance the bioavailability of natural products, nanotechnology and nanocarrier-based approaches have been developed in the delivery of natural product mixtures and their bioactive compounds, which may improve the remedial action and clinical efficacy [149, 150].

6.6 CONCLUSION

A wide range of chemicals have been studied in order to understand the etiology of AD and investigate various probable therapeutic agents for its treatment. Rising evidence has established the neuroprotective abilities of natural products and their active constituents in AD treatment with minimum adverse effects. Most of them act against tau hyper phosphorylation and Aβ formation and inhibit AChE activity. Although previously published reports favor the neuroprotective action of several natural products that are presently available, still advanced investigation is needed to understand the target specific action of the medicinal herbs on different signaling pathways, on the potential to cross the blood brain barrier, and on action mechanisms in various regions of the brain. A comprehensive investigation is also required to understand the mechanism behind the synergistic action of bioactive compounds on the target sites in the wider population.

ABBREVIATIONS

Ach acetylcholine
AChE acetylcholine-esterase
AD Alzheimer's disease

Al	aluminum
AMPA	α-amino-3-hydroxy-5-methyl4-isoxazolepropionic acid
Aβ	amyloid beta
BDNF	brain-derived neurotrophic factor
COX	cyclooxygenase
CRP	C reactive protein
GABA	γ-aminobutyric acid
GPx	glutathione peroxides
GSH	glutathione reductase
GSK-3β	glycogen synthase kinase-3 beta
GST	glutathione-s-transferase
MO	*Moringa olifeira*
ICV	intracerebroventricular
IL	interleukine
LPS	lipopolysaccharide
MDA	malondialdehyde
MDI	Mental Development Index
MWM	Morris water maze
NFTs	neurofibrillary tangles
NGF	anti-nerve growth factor
NMDA	N-methyl-D-aspartate
NO	nitric oxide
OA	okadaic acid
PGG	1,2,3,4,6-Penta-O-galloyl-β-D-glucopyranose
STZ	streptozotocin
PP2A	protein phosphatase 2A
RNS	reactive nitrogen species
ROS	reactive oxygen species
SOD	superoxide dismutase
TEM	transmission electron microscopy

REFERENCES

1. Breijyeh, Z, Karaman, R. Comprehensive review on Alzheimer's disease: Causes and treatment. Molecules. 2020;25(24):5789.
2. Knopman, DS, Amieva, H, Petersen, RC, Chételat, G, Holtzman, DM, Hyman, BT, et al. Alzheimer disease. Nat Rev Dis Primer. 2021;7(1):1–21.
3. Van Dam, D, De Deyn, PP. Animal models in the drug discovery pipeline for Alzheimer's disease. Br J Pharmacol. 2011;164(4):1285–300.
4. Mahdi, O, Baharuldin, MTH, Nor, NHM, Chiroma, SM, Jagadeesan, S, Moklas, MAM. Chemicals used for the induction of Alzheimer's disease-like cognitive dysfunctions in rodents. Biomed Res Ther. 2019;6(11):3460–84.
5. Chakraborty, B, Mukerjee, N, Maitra, S, Zehravi, M, Mukherjee, D, Ghosh, A, et al. Therapeutic potential of different natural products for the treatment of Alzheimer's disease. Oxid Med Cell Longev. 2022; 2022: e6873874.
6. 2020 Alzheimer's disease facts and figures. Alzheimers Dement. 2020;16(3):391–460.
7. Porsteinsson, AP, Isaacson, RS, Knox, S, Sabbagh, MN, Rubino, I. Diagnosis of early Alzheimer's disease: Clinical practice in 2021. J Prev Alzheimers Dis. 2021;8(3):371–86.
8. Peng, Y, Jin, H, hui, Xue Y, Chen, Q, yu, Yao S, qiao, Du M, et al. Current And future therapeutic strategies for Alzheimer's disease: An overview of drug development bottlenecks. Front Aging Neurosci. 2023; 15:1206572
9. Andrade, S, Ramalho, MJ, Loureiro, JA, Pereira M do, C. Natural compounds for Alzheimer's disease therapy: A systematic review of preclinical and clinical studies. Int J Mol Sci. 2019;20(9):2313.

10. Gupta, A, Marquess, AR, Pandey, AK, Bishayee, A. Jackfruit (Artocarpus heterophyllus lam.) In health and disease: A critical review. Crit Rev Food Sci Nutr. 2023;63(23):6344–78.
11. Chen, X, Drew, J, Berney, W, Lei, W. Neuroprotective natural products for Alzheimer's disease. Cells. 2021;10(6):1309.
12. Shal, B, Ding, W, Ali, H, Kim, YS, Khan, S. Anti-neuroinflammatory potential of natural products in attenuation of Alzheimer's disease. Front Pharmacol [Internet]. 2018; 9:548.
13. Mir, MA, Agrewala, JN. Dietary polyphenols in modulation of the immune system. In Polyphenols and Health: New and Recent Advances. 2008. p. 1–27. Nova Science Publisher, ISBN 978-1-60456-349-8.
14. Mir, M. Introduction to costimulation and costimulatory molecules. In Developing Costimulatory Molecules for Immunotherapy of Diseases. 2015. p. 1–44. LAP Publishers: Germany.
15. Gupta, A, Singh, AK, Kumar, R, Jamieson, S, Pandey, AK, Bishayee, A. Neuroprotective potential of ellagic acid: A critical review. Adv Nutr. 2021;12(4):1211–38.
16. Kumar, V. Potential medicinal plants for CNS disorders: An overview. Phytother Res PTR. 2006;20(12): 1023–35.
17. Scott, LJ, Goa, KL. Galantamine: A review of its use in Alzheimer's disease. Drugs. 2000;60(5):1095–122.
18. Shao, R, Xiao, J. Natural products for treatment of Alzheimer's disease and related diseases: Understanding their mechanism of action. Curr Neuropharmacol. 2013;11(4):337.
19. Kennedy, DO, Wightman, EL. Herbal extracts and phytochemicals: Plant secondary metabolites and the enhancement of human brain function. Adv Nutr Bethesda Md. 2011;2(1):32–50.
20. Akram, M, Nawaz, A. Effects of medicinal plants on Alzheimer's disease and memory deficits. Neural Regen Res. 2017;12(4):660–70.
21. Moya-Alvarado, G, Gershoni-Emek, N, Perlson, E, Bronfman, FC. Neurodegeneration and Alzheimer's disease (AD). What can proteomics tell us about the Alzheimer's brain? Mol Cell Proteomics. 2016;15(2): 409–25.
22. O'Brien, RJ, Wong, PC. Amyloid precursor protein processing and Alzheimer's disease. Annu Rev Neurosci. 2011;34:185–204.
23. Essa, MM, Vijayan, RK, Castellano-Gonzalez, G, Memon, MA, Braidy, N, Guillemin, GJ. Neuroprotective effect of natural products against Alzheimer's disease. Neurochem Res. 2012;37(9):1829–42.
24. Islam, F, Shohag, S, Akhter, S, Islam MdR, Sultana, S, Mitra, S, et al. Exposure of metal toxicity in Alzheimer's disease: An extensive review. Front Pharmacol. 2022; 13:903099.
25. Tamagno, E, Guglielmotto, M, Vasciaveo, V, Tabaton, M. Oxidative stress and Beta amyloid in Alzheimer's disease. Which comes first: The chicken or the egg? Antioxidants. 2021;10(9):1479.
26. Medeiros, R, Baglietto-Vargas, D, LaFerla, FM. The role of tau in Alzheimer's disease and related disorders. CNS Neurosci Ther. 2010;17(5):514–24.
27. Marchbanks, RM. Biochemistry of Alzheimer's dementia. J Neurochem. 1982;39(1):9–15.
28. Han, S, Lemire, J, Appanna, VP, Auger, C, Castonguay, Z, Appanna, VD. How aluminum, an intracellular ROS generator promotes hepatic and neurological diseases: The metabolic tale. Cell Biol Toxicol. 2013;29(2):75–84.
29. Chen, X, Zhang, M, Ahmed, M, Surapaneni, KM, Veeraraghavan, VP, Arulselvan, P. Neuroprotective effects of ononin against the aluminium chloride-induced Alzheimer's disease in rats. Saudi J Biol Sci. 2021;28(8):4232–9.
30. Shunan, D, Yu, M, Guan, H, Zhou, Y. Neuroprotective effect of Betalain against AlCl3-induced Alzheimer's disease in Sprague Dawley rats via putative modulation of oxidative stress and nuclear factor kappa b (NF-κB) signaling pathway. Biomed Pharmacother. 2021;137:111369.
31. Chiroma, SM, Mohd Moklas, MA, Mat Taib, CN, Baharuldin, MTH, Amon, Z. d-galactose and aluminium chloride induced rat model with cognitive impairments. Biomed Pharmacother. 2018;103:1602–8.
32. Schattner, A. Colchicine – New horizons for an ancient drug. Review based on the highest hierarchy of evidence. Eur J Intern Med. 2022;96:34–41.
33. Kumar, A, Seghal, N, Naidu, PS, Padi, SS, Goyal, R. Colchicines-induced neurotoxicity as an animal model of sporadic dementia of Alzheimer's type. Pharmacol Rep. 2007;59(3):274–83.
34. Mohamed, AR, Soliman, GY, Ismail, CA, Mannaa HF. Neuroprotective role of vitamin D3 in colchicine-induced Alzheimer's disease in rats. Alex. J. Med. 2015; 51(2):127–136.
35. Kumar, A, Dogra, S, Prakash, A. Protective effect of naringin, a citrus flavonoid, against colchicine-induced cognitive dysfunction and oxidative damage in rats. J Med Food. 2010;13(4):976–84.
36. Saini, N, Singh, D, Sandhir, R. Neuroprotective effects of Bacopa monnieri in experimental model of dementia. Neurochem Res. 2012;37(9):1928–37.
37. Yu, Z, Cheng, G, Hu, B. Mechanism of colchicine impairment on learning and memory, and protective effect of CGP36742 in mice. Brain Res. 1997;750(1–2):53–8.

38. Kumar, A, Seghal, N, Padi, SV, Naidu, PS. Differential effects of cyclooxygenase inhibitors on intracerebroventricular colchicine-induced dysfunction and oxidative stress in rats. Eur J Pharmacol. 2006; 551(1–3):58–66.
39. Spinetta, MJ, Woodlee, MT, Feinberg, LM, Stroud, C, Schallert, K, Cormack, LK, et al. Alcohol-induced retrograde memory impairment in rats: Prevention by caffeine. Psychopharmacology (Berl). 2008;201(3): 361–71
40. Patil, S, Tawari, S, Mundhada, D, Nadeem, S. Protective effect of berberine, an isoquinoline alkaloid ameliorates ethanol-induced oxidative stress and memory dysfunction in rats. Pharmacol Biochem Behav. 2015;136:13–20.
41. Mailliard, WS, Diamond, I. Recent advances in The neurobiology of alcoholism: The role of adenosine. Pharmacol Ther. 2004;101(1):39–46.
42. Cui, SQ, Wang, Q, Zheng, Y, Xiao, B, Sun, HW, Gu, XL, et al. Puerarin protects against damage to spatial learning and memory ability in mice with chronic alcohol poisoning. Braz J Med Biol Res. 2015;48(6):515–22.
43. Tyas, SL. Alcohol use and the risk of developing Alzheimer's disease. Alcohol Res Health. 2001;25(4): 299–306.
44. Maitra, U, Deng, H, Glaros, T, Baker, B, Capelluto, DGS, Li, Z, et al. Molecular mechanisms responsible for the selective and low-grade induction of pro-inflammatory mediators in murine macrophages by lipopolysaccharide. J Immunol Baltim Md 1950. 2012;189(2):1014–23.
45. Sochocka, M, Zwolińska, K, Leszek, J. The infectious etiology of Alzheimer's disease. Curr Neuropharmacol. 2017;15(7):996–1009.
46. Oitzl, MS, van Oers, H, Schöbitz, B, de Kloet, ER. Interleukin-1 beta, but not interleukin-6, impairs spatial navigation learning. Brain Res. 1993;613(1):160–3.
47. Quan, N, Sundar, SK, Weiss, JM. Induction of interleukin-1 in various brain regions after peripheral and central injections of lipopolysaccharide. J Neuroimmunol. 1994;49(1–2):125–34.
48. Hauss-Wegrzyniak, B, Dobrzanski, P, Stoehr, JD, Wenk, GL. Chronic neuroinflammation in rats reproduces components of the neurobiology of Alzheimer's disease. Brain Res. 1998;780(2):294–303.
49. Philippens, IH, Ormel, PR, Baarends, G, Johansson, M, Remarque, EJ, Doverskog, M. Acceleration of amyloidosis by inflammation in the amyloid-Beta marmoset monkey model of Alzheimer's disease. J Alzheimers Dis. 2017; 55(1):101–13.
50. Daulatzai, MA. Fundamental role of pan-inflammation and oxidative-nitrosative pathways in neuropathogenesis of Alzheimer's disease in focal cerebral ischemic rats. Am J Neurodegener Dis. 2016;5(2):102–30.
51. Zakaria, R, Wan Yaacob, WM, Othman, Z, Long, I, Ahmad, AH, Al-Rahbi, B. Lipopolysaccharide-induced memory impairment in rats: A model of Alzheimer's disease. Physiol Res. 2017;66(4):553–65.
52. Zhan, X, Stamova, B, Jin, LW, DeCarli, C, Phinney, B, Sharp, FR. Gram-negative bacterial molecules associate with Alzheimer disease pathology. Neurology. 2016;87(22):2324–32.
53. Zhao, Y, Cong, L, Jaber, V, Lukiw, WJ. Microbiome-derived lipopolysaccharide enriched in the perinuclear region of Alzheimer's disease brain. Front Immunol. 2017;8:1064.
54. Zhao, Y, Jaber, V, Lukiw, WJ. Secretory products of the human GI tract microbiome and their potential impact on Alzheimer's disease (AD): Detection of lipopolysaccharide (LPS) in AD hippocampus. Front Cell Infect Microbiol. 2017;7:318.
55. Kim, C, Lv, G, Lee, JS, Jung, BC, Masuda-Suzukake, M, Hong, CS, et al. Exposure to bacterial endotoxin generates a distinct strain of α-synuclein fibril. Sci Rep. 2016; 6: 30891.
56. Kamat, PK, Tota, S, Saxena, G, Shukla, R, Nath, C. Okadaic acid (ICV) induced memory impairment in rats: A suitable experimental model to test anti-dementia activity. Brain Res. 2010; 1309:66–74.
57. Kamat, PK, Tota, S, Shukla, R, Ali, S, Najmi, AK, Nath, C. Mitochondrial dysfunction: A crucial event in okadaic acid (ICV) induced memory impairment and apoptotic cell death in rat brain. Pharmacol Biochem Behav. 2011;100(2):311–9.
58. Cohen, P, Holmes, CF, Tsukitani, Y. Okadaic acid: A new probe for the study of cellular regulation. Trends Biochem Sci. 1990;15(3):98–102.
59. Ishihara, H, Martin, BL, Brautigan, DL, Karaki, H, Ozaki, H, Kato, Y, et al. Calyculin a and okadaic acid: Inhibitors of protein phosphatase activity. Biochem Biophys Res Commun. 1989;159(3):871–7.
60. Maidana, M, Carlis, V, Galhardi, FG, Yunes, JS, Geracitano, LA, Monserrat, JM, et al. Effects of microcystins over short- and long-term memory and oxidative stress generation in hippocampus of rats. Chem Biol Interact. 2006;159(3):223–34.
61. He, J, Yamada, K, Zou, LB, Nabeshima, T. Spatial memory deficit and neurodegeneration induced by the direct injection of okadaic acid into the hippocampus in rats. J Neural Transm Vienna Austria 1996. 2001;108(12):1435–43.

62. Cagnoli, CM, Kharlamov, E, Atabay, C, Uz, T, Manev, H. Apoptosis induced in neuronal cultures by either the phosphatase inhibitor okadaic acid or the kinase inhibitor staurosporine is attenuated by isoquinolinesulfonamides h-7, h-8, and h-9. J Mol Neurosci MN. 1996;7(1):65–76.
63. Koss, DJ, Hindley, KP, Riedel, G, Platt, B. Modulation of hippocampal calcium signalling and plasticity by serine/threonine protein phosphatases. J Neurochem. 2007;102(4):1009–23.
64. Fernández, MT, Zitko, V, Gascón, S, Torreblanca, A, Novelli, A. Neurotoxic effect of okadaic acid, a seafood-related toxin, on cultured cerebellar neurons. Ann N Y Acad Sci. 1993; 679:260–9.
65. Kamat, PK, Tota, S, Rai, S, Swarnkar, S, Shukla, R, Nath, C. A study on neuroinflammatory marker in brain areas of okadaic acid (ICV) induced memory impaired rats. Life Sci. 2012;90(19–20): 713–20.
66. Costa, AP, Tramontina, AC, Biasibetti, R, Batassini, C, Lopes, MW, Wartchow, KM, et al. Neuroglial alterations in rats submitted to the okadaic acid-induced model of dementia. Behav Brain Res. 2012; 226(2):420–7.
67. Liu, M, Choi, S, Cuny, GD, Ding, K, Dobson, BC, Glicksman, MA, et al. Kinetic studies of Cdk5/p25 kinase: Phosphorylation of tau and complex inhibition by two prototype inhibitors. Biochemistry. 2008;47(32):8367–77.
68. Lucas, JJ, Hernández, F, Gómez-Ramos, P, Morán, MA, Hen, R, Avila, J. Decreased nuclear beta-catenin, tau hyperphosphorylation and neurodegeneration in GSK-3beta conditional transgenic mice. EMBO J. 2001;; 20(1–2):27–39.
69. Arendt, T, Holzer, M, Brückner, MK, Janke, C, Gärtner, U. The use of okadaic acid in vivo and the induction of molecular changes typical for Alzheimer's disease. Neuroscience. 1998;85(4):1337–40.
70. Liu, F, Grundke-Iqbal, I, Iqbal, K, Gong, CX. Contributions of protein phosphatases PP1, PP2A, PP2B and PP5 to the regulation of tau phosphorylation. Eur J Neurosci. 2005;22(8):1942–50.
71. Zhang, Z, Simpkins, JW. Okadaic acid induces tau phosphorylation in SH-SY5Y cells in an estrogen-preventable manner. Brain Res. 2010;1345:176–81.
72. Alvarez-de-la-Rosa, M, Silva, I, Nilsen, J, Pérez, MM, García-Segura, LM, Avila, J, et al. Estradiol prevents neural tau hyperphosphorylation characteristic of Alzheimer's disease. Ann N Y Acad Sci. 2005; 1052:210–24.
73. Zhang, Z, Simpkins, JW. An okadaic acid-induced model of tauopathy and cognitive deficiency. Brain Res. 2010;1359:233–46.
74. Kamat, PK, Rai, S, Nath, C. Okadaic acid induced neurotoxicity: An emerging tool to study Alzheimer's disease pathology. NeuroToxicology. 2013;37:163–72.
75. Clark, D, Tuor, UI, Thompson, R, Institoris, A, Kulynych, A, Zhang, X, et al. protection against recurrent stroke with resveratrol: Endothelial protection. PLoS ONE. 2012; 7(10): e47792.
76. Karthick, C, Periyasamy, S, Jayachandran, KS, Anusuyadevi, M. Intrahippocampal administration of ibotenic acid induced cholinergic dysfunction via NR2A/NR2B expression: Implications of resveratrol against Alzheimer disease pathophysiology. Front Mol Neurosci. 2016;9:28.
77. Rattan, AK, Tejwani, GA. The neurotoxic actions of ibotenic acid on cholinergic and opioid peptidergic systems in the central nervous system of the rat. Brain Res. 1992;571(2):298–305.
78. Antzoulatos, EG, Byrne, JH. Learning insights transmitted by glutamate. Trends Neurosci. 2004;27(9): 555–60.
79. Angelotti, T, Mireles, S, McMahon, D. Anesthetic implications of a near-lethal sodium azide exposure. Anesth Analg. 2007;104(1):229–30.
80. Mattson, MP. Pathways towards and away from Alzheimer's disease. Nature. 2004;430(7000):631–9.
81. Clissold, SP, Heel, RC. Transdermal hyoscine (scopolamine). A preliminary review of its pharmacodynamic properties and therapeutic efficacy. Drugs. 1985;29(3):189–207.
82. Chen, WN, Yeong, KY. Scopolamine, a toxin-induced experimental model, used for research in Alzheimer's disease. CNS Neurol Disord - Drug Targets- CNS Neurol Disord. 2020;19(2):85–93.
83. Riedel, G, Kang, SH, Choi, DY, Platt, B. Scopolamine-induced deficits in social memory in mice: Reversal by donepezil. Behav Brain Res. 2009;204(1):217–25.
84. Klinkenberg, I, Blokland, A. The validity of scopolamine as a pharmacological model for cognitive impairment: A review of animal behavioral studies. Neurosci Biobehav Rev. 2010;34(8):1307–50.
85. Foudah, AI, Devi, S, Alam, A, Salkini, MA, Ross, SA. Anticholinergic effect of resveratrol with vitamin E on scopolamine-induced Alzheimer's disease in rats: Mechanistic approach to prevent inflammation. Front Pharmacol. 2023;14.
86. Khakpai, F, Nasehi, M, Haeri-Rohani, A, Eidi, A, Zarrindast, MR. Scopolamine induced memory impairment; Possible involvement of NMDA receptor mechanisms of dorsal hippocampus and/or septum. Behav Brain Res. 2012;231(1):1–10.

87. Singh, P, Konar, A, Kumar, A, Srivas, S, Thakur, MK. Hippocampal chromatin-modifying enzymes are pivotal for scopolamine-induced synaptic plasticity gene expression changes and memory impairment. J Neurochem. 2015;134(4):642–51.
88. Portero-Tresserra, M, Del Olmo, N, Martí-Nicolovius, M, Guillazo-Blanch, G, Vale-Martínez, A. D-cycloserine prevents relational memory deficits and suppression of long-term potentiation induced by scopolamine in the hippocampus. Eur Neuropsychopharmacol J Eur Coll Neuropsychopharmacol. 2014;24(11):1798–807.
89. Blokland, A, Honig, W, Raaijmakers, WG. Effects of intra-hippocampal scopolamine injections in a repeated spatial acquisition task in the rat. Psychopharmacology (Berl). 1992;109(3):373–6.
90. Elvander, E, Schött, PA, Sandin, J, Bjelke, B, Kehr, J, Yoshitake, T, et al. Intraseptal muscarinic ligands and galanin: Influence on hippocampal acetylcholine and cognition. Neuroscience. 2004;126(3):541–57.
91. Rogers, JL, Kesner, RP. Cholinergic modulation of the hippocampus during encoding and retrieval of tone/shock-induced fear conditioning. Learn Mem Cold Spring Harb N. 2004;11(1):102–7.
92. Wu, YY, Wang, X, Tan, L, Liu, D, Liu, XH, Wang, Q, et al. Lithium attenuates scopolamine-induced memory deficits with inhibition of GSK-3β and preservation of postsynaptic components. J Alzheimers Dis JAD. 2013;37(3):515–27.
93. Damasceno, DC, Netto, AO, Iessi, IL, Gallego, FQ, Corvino, SB, Dallaqua, B, et al. Streptozotocin-induced diabetes models: Pathophysiological mechanisms and fetal outcomes. BioMed Res Int. 2014; 2014:819065.
94. Kamat, PK. Streptozotocin induced Alzheimer's disease like changes and the underlying neural degeneration and regeneration mechanism. Neural Regen Res. 2015;10(7):1050–2.
95. Zhou, S, Yu, G, Chi, L, Zhu, J, Zhang, W, Zhang, Y, et al. Neuroprotective effects of edaravone on cognitive deficit, oxidative stress and tau hyperphosphorylation induced by intracerebroventricular streptozotocin in rats. Neurotoxicology. 2013; 38: 136–45.
96. null, Neha, Sodhi, RK, Jaggi, AS, Singh, N. Animal models of dementia and cognitive dysfunction. Life Sci. 2014;; 109(2):73–86.
97. Yang, S, Zhou, G, Liu, H, Zhang, B, Li, J, Cui, R, et al. Protective effects of p38 MAPK inhibitor SB202190 against hippocampal apoptosis and spatial learning and memory deficits in a rat model of vascular dementia. BioMed Res Int. 2013; 2013:215798.
98. Sodhi, RK, Singh, N, Jaggi, AS. Neuroprotective mechanisms of peroxisome proliferator-activated receptor agonists in Alzheimer's disease. Naunyn Schmiedebergs Arch Pharmacol. 2011;384(2):115–24.
99. Correia, SC, Santos, RX, Santos, MS, Casadesus, G, Lamanna, JC, Perry, G, et al. Mitochondrial abnormalities in a streptozotocin-induced rat model of sporadic Alzheimer's disease. Curr Alzheimer Res. 2013;10(4):406–19.
100. Sharma, M, Briyal, S, Gupta, YK. Effect of alpha lipoic acid, melatonin and trans resveratrol on intracerebroventricular streptozotocin induced spatial memory deficit in rats. Indian J Physiol Pharmacol. 2005;49(4):395–402.
101. Atwal, N, Bindra, CS, Jain, UK. Investigations on molecular mechanism involved in neuroprotective effect of vitamin D against sodium azide induced Alzheimer's disease in rats. World J Pharm Res. 2016; 5(6):1154–1172.
102. Wong-Riley, M, Antuono, P, Ho, KC, Egan, R, Hevner, R, Liebl, W, et al. Cytochrome oxidase in Alzheimer's disease: Biochemical, histochemical, and immunohistochemical analyses of the visual and other systems. Vision Res. 1997;37(24):3593–608.
103. Bennett, MC, Mlady, GW, Kwon, YH, Rose, GM. Chronic in vivo sodium azide infusion induces selective and stable inhibition of cytochrome c oxidase. J Neurochem. 1996;66(6):2606–11.
104. Gao, C, Chang, P, Yang, L, Wang, Y, Zhu, S, Shan, H, et al. Neuroprotective effects of hydrogen sulfide on sodium azide-induced oxidative stress in PC12 cells. Int J Mol Med. 2018;41(1):242–50.
105. Gupta, A, Atkinson, AN, Pandey, AK, Bishayee, A. Health-promoting and disease-mitigating potential of Verbascum thapsus l. (common mullein): A review. Phytother Res. 2022;36(4):1507–22.
106. Shinomol, GK, null, Muralidhara, Bharath, MMS. Exploring the role of "Brahmi" (Bacopa monnieri and Centella asiatica) in brain function and therapy. Recent Pat Endocr Metab Immune Drug Discov. 2011;5(1):33–49.
107. Veerendra Kumar, MH, Gupta, YK. Effect of Centella asiatica on cognition and oxidative stress in an intracerebroventricular streptozotocin model of Alzheimer's disease in rats. Clin Exp Pharmacol Physiol. 2003;30(5–6):336–42.
108. Dhanasekaran, M, Holcomb, LA, Hitt, AR, Tharakan, B, Porter, JW, Young, KA, et al. Centella asiatica extract selectively decreases amyloid beta levels in hippocampus of Alzheimer's disease animal model. Phytother Res PTR. 2009; 23(1):14–9.

109. Chen, CL, Tsai, WH, Chen, CJ, Pan, TM. Centella asiatica extract protects against amyloid β1–40-induced neurotoxicity in neuronal cells by activating the antioxidative defence system. J Tradit Complement Med. 2016;6(4):362–9.
110. Ganguly, R, Hazra, R, Ray, K, Guha, D. Effect of Moringa oleifera in experimental model of Alzheimer's disease : Role of antioxidants. Ann Neurosci. 2010;12(3):33–6.
111. Defeudis, FV. Bilobalide and neuroprotection. Pharmacol Res. 2002;46(6):565–8.
112. Zhou, LJ, Zhu, XZ. Reactive oxygen species-induced apoptosis in PC12 cells and protective effect of bilobalide. J Pharmacol Exp Ther. 2000;293(3):982–8.
113. Tchantchou, F, Lacor, PN, Cao, Z, Lao, L, Hou, Y, Cui, C, et al. Stimulation of neurogenesis and synaptogenesis by bilobalide and quercetin via common final pathway in hippocampal neurons. J Alzheimers Dis JAD. 2009;18(4):787–98.
114. Yakoot, M, Salem, A, Helmy, S. Effect of memo®, a natural formula combination, on mini-mental state examination scores in patients with mild cognitive impairment. Clin Interv Aging. 2013;8:975–81.
115. Chauhan, N, Wang, KC, Wegiel, J, Malik, MN. Walnut extract inhibits the fibrillization of amyloid beta-protein, and also defibrillizes its preformed fibrils. Curr Alzheimer Res. 2004;1(3):183–8.
116. Muthaiyah, B, Essa, MM, Chauhan, V, Chauhan, A. Protective effects of walnut extract against amyloid Beta peptide-induced cell death and oxidative stress in PC12 cells. Neurochem Res. 2011;36(11):2096–103.
117. Borek, C. Antioxidant health effects of aged garlic extract. J Nutr. 2001;131(3s):1010S–5S.
118. Hazzaa, SM, Abdelaziz, SAM, Abd Eldaim, MA, Abdel-Daim, MM, Elgarawany, GE. Neuroprotective potential of Allium sativum against monosodium glutamate-induced excitotoxicity: Impact on short-term memory, gliosis, and oxidative stress. Nutrients. 2020;12(4):1028.
119. Borek, C. Garlic reduces dementia and heart-disease risk. J Nutr. 2006; 136(3): 810–812.
120. Nillert, N, Pannangrong, W, Welbat, JU, Chaijaroonkhanarak, W, Sripanidkulchai, K, Sripanidkulchai, B. Neuroprotective effects of aged garlic extract on cognitive dysfunction and neuroinflammation induced by β-amyloid in rats. Nutrients. 2017;9(1):24.
121. Qu, Z, Mossine, VV, Cui, J, Sun, GY, Gu, Z. Protective effects of AGE and its components on neuroinflammation and neurodegeneration. Neuromolecular Med. 2016;18(3):474–82.
122. Braidy, N, Selvaraju, S, Essa, MM, Vaishnav, R, Al-Adawi, S, Al-Asmi, A, et al. Neuroprotective effects of a variety of pomegranate juice extracts against MPTP-induced cytotoxicity and oxidative stress in human primary neurons. Oxid Med Cell Longev. 2013; 2013.
123. Feng, Y, Yang, S, gao, Du, X, ting, Zhang, X, Sun, X, xia, Zhao, M, et al. Ellagic acid promotes Abeta42 fibrillization and inhibits Abeta42-induced neurotoxicity. Biochem Biophys Res Commun. 2009;390(4):1250–4.
124. ssa MM, Vijayan RK, Castellano-Gonzalez G, Memon MA, Braidy N, Guillemin GJ. Neuroprotective effect of natural products against Alzheimer's disease. Neurochem Res. 2012;37(9):1829–42.
125. Kwak, HM, Jeon, SY, Sohng, BH, Kim, JG, Lee, JM, Lee, KB, et al. β-secretase(BACE1) inhibitors from pomegranate (Punica granatum) husk. Arch Pharm Res. 2005;28(12):1328–32.
126. Hartman, RE, Shah, A, Fagan, AM, Schwetye, KE, Parsadanian, M, Schulman, RN, et al. Pomegranate juice decreases amyloid load and improves behavior in a mouse model of Alzheimer's disease. Neurobiol Dis. 2006;24(3):506–15.
127. Gohil, KJ, Patel, JA. A review on Bacopa monniera: Current research and future prospects. Int J Green Pharm IJGP. 2010;4(1):1–9.
128. Singh, HK, Dhawan, BN. Effect of Bacopa monniera linn. (Brāhmi) extract on avoidance responses in rat. J Ethnopharmacol. 1982;5(2):205–14.
129. Uabundit, N, Wattanathorn, J, Mucimapura, S, Ingkaninan, K. Cognitive enhancement and neuroprotective effects of Bacopa monnieri in Alzheimer's disease model. J Ethnopharmacol. 2010;127(1):26–31.
130. Bhattacharya, SK, Bhattacharya, A, Kumar, A, Ghosal, S. Antioxidant activity of Bacopa monniera in rat frontal cortex, striatum and hippocampus. Phytother Res. 2000;14(3):174–9.
131. Limpeanchob, N, Jaipan, S, Rattanakaruna, S, Phrompittayarat, W, Ingkaninan, K. Neuroprotective effect of Bacopa monnieri on beta-amyloid-induced cell death in primary cortical culture. J Ethnopharmacol. 2008;120(1):112–7.
132. Sadhu, A, Upadhyay, P, Agrawal, A, Ilango, K, Karmakar, D, Singh, GPI, et al. Management of cognitive determinants in senile dementia of Alzheimer's type: Therapeutic potential of a novel polyherbal drug product. Clin Drug Investig. 2014 ;34(12):857–69.
133. Choi, JG, Kim, N, Huh, E, Lee, H, Oh, MH, Park, JD, et al. White Ginseng protects mouse hippocampal cells against amyloid-Beta oligomer toxicity. Phytother Res. 2017; 31(3):497–506.
134. Kim, J, Kim, SH, Lee, DS, Lee, DJ, Kim, SH, Chung, S, et al. Effects of fermented ginseng on memory impairment and β-amyloid reduction in Alzheimer's disease experimental models. J Ginseng Res. 2013;37(1):100–7

135. Lee, MR, Yun, BS, In, OH, Sung, CK. Comparative study of Korean white, red, and black Ginseng extract on cholinesterase inhibitory activity and cholinergic function. J Ginseng Res. 2011;35(4):421–8.
136. Heo, JH, Lee, ST, Oh, MJ, Park, HJ, Shim, JY, Chu, K, et al. Improvement of cognitive deficit in Alzheimer's disease patients by long term treatment with Korean red Ginseng. J Ginseng Res. 2011; 35(4):457–61.
137. Heo, JH, Park, MH, Lee, JH. Effect of Korean red Ginseng on cognitive function and quantitative EEG in patients with Alzheimer's disease: A preliminary study. J Altern Complement Med N Y N. 2016;22(4):280–5.
138. Kim, HJ, Jung, SW, Kim, SY, Cho, IH, Kim, HC, Rhim, H, et al. Panax ginseng as an adjuvant treatment for Alzheimer's disease. J Ginseng Res. 2018;42(4):401–11.
139. Yang, W, Ip, SP, Liu, L, Xian, YF, Lin, ZX. Uncaria rhynchophylla and its major constituents on Central nervous system: A review on their pharmacological actions. Curr Vasc Pharmacol. 2020;18(4):346–57.
140. Hsieh, CL, Tang, NY, Chiang, SY, Hsieh, CT, Lin, JG. Anticonvulsive and free radical scavenging actions of two herbs, Uncaria rhynchophylla (MIQ) jack and Gastrodia elata bl., in kainic acid-treated rats. Life Sci. 1999;65(20):2071–82.
141. Tang, NY, Liu, CH, Su, SY, Jan, YM, Hsieh, CT, Cheng, CY, et al. Uncaria rhynchophylla (miq) jack plays a role in neuronal protection in kainic acid-treated rats. Am J Chin Med. 2010;38(2):251–63.
142. Fujiwara, H, Iwasaki, K, Furukawa, K, Seki, T, He, M, Maruyama, M, et al. Uncaria rhynchophylla, a Chinese medicinal herb, has potent antiaggregation effects on Alzheimer's beta-amyloid proteins. J Neurosci Res. 2006;84(2):427–33.
143. Kang, TH, Murakami, Y, Matsumoto, K, Takayama, H, Kitajima, M, Aimi, N, et al. Rhynchophylline and isorhynchophylline inhibit NMDA receptors expressed in xenopus oocytes. Eur J Pharmacol. 2002;455(1):27–34.
144. Fujiwara, H, Tabuchi, M, Yamaguchi, T, Iwasaki, K, Furukawa, K, Sekiguchi, K, et al. A traditional medicinal herb Paeonia suffruticosa and its active constituent 1,2,3,4,6-penta-o-galloyl-beta-d-glucopyranose have potent anti-aggregation effects on Alzheimer's amyloid beta proteins in vitro and in vivo. J Neurochem. 2009;109(6):1648–57.
145. Singh, M, Arseneault, M, Sanderson, T, Murthy, V, Ramassamy, C. Challenges for research on polyphenols from foods in Alzheimer's disease: Bioavailability, metabolism, and cellular and molecular mechanisms. J Agric Food Chem. 2008;56(13):4855–73.
146. Zhao, D, Simon, JE, Wu, Q. A critical review on grape polyphenols for neuroprotection: Strategies to enhance bioefficacy. Crit Rev Food Sci Nutr. 2020;60(4):597–625.
147. Renaud, J, Martinoli, MG. Considerations for the use of polyphenols as therapies in neurodegenerative diseases. Int J Mol Sci. 2019;20(8):1883.
148. Rein, MJ, Renouf, M, Cruz-Hernandez, C, Actis-Goretta, L, Thakkar, SK, da Silva Pinto, M. Bioavailability of bioactive food compounds: A challenging journey to bioefficacy. Br J Clin Pharmacol. 2013;75(3): 588–602.
149. Ovais, M, Zia, N, Ahmad, I, Khalil, AT, Raza, A, Ayaz, M, et al. Phyto-therapeutic and nanomedicinal approaches to cure Alzheimer's disease: Present status and future opportunities. Front Aging Neurosci. 2018; 10:284.
150. Niu, X, Chen, J, Gao, J. Nanocarriers as a powerful vehicle to overcome blood-brain barrier in treating neurodegenerative diseases: Focus on recent advances. Asian J Pharm Sci. 2019;14(5):480–96.

7 Advancement in the Genetic Model of Alzheimer's Disease

Priti Kanik
Department of Biological Sciences and Engineering, Indian Institute of Technology Gandhinagar, Palaj, Gujarat, India

Rashi Kumar
Department of Neurology, Institute of Human Behaviour and Allied Sciences, Dilshad Garden, Delhi, India

Ritu Varshney
P. D. Patel Institute of Applied Sciences, Charotar University of Science and Technology, CHARUSAT Campus, Anand, Gujarat, India

7.1 INTRODUCTION

Alzheimer's disease (AD), earlier referred to as senile dementia, is the most common neurodegenerative disorder, often characterised by progressive decline and loss of cognitive functions. The global number of patients with AD is estimated to be more than 50 million, and it is the sixth leading cause of death globally [1]. The numbers do not account for subjects with some form of dementia, which is always preceded by a diagnosis of probable AD (presence of positive AD biomarker). Thus, the devastating global burden of AD is grave and requires immense contribution from all sectors of society, including researchers and policy makers, to ensure the implementation of strategies to combat the ever-growing cases of AD.

The history of AD can be traced back more than a hundred years to when Alois Alzheimer, a German psychiatrist, carefully observed a woman named Auguste D. from 1901 until her demise in 1906. During the observation period, he observed symptoms such as untreatable paranoia, sleep disorders, memory disturbances, and behavioral symptoms such as crying and confusion. After her demise, he performed comprehensive investigations on brain morphology and histology, which enabled him to associate the relationship between plaque-like structures in the brain with observed symptoms [2]. Since the first recorded case of AD by Alois Alzheimer, who first described the case in 1906, we have robustly conducted research and evolved our understanding of the disease on a multitude of scientific levels, such as histopathology, genetics, molecular, omics, and clinical. More than 50 years later, we established the role of familial transmission through genetic mechanisms. Another decade later, we identified the disease-predictive genes (i.e. *APP, PSEN1, PSEN2*). To date, other than mutations in these genes, no other disease-associated genes have been discovered. However, these genes contribute to less than 5% of AD cases and are also as called monogenic genes (single gene variants). The remaining 95% of AD cases are due to polygenic genes (multiple gene variants).

With advancements in science and technology, understanding disease pathology and mechanisms have become relatively more straightforward than it was hundreds of years ago. Surprisingly, most of them still remain unclear. To understand the disease dynamics, efforts have been made to study at the genetics and molecular level to determine the dynamics that lead to disease pathology. The Human Genome Project (HGP), which aimed at uncovering the human genome, assisted our efforts in uncovering the genetic

 DOI: 10.1201/9781032620442-7

implications of the disease, such as genome-wide association studies (GWAS). A blast of information from HGP allowed researchers and the scientific community to comprehensively and intricately examine the human genome for identifying disease-associated genetic patterns in the form of polymorphisms or small-nuclear polymorphisms (SNPs). Before the completion of HGP in 2003, most studies related to investigating the underpinning of AD genetics were conducted through Mendelian genetics and linkage analysis. Linkage analysis enabled the researchers to statistically determine if two closely located genes on a chromosome will be inherited together. However, most researchers operated under the assumption that AD can be assessed using a single gene variant, which later proved to be wrong in light of scientific research on sporadic AD and late-onset AD (LOAD). The discovery of the polygenic basis of sporadic AD and LOAD using GWAS further advanced our understanding of AD disease pathology. It also made it complicated to present any single consensus on how the disease actually functions.

AD genetics is governed by the presence of two types of variants: i.e. predictive or mutated genes and susceptible or risk genes. Having monogenic predictive or mutated genes (e.g. *APP, PSEN1, PSEN2*) will very likely contribute to AD development, including early-onset and familial AD. In contrast, having polygenic risk genes (e.g. *APOE, TREM2, PICALM*) will likely increase the vulnerability to the disease, especially late-onset AD, which has been linked to the interaction of lifestyle and environmental exposures with susceptible or risk genes for AD. Thus, the existence of AD due to two contrasting contributing factors (monogenic and polygenic) makes it genetically complicated to study but equally important. Understanding the genetic implications of AD plays a crucial role in clinical research. The detection of AD-associated predictive or susceptible genes would enable clinicians to intervene early and treat AD.

This chapter covers various aspects of AD genetics, including the history of AD genetics, rare variants of AD genes, susceptible genetic variants of AD genes, polygenic risk score estimation, genetic treatments and therapeutics (**Figure 7.1**), interactive omics approach, and gene editing techniques.

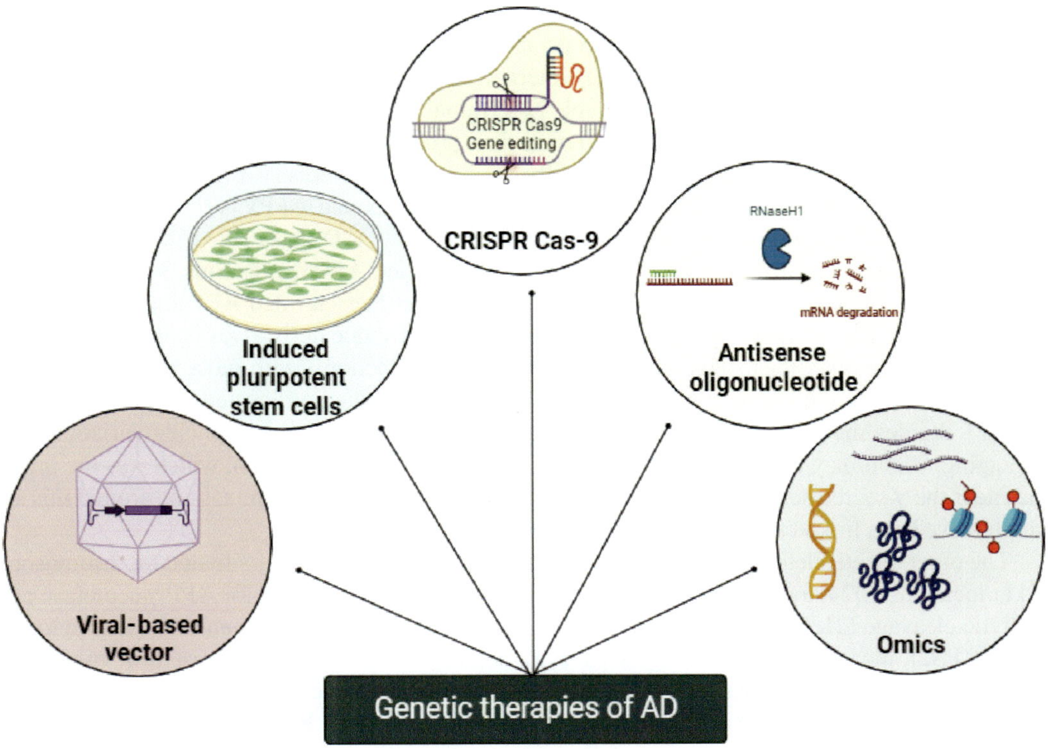

FIGURE 7.1 Overview of genetic therapies in AD.

7.2 EARLY GENETIC DISCOVERIES

In this discussion, we focus on genes related to the familial origins of AD, particularly in relation to APP processing, based on the amyloid hypothesis.

AD is a multifaceted condition. It is delineated into two primary categories: familial AD (fAD) and sporadic AD (sAD). A minority of AD cases (fewer than 5%) are categorised as fAD, characterised by an autosomal dominant, Mendelian inheritance pattern. Onset is often before age 65, thus termed "early-onset", referred to as early-onset familial AD (EO-fAD). The three fully penetrating autosomal dominant mutations – APP, PS1, and PS2 – are determinants of EO-fAD [3, 4]. The majority of AD cases are sporadic (more than 95%) [5], where genetic, environmental, and lifestyle factors interplay. However, genetics still plays a role in predisposition. Onset is often after age 65, thus termed "late-onset, " also referred to as late-onset sporadic AD (LO-sAD). A susceptibility variant, ε4 in the APOE gene, elevates the AD risk, but with incomplete penetrance are determinants of LO-sAD [6, 7].

The key genetic determinant in EO-fAD is APP. In the 1980s, a peptide consisting of 40 amino acids (Aβ) was extracted from amyloid plaques and vascular accumulations. This discovery paved the way to recognise the APP type I integral membrane as the precursor source from which Aβ originates through proteolysis [8]. The APP gene on chromosome 21 at 21q21 has multiple splice variants named by their amino acid length (e.g. APP563, APP695, APP714, APP751, and APP770). While their expression varies by tissue, the three isoforms most pertinent to AD are APP695, APP751, and APP770 [9, 10]. Subsequent mutational studies have revealed that most of these changes are located at the sites where secretase (β and γ) cleaves or within the APP transmembrane region, specifically on exons 16 and 17 [11]. Aβ is a peptide derived from the proteolytic breakdown of a transmembrane APP protein with an expansive extracellular domain. APP is transported to the plasma membrane via the secretory route of protein synthesis. Further, APP is proteolytically cleaved via two paths: The non-amyloidogenic route and the amyloidogenic route. In the former pathway, α-secretase sequentially cleaves APP, producing a soluble N-terminal APPα fragment and a C-terminal CTF-α or C83 fragment. Subsequently, γ-secretase cleaves C83, generating the P3 peptide externally and the APP intracellular domain (AICD) internally. Conversely, in the latter pathway, BACE1 cleaves APP in endosomes, forming a soluble N-terminal APPβ fragment and a membrane-bound C-terminal CTF-β or C99 fragment. γ-secretase then processes C99, releasing Aβ peptides externally and liberating the AICD peptide internally [12, 13]. The "Swedish" APP mutations (APPSW, M671L, APPK670N) and "London" APP mutations (APPLON and APPV717I) are instances that result in elevated Aβ production, subsequently contributing to AD onset [14, 15]. When tested in transgenic mice, these different APP mutations exhibit varied APP expressions. For instance, the mouse model Tg2576 with the "Swedish" mutation demonstrates elevated APP and Aβ levels and cognitive impairments [16]. The role of APP and its various splice derivatives is unclear; their differential tissue expressions suggest potential functional distinctions, with most variants encoding Aβ sequences and the segment responsible for Aβ peptide production located in exons 16 and 17 [17, 18]. Studies show that exons 7 and 8 in the brain experience alternative splicing variations during aging, which may be affected by AD [19, 20]. The V717I APP mutation, which was the initial, lies near the γ-secretase cleavage region [14]. Numerous subsequent APP mutations around this site might have been linked to shifts in Aβ levels.

The other genetic determinants for EOfAD include PS1 and PS2. PS1 is situated on chromosome 14 at location 14q24, a major di-aspartyl protease component responsible for APP γ-secretase proteolytic cleavage [21, 22]. The γ-secretase complex consists of four components: Presenilin (either PS1 or PS2), nicastrin (NCT), anterior pharynx-defective 1 (APH-1), and presenilin enhancer 2 (PEN-2). Endoproteolytic activity in PS1 or PS2 leads to the generation of an N-terminal fragment (NTF) and a C-terminal fragment (CTF) [13]. Mutations in PS1 are the predominant reason for EOfAD as it accounts for a significant proportion of autosomal dominant EOFAD cases. Mutations in PS1 seem to elevate the Aβ42 to Aβ40 ratio, causing a functional shift that diminishes γ-secretase activity [23]. The accumulation of Aβ42 is hypothesised to be an initial, pre-symptomatic occurrence in individuals with PS1 mutations [24]. Different transgenic mouse models have been utilised to explore the influence of the PS1 gene. Conditional knock-out mice hint at PS's association with cognitive memory [25]. Meanwhile, transgenic mice with both PS1 and APP mutations highlight the genes' contribution to Aβ production [26]. These missense mutations lead to amino acid changes in the PS1 protein, resulting in a rise in the Aβ42 to Aβ40 ratio. This can stem from elevated Aβ42 production, reduced Aβ40 production, or a mix of both. The PS2 gene is situated on chromosome 1 at 1q.42, another major component of γ-secretase. Mutations in PS2 are a less frequent cause of EOfAD. Mutations linked to PS2 have been observed to elevate the Aβ42 to Aβ40 ratio in both murine and human subjects, suggesting that PS could influence how γ-secretase processes APP [23, 27]. The roles and significance of presenilin splice variants remain elusive. However, divergent expression of these presenilin isoforms might influence the proteolytic processing of APP. Specifically, PS2 transcripts missing exon 5 seem to enhance Aβ peptide production [28], while native isoforms lacking exons 3, 4, or 8 do not impact Aβ production [29]. These fAD genes, given their critical role in the onset and progression of Alzheimer's disease, emerge as prime therapeutic targets, and manipulating these genes could pave the way for innovative treatments for AD.

7.3 ALZHEIMER'S DISEASE RISK GENES

Apart from FAD and EOAD, 95% of AD cases are due to sporadic AD and late-onset AD (LOAD), which are not governed by Mendelian principles; rather, their action depends on the interaction of numerous independent genetic factors with lifestyle factors (e.g. smoking and physical activity), and environmental factors (e.g. pollution, traumatic brain injury). Sometimes, cases of sporadic AD are also referred to as non-Mendelian AD or polygenic AD, indicating the complex involvement of multiple genetic variants. Saunders et al., 1993, found the association of the APOE4 gene, situated on chromosome 19, with late-onset Alzheimer's disease (LOAD), which contributes significantly as a risk factor for the clinical manifestation of AD [30]. In contrast to predictive or mutated gene variants found in FAD and EOAD, the risk genes in sporadic AD have moderate molecular effects and are easily transmitted. The frequency of such genes increases in the population over time. Once the frequency of such genes is present in 1% or more in a non-diseased population, the DNA sequence of such genes is defined as DNA sequence polymorphism [31]. Such DNA sequence polymorphisms are now capable of forming complex genetic diseases.

Recent advancements in genetics and technology have helped genome-wide associated studies (GWAS) to shed light on AD-associated risk genes. The GWAS method identifies risk genes based on small variations, known as small-nuclear polymorphisms (SNPs), by scanning the whole DNA sequence (the genome) of a large population. A two-stage meta-analysis of 74,000 and more individuals identified 11 novel risk loci associated with AD [32]. In addition to previously identified risk loci *ABCA7* (responsible for cholesterol metabolism and amyloid clearance), *APOE* (responsible for formation of lipoproteins), *BIN1*(responsible endosomal vesicle accumulation), *CLU* (responsible for lipid transport and immune modulation), *CR1* (responsible for encoding transmembrane glycoprotein), *CD2AP* (responsible for signal transduction and cytoskeletal molecules), *EPHA1* (responsible for tyrosine kinase receptor), *MS4A6A-MS4A4E* (responsible for increased concentration contribute to anti-inflammatory neuroprotective cells), and *PICALM* genes (responsible for endocytosis and autophagy) [33–36], the 2013 study under the flagship name International Consortium for Alzheimer's Project found *ZCWPW1* (responsible for proteins regulating epigenetics), *CELF1* (reduced expression responsible for neuronal-viability in AD), *NME8* (responsible for ciliary dyskinesia), *FERMT2* (responsible for angiogenesis, cell matrix adhesion, action assembly), *CASS4* (responsible for action dynamics), *INPP5D* (responsible for modulation metabolism of APP), *MEF2C* (responsible for mental retardation and cerebral malformation), *HLA-DRB5-DRB1* (responsible for immunocompetence and histocompatibility), *SORL1* (responsible for aberrant trafficking of APP), *PTK2B* (responsible for potentiation of hippocampal region CA1), and *SLC24A4* (associated with risk for developing hypertension). In addition to their role in AD pathology, the newly discovered 11 risk loci also contribute to the existing mechanisms identified with previously discovered risk loci.

Post-GWAS analysis involves understanding the GWAS findings and associating probable molecular mechanisms with the disease pathology. Candidate genes are identified within each locus, which vary in length and function. Such genes are identified to be more frequently an intergenic variant (i.e. present outside the coding sequence) than an exonic variant. For example, *APOE* is located at chromosome 19q13.32, consisting of 16 exonic variants spread across nine genes. Quantifying the expression of GWAS-associated loci can be performed using whole transcriptome RNA data, eventually, informing us of how the genes are expressed in the tissues. Such mappings have provided interesting results regarding disease pathology. For example, one would expect AD-risk gene expression more in neuronal cells since AD is a neurodegenerative disease; however, the risk genes are expressed more in microglial cells, a cell type involved in the brain immune response system. The finding is generalised to other parts of the body, in which risk-gene expression is more in non-brain systems such as blood, spleen, and liver; surprisingly, some of them are involved in the body's immune response system. Other pathways affected by risk loci include inflammation, cell migration, lipid transport, endocytosis, hippocampal synaptic function, cytoskeletal function, post-translational protein modification, and microglial cell function. The intricate relationship between genes and the associated mechanisms makes the most common neurodegenerative disease, AD, really complex to study and understand. To understand how such gene sequence polymorphism contributes to the risk of developing AD, polygenic risk scores (PRS) are estimated for individuals.

The best-established and largest risk contributor gene is *APOE,* found on chromosome 19. *APOE* has three alleles, which include *APOE2, APOE3,* and *APOE4,* with *APOE4* being strongly associated with AD neuropathology. The normal function of *APOE* is clearance of amyloid, thus decreasing the concentration. However, in the presence of the *APOE4* variant, the process is known to be less efficient, leading to increased levels of amyloid-beta protein. Eventually, aggregation and deposition of amyloid-beta protein leads to the formation of beta plaques, neurofibrillary tangles, neuroinflammation, neuronal cell death, and ultimately, cognitive impairment and decline. It is important to note that AD neuropathology is preceded by ten years before the first clinical symptoms of AD start to emerge in subjects. Some carriers of *APOE4* do not show any clinical symptoms of AD, which suggests the modulating role of *APOE4* in disease pathology. *APOE* is also a major lipid-protein transporter, contributing to the myelination of neurons in the brain. Dysfunction in the *APOE* transporter could potentially harm the myelination of neurons and lead to disease pathology [37]. *APOE4* exhibits both toxic gain-of-functions and potential loss-of-function characteristics. The toxic gain-of-functions involve tangle formation, amyloid-beta aggregation, and neuronal toxicity. The potential loss of function involves glucose metabolism, neurogenesis, amyloid-beta clearance, lipid metabolism, and vascular function [38].

A complex interaction between polygenic risk genes, lifestyle, and environmental factors determines the diagnosis and prognosis of sporadic or LOAD. To some extent, lifestyle and environmental factors can be treated as possible sites for intervention. According to the Lancet Commission report on dementia [39], early-life factors (i.e. fewer years of education) contribute to 7% to risk of developing dementia, mid-life factors (i.e. hearing loss, traumatic brain injury, hypertension, alcohol, obesity) contribute 15% to risk of developing dementia, and late-life factors (i.e. smoking, depression, social isolation, physical inactivity, air pollution, diabetes) contribute to 19% to risk of developing dementia (**Table 7.1**). These factors, in concert with polygenic risk genes, can increase the vulnerability to the disease. However, most factors are within the control of one's will and choice, unlike genetics; hence, these factors are termed modifiable risk factors. Actively pursuing ways to mitigate the potential modifiable risk factors could theoretically lower one's susceptibility to the disease.

TABLE 7.1
Genetic Risk Factors for Alzheimer's Disease

Gene	Chromosome	Variant	Effect/Other Alleles
Complement Receptor 1 (CR1)	1q32	rs6656401	A/G
Bridging Integrator 1 (BIN1)	2q14.3	rs6733839	T/C
Inositol polyphosphate-5-phosphatase D (INPP5D)	2q37.1	rs35349669	T/C
Myocyte enhancer factor 2C (MEF2C)	5q14.3	rs190982	G/A
Human leukocyte antigen class II histocompatibility, D related beta chain (HLA- DRB1/5)	6p21.32	rs9271192	C/A
CD2 associated protein (CD2AP)	6p12.3	rs10948363	G/A
Triggering receptor expressed on myeloid cells 2 (TREM2)	6p21.1	rs75932628/R47H	T/C
NME/NM23 family member 8 (NME8)	7p14.1	rs2718058	G/A
Zinc finger CW-Type And PWWP domain containing 1 (ZCWPW1)	7q22.1	rs1476679	C/T
Ephrin type-A receptor 1 (EPHA1)	7q34-q35	rs11771145	A/G
Protein tyrosine kinase 2 beta (PTK2B)	8p21. 2	rs28834970	C/T
Clusterin (CLU)	8p21.1	rs9331896	C/T
CUGBP, Elav-like family member 1 (CELF1)	11p11.2	rs10838725	C/T
membrane-spanning 4-domains, subfamily A (MS4A)	11q12	rs983392	G/A
Phosphatidylinositol-binding clathrin assembly protein (PICALM)	11q14.2	rs10792832	A/G
Sirtuin 1 (SORL1)	11q23.2-q24.2	rs11218343	C/T
Solute carrier family 24 (SLC24A4- RIN3)	14q32.12	rs10498633	T/G
Thyroid hormone receptor interactor 4 (TRIP4)	15q22	rs74615166	C/T
Apolipoprotein E (APOE)	19q13	rs429358/ C112R rs7412/ C158R	ε4
TP-binding cassette, sub-family A, member 7 (ABCA7)	19p13.3	rs4147929	A/G
Siglec-3/CD33	19q13. 33	rs3865444	A/C
Cas scaffold protein family member 4 (CASS4)	20q13. 31	rs7274581	C/T

7.4 RARE VARIANT AND MENDELIAN AD GENES

Early-onset Alzheimer's disease (EOAD) or familial AD (FAD) is an inheritable form of AD that occurs between 40 to 65 years of age and contributes to about 5% of AD cases. The mode of genetic transmission follows the autosomal dominant characteristics of Mendelian genetics. Large-scale and in-depth generational pedigree analysis was the source of identification of genetic cause and effect of familial AD. The associated genes are rare in the population due to their high-penetrant molecular effect. As a result, these genes are selected against during genetic transmission, hence known as the rare variant.

In the early 1980s, 125 probable autopsy-confirmed AD cases were investigated to determine underlying molecular and genetic mechanisms [40]. The study reported that the disease exhibits familial aggregation and genetic transmission characteristics. A decade later, in the early 1990s, genetic studies based on linkage analysis in families with early-onset Alzheimer's disease (EOAD) found AD-specific mutated genes, which include: *APP* (Amyloid-beta precursor protein), *PSEN1* (Presenilin 1), and *PSEN2* (Presenilin 2) [14] found a locus situated close to *APP*, situated on chromosome 21, causing a subset of familial AD cases, which suggests the important role of point mutation in the APP gene (amino-acid substitution of valine for isoleucine). Sherrington et al. (1995) found a transcript (S182), which later came to be named *PSEN1*, situated in the *AD3* gene region, linked to chromosome 14q24.3, causing early-onset AD [41] (**Table 7.2**). Levy-Lahad et al. (1995) found another autosomal dominant familial AD gene called *PSEN2* and linked to chromosome 1q31.42. Since the 1990s, no further predictive genes have been established [42].

TABLE 7.2
Variability in Alzheimer's Disease Genetics: Causal Genes for Early-onset Alzheimer's Disease

Gene	Chromosome	Inheritance	Gene Identification	Disease Onset Age of Mutation Carrier
Amyloid precursor protein (APP)	21q21.1–21q21.3	Autosomal dominant & autosomal recessive	Linkage analysis	45–60 years
Presenilins 1 (PSEN1)	14q24.3	Autosomal dominant	Linkage analysis	30–50 years
Presenilins 2 (PSEN2)	1q31–q42	Autosomal dominant	Linkage analysis and homology mapping	40–70 years

The *APP* gene contributes to encoding amyloid-beta precursor protein and plays a crucial role in synaptic development [43] and neuronal migration [44]. Missense point mutation is common and occurs at least four times more frequently than *APP* gene duplication. Almost 54 mutations have been identified in the *APP* gene, which directly or indirectly controls the cleavage of the amyloid-beta protein. Over-expression of the *APP* gene promotes the synthesis of pathogenic amyloid-beta 42 [45]. Also, depending on the type of mutation, the levels of amyloid-beta 42 protein vary, hinting at the heterozygous etiology of EOAD. The most popular types of *APP* mutations are the Swedish, Artic, and London mutations. The Swedish mutation is a double mutation that results in increased AB42 production and altered localisation of intercellular[46]. The Artic mutation lowers AB42 and AB40 levels with full disease penetration at 57 years of mean age [47]. The London mutation results in increased AB42 levels [48]. Brain pathologic investigation of subjects with Down syndrome (DS) revealed the presence of plaque-like structures, which allowed the researchers to associate the extra copy of chromosome 21 with amyloid-beta plaques, hinting at the role of genes present in chromosome 21. Since the *APP* gene is found on chromosome 21, individuals with DS (trisomy of chromosome 21) are more susceptible to AD than others. *PSEN1* is a transmembrane protein that shares 66% homology with *PSEN2* and plays a critical role in *APP* gene processing [49]. To date, more than 300 mutations have been identified that can cause impairment in critical biological pathways, such as a higher AB42/AB40 ratio [50] and Parkinson's disease (PD) with non-amnestic mild cognitive impairment (MCI) [51]. More frequent genetic mutations occur in *PSEN1* than in other predictive genes, accounting for 60%–70% of EOAD. However, the gene is not fully penetrant like *APP*. The pathology resulting from *PSEN1* can be divided into two groups. Mutations occurring before position 200 share a similar disease construct like sporadic AD, and mutations occurring after position 200 share a similar disease construct like amyloid angiopathy [52]. To date, more than 30 mutations have been identified in the *PSEN2* gene. Mutations in *PSEN2* have been known to increase the concentration of pathogenic amyloid-beta protein [23]. However, *PSEN2* has been studied much less than *PSEN1*.

With linkage analysis studies, there are many drawbacks, including a requirement for a large sample consisting of families affected by AD, in-depth generational family pedigree charts to find the affected and unaffected members of the family, and a long waiting period for the study, which could sometimes stretch to several months or years. However, Mendelian genetics have proved to be extremely helpful in providing insights into autosomal dominant characteristics and genetic mechanisms of EOAD and FAD, also referred to as Mendelian-AD or monogenic-AD (caused by a single genetic variant).

7.5 POLYGENIC RISK SCORE

Multiple risk gene variants are used in polygenic disorders to assess the disease risk. Polygenic risk score (PRS) provides a single number denoting an individual's disease vulnerability by summing the risk alleles for each variant and weighing by its effect size derived from GWAS. The number of risk alleles for a variant ranges from 0 to 2. The effect size derived from GWAS is usually the odds ratio of developing the disease for binary traits or the beta-coefficient value for continuous traits. Some studies have argued the validity of PRS in assessing disk risk and its potential application in clinical trials. The additive model of PRS is often argued to be an oversimplified genetic mechanism, and alternately, multiplicative models have been fit to assess the disease risk, which performs better than PRS with a better area under the curve (AUC) [53]. Polygenic hazard score (PHS) has also been shown to perform better than PRS for predicting age-specific disease risk based on SNP P-value disease threshold [54]. R. S. Desikan et al. carried out one of the first assessments of AD risk calculation using polygenic hazard score (PHS), which used the Cox regression model as opposed to linear (for continuous traits) or logistic (for binary traits) models used in PRS [55]. The PRS estimation allows the calculation of the disease risk without considering lifestyle and environmental factors, which has been argued to be counter-intuitive in light of the polygenic nature of sporadic AD. Another big drawback of PRS is the applicability of the score to non-European families. As most GWAS are conducted on European family lineage, many researchers and clinicians have criticised the applicability in other races or descendants, such as African Americans. However, many studies report PRS showing convincing predictive capabilities in population-based and electronic-record-based cohorts.

In clinical research, PRS is used extensively to identify subjects with increased AD, hence allowing clinical researchers to make sound designs for early intervention and early treatment strategies. As PRS is subject-specific, the intervention and treatment could be tailored specifically to the subject, taking into account the risk genes and the impaired biological pathways. Subject-specific PRS opens endless opportunities for exploring the implications of precision medicine in AD. The generation of PRS also allows for more structured and informed clinical trial stratification, ensuring better efficacy and generalisable results. Since PRS can predict risks, disease trajectory could be estimated, which could guide treatment.

7.6 GENE EDITING TECHNOLOGIES

In this topic, based on the amyloid hypothesis, an overview of various gene editing tools, the rationale behind the use of CRISPR Cas9 for AD therapy, and an account of targeting genes under study that are modified using CRISPR will be discussed.

AD's pathophysiology is known to be associated with genetic mutations in both early and late onset of disease. Over the past decade, more than 400 clinical trials for treating AD have been unsuccessful, with a failure rate of 99.5% [56]. These failures have increased the need for more potent gene editing approaches targeting specific genes linked with AD. At present, the engineered nucleases used as primary gene editing tools in experimental and clinical studies are zinc finger nucleases (ZFN), transcription activator-like endonuclease (TALEN), clustered regularly interspaced short palindromic repeats (CRISPR)-associated protein (Cas9) (CRISPR-Cas9) [57]. ZFNs are engineered proteins that combine the DNA-binding specificity of zinc finger domains with the cleavage activity of the FokI endonuclease, enabling targeted genomic modifications. The pairs of ZFNs recognise specific DNA sequences on opposite DNA strands; and the tailored zinc finger domains confer a specificity. Upon binding, the consequent domains (folk domain) undergo activation, facilitating a double-stranded DNA cleavage. After this disruption, cellular repair processes utilise either a non-homologous end joining (NHEJ) or homologous recombination (HR) mechanism. Despite their specificity, ZFNs can sometimes bind to and cleave non-target DNA sequences, leading to unintended genetic modifications [58]. TALENs are engineered proteins that merge the sequence-specific DNA-binding properties of TALE domains with the DNA-cleaving capability of the FokI nuclease, facilitating targeted genome alterations. Through their TALE domains, TALENs are capable of identifying distinct DNA sequences with each repeat in the domain being specific to a single DNA base, ensuring accurate targeting. Upon binding, the FokI nuclease domains of nearby TALENs come together in a dimer, resulting in a double-stranded DNA cut. Similar to ZFNs, following this incision, the cellular response to this break can be NHEJ or HR, resulting in off-target modifications [59]. TALENs, like ZFNs, played a pioneering role in genome editing before the emergence of CRISPR. Each method has its pros and cons, but CRISPR is a more versatile tool in terms of delivery system, stability, and specificity as compared to the other counterparts [56, 60]. CRISPR-Cas gene editing is notable for enabling inheritable genome modifications using short guide RNAs targeting coding genes, bypassing the need for modified proteins, a primary limitation with ZFN and TALEN [61]. The CRISPR/Cas9 genome editing system provides an efficacious alternative to traditional techniques, possessing potential benefits for gene therapy applications [62]. CRISPR Cas9 is a bacterial defense mechanism that targets invasive viral DNA. By incorporating viral DNA fragments into their own DNA, bacteria form sequences known as CRISPR loci. These loci are transformed into CRISPR RNAs, directing the Cas9 to slice the desired DNA [63]. The system primarily consists of the Cas9 enzyme, which cuts double-stranded DNA, and the guide RNA (gRNA) that recognises and leads the Cas9 to the exact DNA location. The DNA break is repaired either by NHEJ or HR mechanisms [64, 65].

In the realm of AD research, cellular models are the cornerstone due to their efficient circumvention of ethical constraints and shorter experimental timelines. To date, in vitro AD research features cell lines such as SH-SY5Y and SK-N-SH from the neuroblastoma category, HT22 cells from mouse hippocampal neurons, BV2 from the glial cell group, and N2a representing mouse glioblastoma [66]. Research with N2a cells demonstrated that CRISPR/Cas9-mediated (**Figure 7.2**) knockout of the PS1 genes eliminates the backdrop of innate γ-secretase activity. Subsequent experiments with recombinant proteins from PS1 mutations noted a decline in Aβ42 and Aβ40 production [67]. Furthermore, a modified SH-SY5Y cell line used to probe the pathogenic implications of the PS1 gene mutation at codon 97 (V97L) unveiled a notable rise in Aβ42 levels [68]. Concurrent studies emphasised the regulatory role of γ-secretase activating protein (GSAP) over γ-secretase activity.

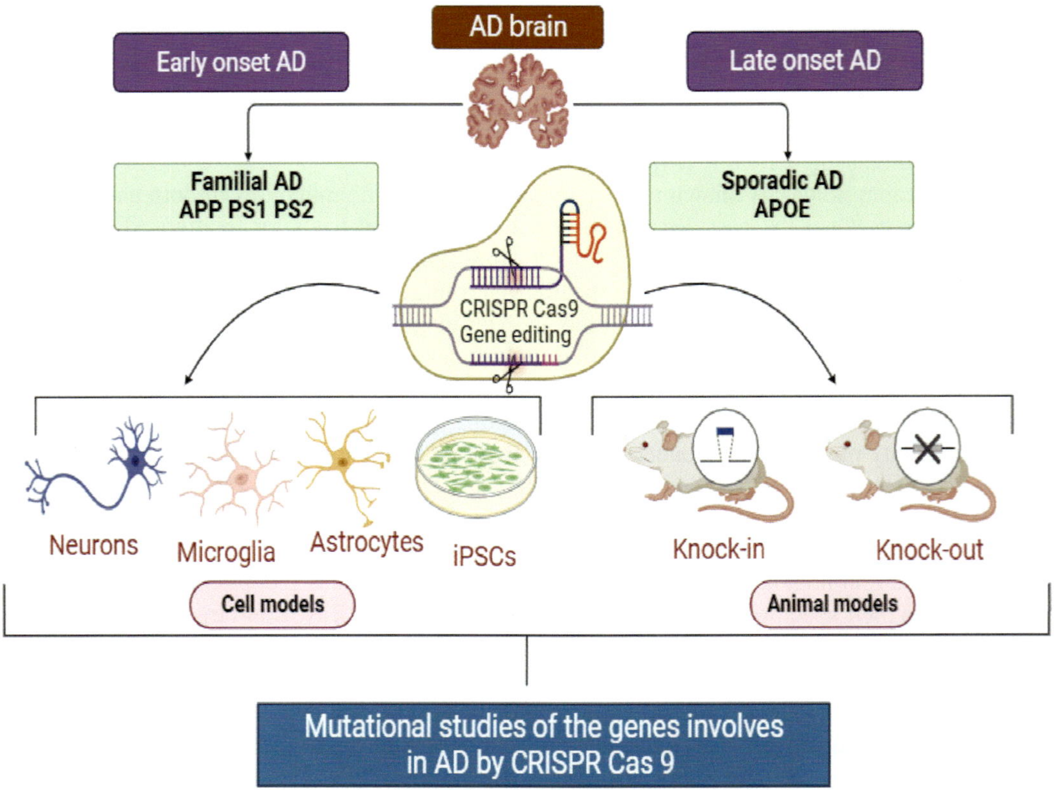

FIGURE 7.2 CRISPR/Cas9-mediated gene editing approach in AD.

Utilising CRISPR-Cas9 to eliminate GSAP in HEK-APP cells led to a marked reduction in both Aβ secretion and γ-secretase function [69]. Additionally, the Glia maturation factor (GMF) identified in AD-affected brain regions' glial cells is understood to instigate neuronal death. CRISPR/Cas9 interventions in BV2 cell lines inhibited pp38 MAPK and curtailed GMF expression [70]. A particularly compelling study underscored the TREM2 gene's role in AD, revealing that its knock-out in iPSCs diminished microglia survival and APOE clearance, further impairing SDF-1α/CXCR4-mediated chemotaxis, which is crucial for addressing beta-amyloid plaques [71]. On a different note, a specific CRISPR/Cas9 application on 22 AD-linked genes in HEK293T cells indicated no alterations in tau metabolism [72].

In addition to cell models, CRISPR Cas9 is employed in creating AD animal models. For instance, in the Tg2676 mice model, disruption of the APP gene using CRISPR/Cas9 technology led to diminished expression of Aβ [73]. In another notable study, researchers employed CRISPR Cas9 to target the APP gene in mice and rats, converting it to its human counterpart by introducing the point mutations G676R, F681Y, and R684H, creating a humanised animal model. This results in an increased level of Aβ as compared to the wild-type strain and reveals new mechanisms of disease pathogenesis [74]. Using CRISPR Cas9 in an AD mouse model targeting APP deletions led to reduced Aβ pathology and unveiled protective deletions in APP's 30-UTR [75]. A novel tau mouse model, post CRISPR-induced deletions in the Mapt gene encoding tau. This knock-out tau mouse model, on a C57Bl/6J background, shows resistance to excitotoxic effects and no memory deficit [76]. Moreover, a knock-in mouse model with a Plcγ2-P522R was created utilising CRISPR Cas9, showing enhanced microglial functions, suggesting a potential AD treatment strategy [73, 77]. The potential of BACE 1 as an AD therapeutic target was also explored with promising results using Cas9 nano complexes, where its expression was effectively reduced by using Cas9 nano complexes, which were produced by incorporating amphiphilic R7L10 peptide into Cas9-sgRNA [78]. These cumulative endeavors highlight the transformative potential of CRISPR/Cas9 in advancing our understanding and treatment approaches for AD.

7.7 INDUCED PLURIPOTENT STEM CELLS (iPSCs)

Based on familial and sporadic AD, an overview of iPSC, the rationale behind the use, and the various AD model systems studied will be discussed in this topic.

Presently, no therapeutic intervention exists that halts the progression of multifaceted AD. So, the first step to understanding the underlying mechanism of such a multifaceted disease is to develop an effective cell model. Traditionally, AD has been studied in animal models or immortalised cell lines, both of which have limitations in faithfully replicating human disease processes. Replicating AD in a laboratory setting is difficult due to the limited accessibility of pertinent living tissue, the intricate interplay between genetic factors and environmental variables, and the distinct disparities observed between animal models and human subjects. For example, post-mortem tissues typically reflect end-stage manifestations of complex diseases, potentially masking early pathological indicators and unique disease stage-specific characteristics [79]. Mice vary widely from humans in physiology, lifespan, and drug response [80]. Thus, many researchers and pharmaceutical firms deem data from human cell lines more applicable therapeutically than mouse model data [81, 82]. These challenges significantly constrain efforts to model the complex AD. However, the prospect of utilising neurons derived from human iPSCs for *in vitro* investigations has profoundly transformed the realm of intricate disease modeling, allowing *in vitro* studies of patient-specific genetic profiles and selective differentiation of neurons into the central nervous system in AD [83]. iPSC technology enables researchers to emulate the pathology and study genes linked to NDDs, providing a "disease-in-a-dish" model to examine cellular abnormalities in AD (**Figure 7.3**). iPSCs are a stem cell derived from mature human cells such as skin fibroblasts or blood cells. These cells are reprogrammed to reach a pluripotent stage, mainly driven by the exogenous expression of four transcription (commonly Oct3/4, Sox2, Klf4, and c-Myc) factors [84]. The neurons derived from iPSCs of AD

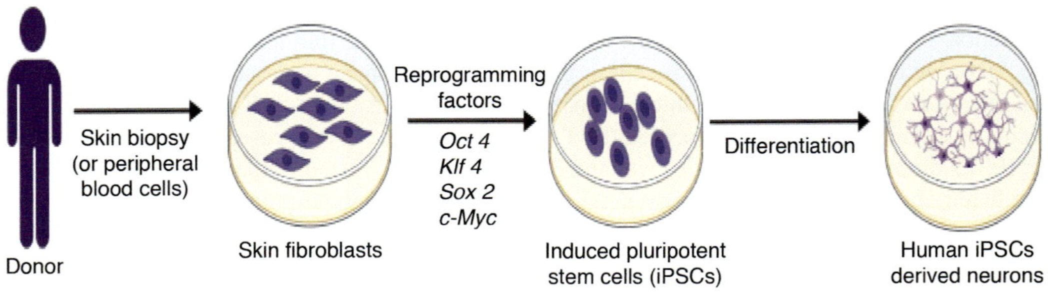

FIGURE 7.3 Schematic representation to generate induced pluripotent stem cells (iPSCs) from human somatic cells as skin fibroblasts or blood cells.

patients retain the donor's genetic makeup, which facilitates disease study within a human genetic framework, offering insights into how genetic changes drive AD pathogenesis. The fibroblasts have the capability to be directly reprogrammed, producing differentiated cells like neurons [85].

In the case of FAD, which accounts for 5% and is closely associated with mutations in the APP, PS1, or PS2 genes, a seminal model for understanding FAD was derived from patients' fibroblasts harboring mutations in PS1 (A246E) and PS2 (N141I). This AD model exhibited an elevated Aβ42 production and Aβ42/Aβ40 ratio, serving as a vital tool to recapitulate the molecular pathogenesis associated with the PS mutation [86]. Likewise, neurons derived from iPSCs, sourced from the fibroblasts of AD and Down syndrome patients, displayed hallmark AD features, notably Aβ42 accumulation and tau hyperphosphorylation. This underlines their potential utility in evaluating prospective treatment [87, 88]. Additionally, neural progenitor cells (NPCs) derived from AD-specific iPSCs, with either PS1 (A246E) or (M146L) mutations, manifested an augmented Aβ42/Aβ40 ratio. Such NPCs contribute to unveiling novel genes implicated in AD pathology [89]. Researchers created AD-specific iPSC lines from a patient exhibiting the APP E693 deletion (APP E693D) and another patient with the APP (V717L) mutation. These lines revealed heightened Aβ production and oxidative stress, exemplifying the value of patient-specific iPSCs in probing AD pathogenesis [90].

Over 95% of AD cases are represented by sAD [91]. Given its uncertain onset and etiology, creating a cellular model for sAD is challenging and leads to inconsistencies among cell lines from different patients. Remarkably, the APOE gene is the only one found to have genetic variations in sAD. Utilising iPSCs sourced from AD patients, the CRISPR/Cas9 mechanism was used to transform the E4 allele into an E3/E3 genotype. Neurons differentiated from these iPSCs with the E3 genotype exhibited increased resilience to ionomycin-induced cytotoxic effects. Conversely, cells expressing the E4 allele displayed elevated levels of tau phosphorylation, culminating in calcium dysregulation and consequent cellular apoptosis. This research enhances our comprehension of the link between the APOE4 allele and sAD development [92]. Neuronal models derived from iPSCs utilising sAD cell lines shed light on additional AD-linked phenotypic manifestations such as endoplasmic reticulum stress, inflammatory responses, oxidative stress, and environmental factors. For instance, when iPSC-derived neurons were exposed to H2O2 and NO – NO-mimicking oxidative stress and inflammation – they displayed reduced viability due to an increase in Aβ42 species, a pattern seen especially in sAD and familial AD (fAD) lines. Interestingly, the compound apigenin, known to possibly inhibit NF-κB, was found to protect against this inflammation-induced mortality [93]. Another insightful study by Birnbaum in 2018 used iPSC-derived neurons to delve deeper into sAD's onset and progression, pinpointing oxidative stress and alterations in mitochondrial protein expression, even in the absence of the hallmark Aβ and tau pathology [94] (**Figure 7.4**). Various phenotypic alterations were noted in cholinergic neurons derived from sAD patients carrying the APOE4 allele through iPSCs. The APOE3/E4 genetic makeup of these sAD patients led to neurons more vulnerable to harmful stimuli and heightened calcium levels, leading to enhanced neuronal

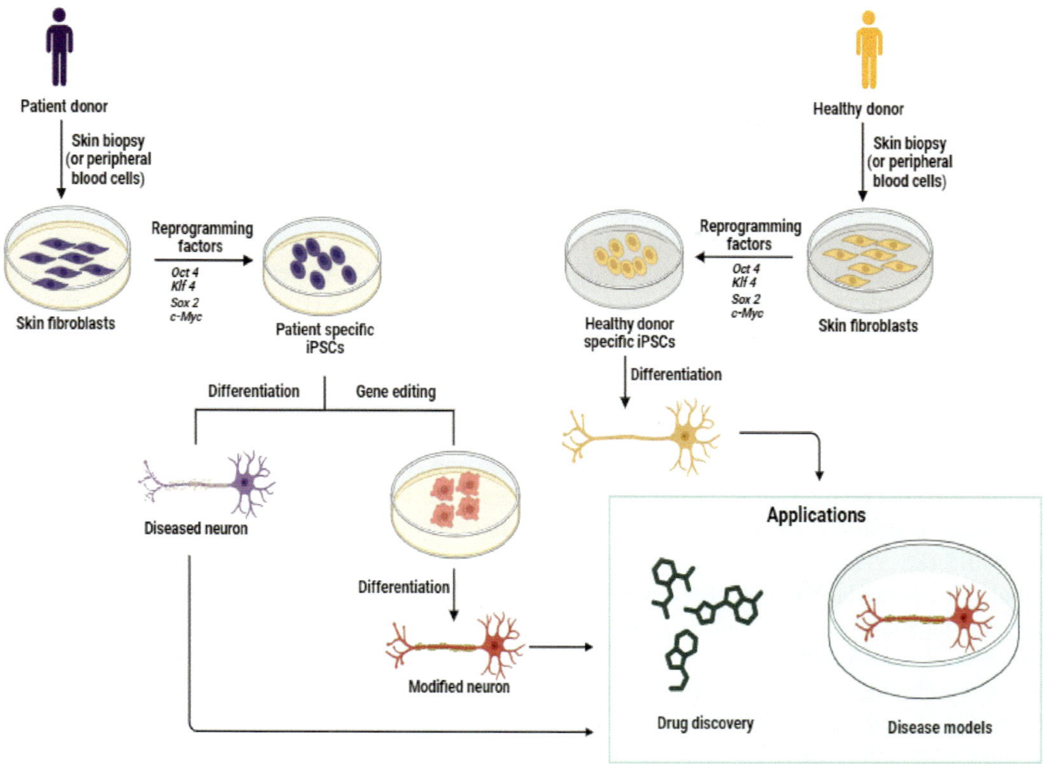

FIGURE 7.4 Schematic of iPSC technology and applications of iPSCs-derived diseased specific neurons in drug discovery and disease models for studying AD pathology.

excitotoxicity relative to standard neurons [95]. In another study, iPSC-derived neurons and astrocytes were used to highlight pathological manifestations in the sAD model, such as increased endoplasmic reticulum stress, oxidative damage, and Aβ oligomer buildup [90]. These rigorous research initiatives utilising iPSCs, the application of iPSC stands out as a beacon, heralding the future of innovative therapeutic strategies and more profound comprehension of this complex ailment.

7.8 GENETIC THERAPIES

In this topic, two major genetic therapies, antisense oligonucleotide (ASO) and adeno-associated viral vectors (AAV), and their potential target genes under study that are modified using these approaches will be discussed.

7.8.1 Antisense Oligonucleotide (ASO)

For AD treatments, therapies range from cell-based solutions like stem cells to molecular interventions targeting proteins or nucleic acids. Proteins are frequently modified using antibodies or substituted with synthetic versions, whereas gene therapies address DNA mutations through vectors. However, the standout approach focuses on RNA using synthetic antisense oligonucleotides (ASOs), which regulate protein synthesis, underscoring their significant therapeutic potential. Treatments should be specific to reduce side effects, tailored to the disease's stage, and encompass its multifaceted nature for optimal outcomes. Efficiently delivering therapeutic agents while navigating challenges like the blood-brain barrier remains a critical objective. Utilising short, synthetic ASO to target mRNA could potentially address the aforementioned needs. ASO is a short, synthetic strand of DNA or RNA designed to bind specifically to a target gene's messenger RNA (mRNA), thereby regulating its expression. This can lead to modifying, reducing, or eliminating protein production, making ASOs a valuable tool for therapeutic interventions. The highly specific complementary binding of ASOs to RNA, grounded in Watson-Crick base pairing, is noteworthy. Additionally, the rise of RNA interference techniques and the exploration of non-coding microRNAs (miRNA) have enhanced the interest in miRNAs as potential therapeutic avenues for ASO applications. Furthermore, advancements in techniques for delivering ASOs to the brain have enhanced their potential as treatments for neurological disorders [96].

Several explorations have employed ASOs to address hallmark AD features, such as the Aβ level in models like Tg2576 (APPswe), tau protein regulation in tauP301S (PS19) mice as well as the critical ApoE4 isoform as a major genetic AD risk factor. ASOs, designed to reduce toxic Aβ levels, target the APP mRNA and its enzymes. The OL-1 ASO focused on a specific amino acid segment of Aβ and dropped APP expression in Tg2576 (APPswe) and aging SAMP8 mouse models. Despite increased soluble Aβ, OL-1–enhanced cognition and reduced inflammation in these models [97, 98], in a separate study on Tg2576 (APPswe) mice, while one ASO reduced the cerebral Aβ40/42 ratio by targeting the β-site, another targeting the γ-secretase site had no evident effect [99]. Another ASO designed for RNase H-mediated tau mRNA degradation reduced tau protein in the brain, benefiting tauP301S (PS19) mice by preventing neuronal loss, limiting tau aggregate spread, and extending survival. This tau reduction was also observed in Cynomolgus monkeys [100]. Another ASO targeting the enzyme GSK-3b, crucial for tau phosphorylation, showed reduced tau phosphorylation and oxidative stress in SAMP8 mice, leading to improved cognitive function [97]. The ApoE4 isoform significantly elevates AD risk, especially for homozygotes. In the APP/PS1 mouse model with ApoE4, an ASO reduced brain APOE proteins. A decrease in Aβ plaque was observed only when treatment commenced shortly after birth, not in mature mice [101]. Future advancements in ASO chemistry and targeted delivery are poised to enhance antisense drug efficacy. The key breakthrough will be effectively and safely transporting ASOs across the blood-brain barrier for systemic administration.

7.8.2 Adeno-Associated Viral Vectors (AAV)

To delve into the intricacies of neurological disorders, scientists often rely on small animals, like mice and rats, and larger species, such as rabbits, dogs, and non-human primates, to understand complex neurological disorders. However, modeling disorders in laboratory animals, particularly through transgenesis or genome editing, poses substantial challenges. One significant impediment includes embryonic lethality due to certain genetic modifications [102]. To circumvent the challenges posed by traditional methods, the field has witnessed a growing inclination toward using viral vectors for precise transgene delivery. Notably, adeno-associated virus (AAV) vectors have gained prominence owing to their ability to efficiently introduce genes into various animal tissues, offering a flexible approach for accurately modeling genetic neurological disorders [103]. These vectors, derived from non-pathogenic viruses, are engineered to introduce therapeutic genes into cells. Once inside a cell, the AAV vector's genetic material is released in the nucleus, instructing the cell to produce therapeutic proteins. This method ensures long-term gene expression, reducing the risk of eliciting strong immune responses or disrupting the host DNA. Diverse AAV serotypes and administration techniques permit targeted interventions in distinct neuronal cell types and brain regions [103]. Other viral vectors such as adenovirus, herpesvirus, and lentivirus are also used in neuroscience for gene delivery [104]. Once the AAV vector binds to a cell, it is taken up and transported to the nucleus, where it releases its genetic payload. This introduced genetic material then instructs the cell to produce specific proteins based on the therapeutic intent. AAV vectors are favored in gene therapy because they can lead to long-term gene expression, elicit minimal immune responses, and have a low risk of disrupting the host cell's natural DNA.

AAV vectors can be employed to induce the overexpression of causative genes (APP or PS) in neurons, thereby establishing AD pathological conditions within a healthy animal model [105].For modeling neurodegenerative conditions in adult animals, intraparenchymal injections of AAV vectors target the primary affected brain areas. For instance, ADs can be modeled by direct injections into the hippocampus, a region, along with the associative cortex and subcortical structures, that is notably impacted in AD [106–108]. Furthermore, to address the prominent neuroinflammation seen in AD, an AD mouse model was designed to leverage the anti-inflammatory properties of interleukin-2 by delivering it encapsulated within an AAV vector, which demonstrated that the AAV-interleukin-2 not only reshaped the hippocampal structure by augmenting spine density but also reinforced synaptic plasticity. Such interventions led to significant enhancements in memory capabilities while exhibiting a limited spectrum of adverse effects [109, 110]. Additionally, in late-onset AD, where the APOE4 allele contributes to Aβ buildup, the other allele APOE2 offers protection. AAV-mediated delivery of the APOE2 allele has been shown to alleviate AD symptoms [111]. While rodents don't naturally develop AD, AAV vectors encoding mutations like APP [105], tau [107], and PS1 [106] have been injected into regions like the hippocampus to establish AD models. In a recent advancement, an AD model was established in non-human primates by injecting recombinant AAV1 with two AD-associated tau mutations. This AAV-driven expression led to the spread of misfolded tau protein, triggering a neuroinflammatory response and significant axonal damage [112]. The advent of innovative animal models using AAV, which closely mimic human diseases, enhances our arsenal for drug evaluation and the testing of cutting-edge therapeutic products. With these advancements, we optimistically anticipate a faster drug discovery journey for conditions that remain without cures.

7.9 CONCLUSION

AD is a progressive neurodegenerative disorder with brain pathologic changes taking place prior to the first presentation of clinical symptoms, which includes memory disturbance and behavioral symptoms. With over 50 million people globally afflicted by AD, understanding its genetic factors is pivotal for improved treatments and patient care. Efforts have been made in the field of genetics to develop an understanding of the disease, but most of these results remain unclear and debated. German psychiatrist Alois Alzheimer identified AD's clinical and neurological features in the early 20th century, paving the way for advanced genetic research into the disease. Predictive or mutated genes cause familial or early-onset AD. Although familial AD is very rare in the population, the genes show complete to near-complete disease penetrance and follow principles of Mendelian genetics. To date, only three disease-predictive genes have been discovered, which include APP, PSEN1, and PSEN2. Having an allele of the predictive genes will very likely contribute to AD development before 65 years of age. Intriguingly, mutations in these genes account for only about 5% of AD cases. Most (95%) of AD cases are associated with multiple gene variants, a condition often termed polygenic AD. Susceptible or risk genes cause sporadic or late-onset AD, which is the most common form of AD in which genes show moderate disease penetrance. Such genes do not follow the principles of Mendelian genetics. Having an allele of the risk genes will contribute to vulnerability or susceptibility to the disease, usually manifesting in clinical symptoms after 65 years of age. Due to the presence of polygenic gene variants and interaction with lifestyle and environmental factors, sporadic AD or LOAD is really complex to study.

Following the 2003 Human Genome Project, advancements have enabled advanced genetic studies like genome-wide association studies, deepening our understanding of AD's genetic intricacies, notably distinguishing familial from sporadic forms. Familial AD is tied to mutations like APP, PS1, and PS2, influencing amyloid-beta peptide aggregation. In contrast, sporadic AD intertwines genetics with environmental factors, with APOE4 being a key risk gene, yet many genes identified via GWAS contribute alongside modifiable lifestyle factors.

Moreover, while the primary focus has been on common genes and their variants, rarer genetic variants also play a crucial role, especially in early-onset Alzheimer's disease or familial AD. These are characterised by their autosomal dominant Mendelian inheritance pattern and have provided profound insights into AD's genetic mechanics.

In this expansive genetic landscape, tools like the polygenic risk score are transformative, gauging an individual's AD susceptibility by assessing multiple gene variants. Despite its prowess, PRS is clouded by debates on its broad applicability, especially outside European lineages. Yet, its value is evident in clinical settings where it shapes early intervention strategies, thereby heralding a new era of precision medicine and refined clinical trial designs. Simultaneously, the synergy of genetic data with state-of-the-art omics technologies has unearthed novel AD-associated pathways and promising drug targets. At the forefront of therapeutic modalities, gene editing technologies, particularly CRISPR/Cas9, are revolutionising our approach to AD by offering precise genomic interventions, both in cell and animal models.

Meanwhile, the power of iPSCs is being harnessed to replicate AD in a lab setting, providing a human-centric lens to view and understand disease pathogenesis, given the shortcomings of traditional animal and immortalised cell line models. iPSCs have emerged as a transformative tool in AD research, providing a "disease-in-a-dish" platform. They facilitate the exploration of patient-specific genetic intricacies, thereby offering invaluable insights into familial and sporadic AD pathogenesis. The ability of iPSCs to replicate the genetic makeup of AD patients has enabled researchers to better understand gene interactions, pathological manifestations, and potential treatments. Recent advancements in CRISPR/Cas9 techniques combined with iPSC technology further underscore the potential of this approach. The collective efforts and findings in the field enhance our understanding of AD and illuminate the path forward, with iPSCs playing a pivotal role in driving innovative therapeutic interventions and strategies. Finally, the realm of genetic therapies beckons with promise. The landscape of genetic therapies has seen remarkable advancements, particularly in the realms of antisense oligonucleotides (ASOs) and adeno-associated viral vectors (AAV), each presenting a unique approach to address the multifaceted nature of Alzheimer's disease (AD). ASOs, with their ability to target specific mRNAs, offer a promising therapeutic potential, enabling the regulation of protein synthesis in a manner that's tailored to the disease's complexities. Their specificity and capability to penetrate the blood-brain barrier herald a future where RNA-targeting interventions could redefine AD treatments. On the other hand, AAV vectors, derived from non-pathogenic viruses, have proven instrumental in delivering therapeutic genes to cells, thereby offering a mechanism to model and potentially treat neurological disorders. By harnessing these vectors, researchers can induce precise genetic changes in targeted brain regions, paving the way for better animal models and therapeutic interventions for AD. As both ASOs and AAV vectors continue to evolve, their combined potential underscores the strides in genetic therapies. These advancements, rooted in rigorous research and technological innovations, offer hope for more effective treatments and deeper insights into the enigma of AD. Collectively, these groundbreaking strides offer renewed hope in understanding and eventually conquering AD.

REFERENCES

1. Nichols E, Steinmetz JD, Vollset SE, Fukutaki K, Chalek J, Abd-Allah F, et al. Estimation of the global prevalence of dementia in 2019 and forecasted prevalence in 2050: an analysis for the Global Burden of Disease Study 2019. Lancet Public Health 2022;7:e105–25.
2. Hippius H, Neundörfer G. The discovery of Alzheimer's disease. Dialogues Clin Neurosci 2003;5:101–8.
3. Ott A, Slooter AJC, Hofman A, van Harskamp F, Witteman JCM, Van Broeckhoven C, et al. Smoking and risk of dementia and Alzheimer's disease in a population-based cohort study: the Rotterdam Study. The Lancet 1998;351:1840–3.
4. Tanzi RE. A genetic dichotomy model for the inheritance of Alzheimer's disease and common age-related disorders. J Clin Invest 1999;104:1175–9.
5. Bertram L, Tanzi RE. Alzheimer's disease: one disorder, too many genes? Hum Mol Genet 2004;13: R135–41.
6. Bertram L, Tanzi RE. The genetic epidemiology of neurodegenerative disease. J Clin Invest 2005;115: 1449–57.
7. Tanzi RE, Bertram L. Twenty years of the Alzheimer's disease amyloid hypothesis: a genetic perspective. Cell 2005;120:545–55.

8. Kang J, Lemaire H-G, Unterbeck A, Salbaum JM, Masters CL, Grzeschik K-H, et al. The precursor of Alzheimer's disease amyloid A4 protein resembles a cell-surface receptor. Nature 1987;325:733–6.
9. Tanzi RE, McClatchey AI, Lamperti ED, Villa-Komaroff L, Gusella JF, Neve RL. Protease inhibitor domain encoded by an amyloid protein precursor mRNA associated with Alzheimer's disease. Nature 1988;331:528–30.
10. Kitaguchi N, Takahashi Y, Tokushima Y, Shiojiri S, Ito H. Novel precursor of Alzheimer's disease amyloid protein shows protease inhibitory activity. Nature 1988;331:530–2.
11. Cruts M, vanBroeckhoven C. Molecular genetics of Alzheimer's disease. Ann Med 1998;30:560–5.
12. Thinakaran G, Koo EH. Amyloid precursor protein trafficking, processing, and function. J Biol Chem 2008;283:29615–9.
13. Zhao J, Liu X, Xia W, Zhang Y, Wang C. Targeting amyloidogenic processing of APP in Alzheimer's disease. Front Mol Neurosci 2020;13:137.
14. Goate A, Chartier-Harlin M-C, Mullan M, Brown J, Crawford F, Fidani L, et al. Segregation of a missense mutation in the amyloid precursor protein gene with familial Alzheimer's disease. Nature 1991;349:704–6.
15. Mullan M. Familial Alzheimer's disease: second gene locus located. BMJ 1992;305:1108.
16. Westerman MA, Cooper-Blacketer D, Mariash A, Kotilinek L, Kawarabayashi T, Younkin LH, et al. The relationship between Aβ and memory in the Tg2576 mouse model of Alzheimer's disease. J Neurosci 2002;22:1858–67.
17. Lemaire HG, Salbaum JM, Multhaup G, Kang J, Bayney RM, Unterbeck A, et al. The PreA4 695 precursor protein of Alzheimer's disease A4 amyloid is encoded by 16 exons. Nucleic Acids Res 1989;17: 517–22.
18. König G, Salbaum M, Wiestler O, Lang W, Schmitt HP, Masters CL, et al. Alternative splicing of the βA4 amyloid gene of Alzheimer's disease in cortex of control and Alzheimer's disease patients. Mol Brain Res 1991;9:259–62.
19. Neve RL, Finch EA, Dawes LR. Expression of the Alzheimer amyloid precursor gene transcripts in the human brain. Neuron 1988;1:669–77.
20. Sisodia SS, Koo EH, Beyreuther K, Unterbeck A, Price DL. Evidence that β-amyloid protein in Alzheimer's disease is not derived by normal processing. Science (1979) 1990;248:492–5.
21. De Strooper B, Saftig P, Craessaerts K, Vanderstichele H, Guhde G, Annaert W, et al. Deficiency of presenilin-1 inhibits the normal cleavage of amyloid precursor protein. Nature 1998;391:387–90.
22. Wolfe MS, Xia W, Ostaszewski BL, Diehl TS, Kimberly WT, Selkoe DJ. Two transmembrane aspartates in presenilin-1 required for presenilin endoproteolysis and γ-secretase activity. Nature 1999;398:513–7.
23. Citron M, Westaway D, Xia W, Carlson G, Diehl T, Levesque G, et al. Mutant presenilins of Alzheimer's disease increase production of 42-residue amyloid β-protein in both transfected cells and transgenic mice. Nat Med 1997;3:67–72.
24. Lippa CF, Nee LE, Mori H, George-Hyslop Ps. Aβ– 42 deposition precedes other changes in PS-1 Alzheimer's disease. The Lancet 1998;352:1117–8.
25. Yu H, Saura CA, Choi S-Y, Sun LD, Yang X, Handler M, et al. APP processing and synaptic plasticity in presenilin-1 conditional knockout mice. Neuron 2001;31:713–26.
26. Holcomb L, Gordon MN, McGowan E, Yu X, Benkovic S, Jantzen P, et al. Accelerated Alzheimer-type phenotype in transgenic mice carrying both mutant amyloid precursor protein and presenilin 1 transgenes. Nat Med 1998;4:97–100.
27. Scheuner D, Eckman C, Jensen M, Song X, Citron M, Suzuki N, et al. Secreted amyloid β–protein similar to that in the senile plaques of Alzheimer's disease is increased in vivo by the presenilin 1 and 2 and APP mutations linked to familial Alzheimer's disease. Nat Med 1996;2:864–70.
28. Sato N, Imaizumi K, Manabe T, Taniguchi M, Hitomi J, Katayama T, et al. Increased production of β-amyloid and vulnerability to endoplasmic reticulum stress by an aberrant spliced form of presenilin 2. J Biol Chem 2001;276:2108–14.
29. Grünberg J, Walter J, Eckman C, Capell A, Schindzielorz A, Younkin S, et al. Truncated presenilin 2 derived from differentially spliced mRNAs does not affect the ratio of amyloid β-peptide 1-42/1-40. Neuroreport 1998;9:3293–9.
30. Saunders AM, Strittmatter WJ, Schmechel D, St. George-Hyslop PH, Pericak-Vance MA, Joo SH, et al. Association of apolipoprotein E allele ε4 with late-onset familial and sporadic Alzheimer's disease. Neurology 1993;43:1467.
31. Bertram L, Tanzi RE. Genomic mechanisms in Alzheimer's disease. Brain Pathol 2020;30:966–77.
32. Lambert J-C, Ibrahim-Verbaas CA, Harold D, Naj AC, Sims R, Bellenguez C, et al. Meta-analysis of 74,046 individuals identifies 11 new susceptibility loci for Alzheimer's disease. Nat Genet 2013;45:1452–8.

33. Naj AC, Jun G, Beecham GW, Wang L-S, Vardarajan BN, Buros J, et al. Common variants at MS4A4/MS4A6E, CD2AP, CD33 and EPHA1 are associated with late-onset Alzheimer's disease. Nat Genet 2011;43:436–41.
34. Lambert J-C, Heath S, Even G, Campion D, Sleegers K, Hiltunen M, et al. Genome-wide association study identifies variants at CLU and CR1 associated with Alzheimer's disease. Nat Genet 2009;41:1094–9.
35. Hollingworth P, Harold D, Sims R, Gerrish A, Lambert J-C, Carrasquillo MM, et al. Common variants at ABCA7, MS4A6A/MS4A4E, EPHA1, CD33 and CD2AP are associated with Alzheimer's disease. Nat Genet 2011;43:429–35.
36. Seshadri S, Fitzpatrick AL, Ikram MA, DeStefano AL, Gudnason V, Boada M, et al. Genome-wide analysis of genetic loci associated with Alzheimer disease. JAMA 2010;303:1832–40.
37. Jeong W, Lee H, Cho S, Seo J. ApoE4-induced cholesterol dysregulation and its brain cell type-specific implications in the pathogenesis of Alzheimer's disease. Mol Cells 2019;42:739–46.
38. Liu C-C, Kanekiyo T, Xu H, Bu G. Apolipoprotein E and Alzheimer disease: risk, mechanisms and therapy. Nat Rev Neurol 2013;9:106–18.
39. Livingston G, Huntley J, Sommerlad A, Ames D, Ballard C, Banerjee S, et al. Dementia prevention, intervention, and care: 2020 report of the Lancet Commission. The Lancet 2020;396:413–46.
40. Heston LL, Mastri AR, Anderson VE, White J. Dementia of the Alzheimer type: clinical genetics, natural history, and associated conditions. Arch Gen Psychiatry 1981;38:1085–90.
41. Sherrington R, Rogaev EI, Liang Y al, Rogaeva EA, Levesque G, Ikeda M, et al. Cloning of a gene bearing missense mutations in early-onset familial Alzheimer's disease. Nature 1995;375:754–60.
42. Levy-Lahad E, Wasco W, Poorkaj P, Romano DM, Oshima J, Pettingell WH, et al. Candidate gene for the chromosome 1 familial Alzheimer's disease locus. Science (1979) 1995;269:973–7.
43. Priller C, Bauer T, Mitteregger G, Krebs B, Kretzschmar HA, Herms J. Synapse formation and function is modulated by the amyloid precursor protein. Journal of Neuroscience 2006;26:7212–21.
44. Young-Pearse TL, Bai J, Chang R, Zheng JB, LoTurco JJ, Selkoe DJ. A critical function for β-amyloid precursor protein in neuronal migration revealed by in utero RNA interference. Journal of Neuroscience 2007;27:14459–69.
45. Rovelet-Lecrux A, Hannequin D, Raux G, Meur N Le, Laquerrière A, Vital A, et al. APP locus duplication causes autosomal dominant early-onset Alzheimer disease with cerebral amyloid angiopathy. Nat Genet 2006;38:24–6.
46. Haass C, Lemere CA, Capell A, Citron M, Seubert P, Schenk D, et al. The Swedish mutation causes early-onset Alzheimer's disease by β-secretase cleavage within the secretory pathway. Nat Med 1995;1:1291–6.
47. Nilsberth C, Westlind-Danielsson A, Eckman CB, Condron MM, Axelman K, Forsell C, et al. The 'Arctic' APP mutation (E693G) causes Alzheimer's disease by enhanced Aβ protofibril formation. Nat Neurosci 2001;4:887–93.
48. Van Dorpe J, Smeijers L, Dewachter I, Nuyens D, Spittaels K, Van den Haute C, et al. Prominent cerebral amyloid angiopathy in transgenic mice overexpressing the London mutant of human APP in neurons. Am J Pathol 2000;157:1283–98.
49. Bagaria J, Bagyinszky E, An SSA. Genetics, functions, and clinical impact of presenilin-1 (PSEN1) gene. Int J Mol Sci 2022;23:10970.
50. Kumar-Singh S, Theuns J, Van Broeck B, Pirici D, Vennekens K, Corsmit E, et al. Mean age-of-onset of familial Alzheimer disease caused by presenilin mutations correlates with both increased Aβ42 and decreased Aβ40. Hum Mutat 2006;27:686–95.
51. Gatto EM, Rojas GJ, Nemirovsky SI, Da Prat G, Persi G, Cesarini M, et al. A novel mutation in PSEN1 (p. Arg41Ser) in an Argentinian woman with early onset Parkinsonism. Parkinsonism Relat Disord 2020;77:21–5.
52. Ryan NS, Rossor MN. Correlating familial Alzheimer's disease gene mutations with clinical phenotype. Biomark Med 2010;4:99–112.
53. Moonesinghe R, Khoury MJ, Liu T, Janssens ACJW. Discriminative accuracy of genomic profiling comparing multiplicative and additive risk models. EJHG 2011;19:180–5.
54. Leonenko G, Sims R, Shoai M, Frizzati A, Bossù P, Spalletta G, et al. Polygenic risk and hazard scores for Alzheimer's disease prediction. Ann Clin Transl Neurol 2019;6:456–65.
55. Desikan RS, Fan CC, Wang Y, Schork AJ, Cabral HJ, Cupples LA, et al. Genetic assessment of age-associated Alzheimer disease risk: Development and validation of a polygenic hazard score. PLoS Med 2017;14:e1002258–e1002258.
56. Sander JD, Joung JK. CRISPR-Cas systems for editing, regulating and targeting genomes. Nat Biotechnol 2014;32:347–55.
57. Doudna JA. The promise and challenge of therapeutic genome editing. Nature 2020;578:229–36.

58. Carroll D. Genome engineering with zinc-finger nucleases. Genetics 2011;188:773–82.
59. Bhardwaj A, Nain V. TALENs—an indispensable tool in the era of CRISPR: a mini review. J Genet Eng & Biotechnol 2021;19:125.
60. Khadempar S, Familghadakchi S, Motlagh RA, Farahani N, Dashtiahangar M, Rezaei H, et al. CRISPR–Cas9 in genome editing: Its function and medical applications. J Cell Physiol 2019;234:5751–61.
61. Eid A, Mahfouz MM. Genome editing: the road of CRISPR/Cas9 from bench to clinic. Exp Mol Med 2016;48:e265–e265.
62. Gopinath C, Nathar TJ, Ghosh A, Hickstein DD, Remington Nelson EJ. Contemporary animal models for human gene therapy applications. Curr Gene Ther 2015;15:531–40.
63. Barrangou R, Horvath P. A decade of discovery: CRISPR functions and applications. Nat Microbiol 2017;2:1–9.
64. Zaboikin M, Zaboikina T, Freter C, Srinivasakumar N. Non-homologous end joining and homology directed DNA repair frequency of double-stranded breaks introduced by genome editing reagents. PLoS One 2017;12:e0169931–e0169931.
65. Mir A, Edraki A, Lee J, Sontheimer EJ. Type II-C CRISPR-Cas9 biology, mechanism, and application. ACS Chem Biol 2018;13:357–65.
66. Lu L, Yu X, Cai Y, Sun M, Yang H. Application of CRISPR/Cas9 in Alzheimer's disease. Front Neurosci 2021;15:803894.
67. Sun L, Zhou R, Yang G, Shi Y. Analysis of 138 pathogenic mutations in presenilin-1 on the in vitro production of Aβ42 and Aβ40 peptides by γ-secretase. Proc Natl Acad Sci 2017;114:E476–85.
68. Fang B, Jia L, Jia J. Chinese Presenilin-1 V97L mutation enhanced Aβ42 levels in SH-SY5Y neuroblastoma cells. Neurosci Lett 2006;406:33–7.
69. Wong E, Liao GP, Chang JC, Xu P, Li Y-M, Greengard P. GSAP modulates γ-secretase specificity by inducing conformational change in PS1. Proc Natl Acad Sci 2019;116:6385–90.
70. Raikwar SP, Thangavel R, Dubova I, Selvakumar GP, Ahmed ME, Kempuraj D, et al. Targeted gene editing of glia maturation factor in microglia: a novel Alzheimer's disease therapeutic target. Mol Neurobiol 2019;56:378–93.
71. McQuade A, Kang YJ, Hasselmann J, Jairaman A, Sotelo A, Coburn M, et al. Gene expression and functional deficits underlie TREM2-knockout microglia responses in human models of Alzheimer's disease. Nat Commun 2020;11:5370.
72. Kolay S, Diamond MI. Alzheimer's disease risk modifier genes do not affect tau aggregate uptake, seeding or maintenance in cell models. FEBS Open Bio 2020;10:1912–20.
73. György B, Lööv C, Zaborowski MP, Takeda S, Kleinstiver BP, Commins C, et al. CRISPR/Cas9 mediated disruption of the Swedish APP allele as a therapeutic approach for early-onset Alzheimer's disease. Mol Ther-Nucleic Acids 2018;11:429–40.
74. Serneels L, T'Syen D, Perez-Benito L, Theys T, Holt MG, De Strooper B. Modeling the β-secretase cleavage site and humanizing amyloid-beta precursor protein in rat and mouse to study Alzheimer's Mol Neurodegener. 2020 Oct;15(1):60. doi: 10.1186/s13024-020-00399-z. PMID: 33076948; PMCID: PMC7574558.
75. Nagata K, Takahashi M, Matsuba Y, Okuyama-Uchimura F, Sato K, Hashimoto S, et al. Generation of App knock-in mice reveals deletion mutations protective against Alzheimer's disease-like pathology. Nat Commun 2018;9:1800.
76. Tan D, Yao S, Ittner A, Bertz J, Ke YD, Ittner LM, et al. Generation of a new tau knockout (tau Δex1) line using CRISPR/Cas9 genome editing in mice. J Alzheimer's Dis 2018;62:571–8.
77. Takalo M, Wittrahm R, Wefers B, Parhizkar S, Jokivarsi K, Kuulasmaa T, et al. The Alzheimer's disease-associated protective Plcγ2-P522R variant promotes immune functions. Mol Neurodegener 2020;15:1–14.
78. Park H, Oh J, Shim G, Cho B, Chang Y, Kim S, et al. In vivo neuronal gene editing via CRISPR–Cas9 amphiphilic nanocomplexes alleviates deficits in mouse models of Alzheimer's disease. Nat Neurosci 2019;22:524–8.
79. Siller R, Greenhough S, Park I-H, J Sullivan G. Modelling human disease with pluripotent stem cells. Curr Gene Ther 2013;13:99–110.
80. Haston KM, Finkbeiner S. Clinical trials in a dish: the potential of pluripotent stem cells to develop therapies for neurodegenerative diseases. Annu Rev Pharmacol Toxicol 2016;56:489–510.
81. Benam KH, Dauth S, Hassell B, Herland A, Jain A, Jang K-J, et al. Engineered in vitro disease models. Annual Review of Pathology: Mechanisms of Disease 2015;10:195–262.
82. Donowitz M, Turner JR, Verkman AS, Zachos NC. Current and potential future applications of human stem cell models in drug development. J Clin Invest 2020;130:3342–4.
83. Young JE, Delgado-Morales R. Human-induced pluripotent stem cell-derived neurons to model and gain insights into Alzheimer's disease pathogenesis. In Stem Cell Genetics for Biomedical Research: Past, Present, and Future (pp. 3–12). 2018.

84. Takahashi K, Yamanaka S. Induction of pluripotent stem cells from mouse embryonic and adult fibroblast cultures by defined factors. Cell 2006;126:663–76.
85. Vierbuchen T, Ostermeier A, Pang ZP, Kokubu Y, Südhof TC, Wernig M. Direct conversion of fibroblasts to functional neurons by defined factors. Nature 2010;463:1035–41.
86. Yagi T, Ito D, Okada Y, Akamatsu W, Nihei Y, Yoshizaki T, et al. Modeling familial Alzheimer's disease with induced pluripotent stem cells. Hum Mol Genet 2011;20:4530–9.
87. Jang J, Yoo J-E, Lee J-A, Lee DR, Kim JY, Huh YJ, et al. Disease-specific induced pluripotent stem cells: a platform for human disease modeling and drug discovery. Exp Mol Med 2012;44:202–13.
88. Shi Y, Kirwan P, Smith J, MacLean G, Orkin SH, Livesey FJ. A human stem cell model of early Alzheimer's disease pathology in Down syndrome. Sci Transl Med 2012;4:124ra29-124ra29.
89. Sproul AA, Jacob S, Pre D, Kim SH, Nestor MW, Navarro-Sobrino M, et al. Characterization and molecular profiling of PSEN1 familial Alzheimer's disease iPSC-derived neural progenitors. PLoS One 2014;9:e84547–e84547.
90. Kondo T, Asai M, Tsukita K, Kutoku Y, Ohsawa Y, Sunada Y, et al. Modeling Alzheimer's disease with iPSCs reveals stress phenotypes associated with intracellular Aβ and differential drug responsiveness. Cell Stem Cell 2013;12:487–96.
91. Zhang L, Chen C, Mak MSH, Lu J, Wu Z, Chen Q, et al. Advance of sporadic Alzheimer's disease animal models. Med Res Rev 2020;40:431–58.
92. Wadhwani AR, Affaneh A, Van Gulden S, Kessler JA. Neuronal apolipoprotein E4 increases cell death and phosphorylated tau release in Alzheimer disease. Ann Neurol 2019;85:726–39.
93. Balez R, Steiner N, Engel M, Muñoz SS, Lum JS, Wu Y, et al. Neuroprotective effects of apigenin against inflammation, neuronal excitability and apoptosis in an induced pluripotent stem cell model of Alzheimer's disease. Sci Rep 2016;6:31450.
94. Birnbaum JH, Wanner D, Gietl AF, Saake A, Kündig TM, Hock C, et al. Oxidative stress and altered mitochondrial protein expression in the absence of amyloid-β and tau pathology in iPSC-derived neurons from sporadic Alzheimer's disease patients. Stem Cell Res 2018;27:121–30.
95. Duan L, Bhattacharyya BJ, Belmadani A, Pan L, Miller RJ, Kessler JA. Stem cell derived basal forebrain cholinergic neurons from Alzheimer's disease patients are more susceptible to cell death. Mol Neurodegener 2014;9:1–14.
96. Grabowska-Pyrzewicz W, Want A, Leszek J, Wojda U. Antisense oligonucleotides for Alzheimer's disease therapy: from the mRNA to miRNA paradigm. EBioMedicine 2021;74.
97. Farr SA, Erickson MA, Niehoff ML, Banks WA, Morley JE. Central and peripheral administration of antisense oligonucleotide targeting amyloid precursor protein improves learning and memory and reduces neuroinflammatory cytokines in Tg2576 (APPswe) mice. J Alzheimer's Dis 2014;40:1005.
98. Erickson MA, Niehoff ML, Farr SA, Morley JE, Dillman LA, Lynch KM, et al. Peripheral administration of antisense oligonucleotides targeting the amyloid-β protein precursor reverses AβPP and LRP-1 overexpression in the aged SAMP8 mouse brain. J Alzheimer's Dis 2012;28:951–60.
99. Chauhan NB, Siegel GJ. Antisense inhibition at the β-secretase-site of β-amyloid precursor protein reduces cerebral amyloid and acetyl cholinesterase activity in Tg2576. Neuroscience 2007;146:143–51.
100. DeVos SL, Miller RL, Schoch KM, Holmes BB, Kebodeaux CS, Wegener AJ, et al. Tau reduction prevents neuronal loss and reverses pathological tau deposition and seeding in mice with tauopathy. Sci Transl Med 2017;9:eaag0481–eaag0481.
101. Huynh T-P V, Liao F, Francis CM, Robinson GO, Serrano JR, Jiang H, et al. Age-dependent effects of apoE reduction using antisense oligonucleotides in a model of β-amyloidosis. Neuron 2017;96:1013–23.
102. Gurumurthy CB, Lloyd KCK. Generating mouse models for biomedical research: technological advances. Dis Model Mech 2019;12:dmm029462–dmm029462.
103. Lunev E, Karan A, Egorova T, Bardina M. Adeno-associated viruses for modeling neurological diseases in animals: achievements and prospects. Biomedicines 2022;10:1140.
104. Nectow AR, Nestler EJ. Viral tools for neuroscience. Nat Rev Neurosci 2020;21:669–81.
105. Lawlor PA, Bland RJ, Das P, Price RW, Holloway V, Smithson L, et al. Novel rat Alzheimer's disease models based on AAV-mediated gene transfer to selectively increase hippocampal Aβ levels. Mol Neurodegener 2007;2:1–13.
106. Audrain M, Fol R, Dutar P, Potier B, Billard J-M, Flament J, et al. Alzheimer's disease-like APP processing in wild-type mice identifies synaptic defects as initial steps of disease progression. Mol Neurodegener 2016;11:1–17.
107. Jaworski T, Dewachter I, Lechat B, Croes S, Termont A, Demedts D, et al. AAV-tau mediates pyramidal neurodegeneration by cell-cycle re-entry without neurofibrillary tangle formation in wild-type mice. PLoS One 2009;4:e7280–e7280.

108. Braak H, Thal DR, Ghebremedhin E, Del Tredici K. Stages of the pathologic process in Alzheimer disease: age categories from 1 to 100 years. J Neuropathol Exp Neurol 2011;70:960–9.
109. Saadoun D, Rosenzwajg M, Joly F, Six A, Carrat F, Thibault V, et al. Regulatory T-cell responses to low-dose interleukin-2 in HCV-induced vasculitis. N Engl J Med 2011;365:2067–77.
110. Alves S, Churlaud G, Audrain M, Michaelsen-Preusse K, Fol R, Souchet B, et al. Interleukin-2 improves amyloid pathology, synaptic failure and memory in Alzheimer's disease mice. Brain 2017;140:826–42.
111. Zhao L, Gottesdiener AJ, Parmar M, Li M, Kaminsky SM, Chiuchiolo MJ, et al. Intracerebral adeno-associated virus gene delivery of apolipoprotein E2 markedly reduces brain amyloid pathology in Alzheimer's disease mouse models. Neurobiol Aging 2016;44:159–72.
112. Beckman D, Chakrabarty P, Ott S, Dao A, Zhou E, Janssen WG, et al. A novel tau-based rhesus monkey model of Alzheimer's pathogenesis. Alzheimer's Dement 2021;17:933–45.

8 Organoids and Alzheimer's Disease

Payal Singh
MMV, Department of Zoology, Banaras Hindu University, Varanasi, India

Shashank Mishra
Center of Experimental Medicine & Surgery, Institute of Medical Sciences, Banaras Hindu University, Varanasi, India

Pradeep Kumar
Department of Neurology, Faculty of Medicine, Hadassah Medical Center, Hebrew University of Jerusalem, Hadassah, Jerusalem, Israel

8.1 INTRODUCTION

The prevalence of dementia is rising as the world's population ages quickly. An estimated 35×10^6 persons globally are thought to have Alzheimer's disease (AD) or other types of dementia. By 2030, there will be about 82 million cases of dementia, and by 2050, there will be 150 million cases [1]. The most common neurodegenerative illness and the main cause of dementia in the elderly is Alzheimer's disease. Alzheimer's disease is characterised by a progressive decline in cognitive functioning, including memory loss and impaired motor function. The widespread loss of neurons and synaptic connections in the cortex, hippocampus, amygdala, and basal forebrain, among other brain regions, is thought to be the cause of these consequences [2–4].

The Prince et al. states that people with AD require long-term care, financial resources, and goods and services. Additionally, the study predicts that by 2030, the socioeconomic cost of AD will increase from approximately $1 trillion in 2018 to $2 trillion. The international medical community urgently needs to focus on prevention and therapy for Alzheimer's disease since there is currently no known cure for it [5]. It is challenging to assess the effectiveness and mechanisms of Alzheimer's disease treatments since current animal models of AD are unable to accurately mimic the pathophysiology in human patients. Despite their strong therapeutic efficacy in early animal-model research, this is the main cause of new AD drugs' failure in phase III clinical trials.

Compared to animal and cellular models, 3D cerebral organoid cultures created with stem cell technologies have demonstrated notable advances in modeling nervous system disorders in recent years. The increase in modeling accuracy has established new standards for the study of nervous system illnesses, such as Alzheimer's disease. While assessments of the cerebral organoid's technology have been published, an overview of the AD mini-brain has not yet been published. The advantages and disadvantages of AD cerebral organoids are discussed here, along with an overview of new technologies that can be used to customise cerebral organoids. To provide a new angle on AD research, prospective methods or approaches for improving the existing brain organoid models are also suggested.

Organoids are essentially structures that resemble organs that are created from primary tissue or pluripotent stem cells [6]. Every cell type has different benefits and drawbacks when it comes to being the main source of organoids. As the main source of cells, ESCs or iPSCs, for instance, might

 DOI: 10.1201/9781032620442-8

be used since they are frequently reprogrammed from fibroblasts and are simple to obtain in large quantities from patients. These pluripotent stem cells have the ability to mimic the gene expression patterns of fetuses [7]. However, in more complex 3D settings, where a lack of knowledge about differentiation signals may have unintended consequences in the form of less developed organoids, accidental heterogeneity of such pluripotent stem cells is seen [8, 9]. Conversely, although adult human stem cells are well-defined homogenous cells, they are rarely separated from a particular tissue, limiting their differentiation potential [10]. Organoids are powerful systems that enable the study of certain diseases and the creation of customised treatments. The possibility that pharmaceuticals can be applied to preclinical therapies and that organ complexity can be simulated has increased due to this emerging technology. This suggests a variety of approaches for modeling human disease, tissue engineering, drug research, diagnosis, and regenerative medicine. In this paper, we first examine the development of organoid technology and its uses, and then we talk about the challenges and limitations faced by three-dimensional organoids. Finally, we suggest that a basic mechanical underpinning for 'human modeling' systems that recommend tailored drugs can be found in human organoids.

8.2 HISTORICAL BACKGROUND

German doctor Alois Alzheimer initially reported the symptoms of Alzheimer's disease in 1907 [4]. In 1901, he laid eyes on 51-year-old Auguste Deter for the first time. When Auguste began acting strangely hiding things, threatening neighbors, accusing her husband of adultery – her husband, Karl, sent her to a mental health facility. She also lost the ability to cook and clean the house on a daily basis. Auguste was admitted to a Frankfurt psychiatric hospital to receive treatment. He observed and documented her actions there: She was able to talk but not write her name, identify items like a pencil but not the food she was eating, and alternate between being polite and annoying. Alzheimer diagnosed Auguste with 'presenile dementia' [5]. An Alzheimer's brain biopsy performed after her death in 1906 identified 'particular abnormalities in cortical cell clusters' in addition to widespread cortical atrophy [6]. Alzheimer's disease was associated with nerve fiber plaques and tangles, which scientists would later identify as beta amyloid plaques and tau neurofibrillary tangles in the 1980s [7, 8]. At a German psychiatric conference that year, Alzheimer gave a presentation on Auguste and claimed that these lesions in her brain were the cause of her symptoms. The following year, he released a study report, and in 1910, he named the condition 'Alzheimer's disease' in a psychiatric textbook. In the United States, the clinical diagnostic standards for AD were established in 1984 [9]. They were changed in 2011 and 2018 to emphasise the importance of biomarkers in the diagnosis of AD and to specify distinct diagnoses for the preclinical, dementia, and moderate cognitive impairment (MCI) phases of the disease [10, 11].

8.3 PATHOGENESIS

AD is a multifaceted, complex neurological illness brought on by a complex web of interplay between environment, age, education, and heredity. One of the earliest theories that laid the foundation for comprehending the disease's genesis is the cholinergic theory. The finding that patients with Alzheimer's disease had lower levels of acetylcholinesterase and choline acetyltransferase in their cerebral cortex when compared to healthy brains lends credence to this theory [11]. Reduced neurotransmitter pathway activity was shown in postmortem brain tissue from AD patients, suggesting that cholinergic neuron degeneration and neurotransmission loss are major contributors to the cognitive impairment seen in AD [11]. Given that AD histology shows tau protein-based neurofibrillary intraneuronal lesions, the tau theory has also been proposed. The creation and stability of the neuronal microtubule network are mediated by tau proteins, which are mostly found in neurons. Tau protein turns pathogenic when phosphorylation regulation is disrupted, causing hyperphosphorylated tau proteins to polymerise into filaments and create neurofibrillary tangles.

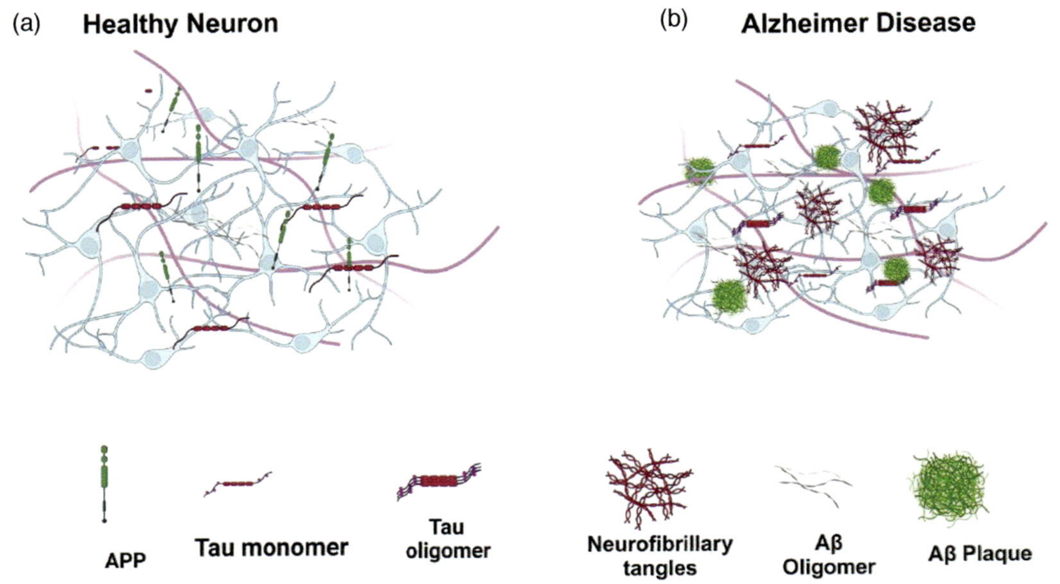

FIGURE 8.1 Histopathological alterations observed in the brain of Alzheimer's disease patients. (a) Healthy brain. (b) AD brain with neurofibrillary tangles inside of neurons and amyloid plaques outside.

This leads to abnormal axonal transport, synaptic function, and neuronal morphology as well as the cytoskeleton's defective structural and regulatory roles [11], causing the degeneration of neurons. Two main traits associated with Alzheimer's disease (AD) have been identified from post-mortem brain examinations of affected individuals. The accumulation of misfolded proteins, such as amyloid plaques and neurofibrillary tangles (NFTs), is one of these characteristics (**Figure 8.1**) [12].

The abnormal build-up of hyperphosphorylated tau protein inside the intracellular milieu is what causes the formation of NFTs. The fibrillar deposits of amyloid-β (Aβ) peptide, which is misfolded and comprises 40 or 42 amino acids (Aβ40, Aβ42), are what make up the amyloid plaques.

These peptides are generated as byproducts of the amyloid precursor protein's (APP) metabolism [3, 12–15]. Alzheimer's disease has no known cause, but becoming older is considered the greatest risk factor because the disease doubles in frequency every five years after age 65. Alzheimer's disease with sporadic or late-onset can be applied to about 95% of cases.

8.4 GENETICS

The genetic heritability of Alzheimer's disease (including memory components) ranges from 49%–79%, according to analyses of twin and family studies. About 0.1% of cases occur before the age of 65 and have autosomal dominant inheritance, which is not sex-linked. This type of Alzheimer's is known as early-onset familial Alzheimer's disease. The etiology of Alzheimer's disease should include the genetics of the disease in addition to inflammation, apoptosis, and plaque deposition. The first gene identified as causative for AD was the APP gene on 21q21 [16]. Advances in genetic research have identified two distinct types of Alzheimer's disease (AD): Sporadic Alzheimer's disease (SAD), which accounts for the majority of cases, and familial Alzheimer's disease (FAD). Significant discoveries in the 1990s and early 2000s demonstrated that APP, PSEN1, and PSEN2 gene mutations, which are found on chromosomes 21, 14, and 1, respectively, are the cause of FAD [17, 18]. The amino acid residues required for the catalytic active site of gamma-secretase are specifically present in PSEN1 and PSEN2. Aβ peptide production and neurodegeneration are enhanced by specific mutations in these genes [11]. Much more commonly, a genetic risk factor for SAD was found to be the type e4 allele on chromosome 19, which is home to the low-density

TABLE 8.1
Genes that Significantly Influence the Development of Sporadic Alzheimer's Disease (SAD) [20]

Genes Involved in Pathogenesis of AD	Abbreviations
Amyloid precursor protein gene	APP
Presenilin gene	PS
Apolipoprotein E gene	APOE
Clusterin gen	n/a
Complement Receptor 1 gene	n/a
Phospholipids bind to clathrin protein gene	PICALM
Cholesterol metabolism gene	CH25H, ABCAL, and CH24H
Sterol O-acyltransferase gene	SOAT1
Prostaglandin-endoperoxide Synthase 2 gen	Ptgs2
Angiotensin-converting enzyme gene	n/a
SLC26A38 gene	n/a

lipoprotein carrier gene apolipoprotein E (APOE). Compared to 20%–25% of healthy older people without a family history of the disease, 50%–60% of Alzheimer's patients carry APOE. An eight-fold increased risk of Alzheimer's disease is linked to two copies of APOE, compared to a three-fold increased risk with one copy [19]. **Tables 8.1** and **8.2** list these as well as additional important SAD-influencing genes.

In recent decades, a number of pharmaceutical treatments have been approved, such as N-methyl D-aspartate receptor antagonists and acetylcholinesterase inhibitors. While some symptoms are alleviated by these therapies, the neurodegenerative process is not stopped [35, 36]. Although the clinical features of the disease are well understood, the exact mechanism causing or exacerbating the beginning of Alzheimer's disease remains poorly understood [37].

The lack of suitable in vivo and in vitro models that faithfully reproduce the illness and aid in understanding the molecular pathways behind Alzheimer's disease is the main contributing factor. These models are also necessary to assess the effectiveness of recently created medicinal therapies.

TABLE 8.2
Important Genes Implicated in Alzheimer's Disease Etiology

Gene	Proposed Mechanism of Dysfunction Contributing to Pathogenesis in Alzheimer's Disease
APP	Defects in synaptic development and neuronal migration [21] A-beta peptide synthesis defects [22]
PSEN 1 / 2	Altered gamma-secretase activity resulting in elevated amyloid beta-42 levels [23]
APOE – epsilon n 4	Lipid and cholesterol dysmetabolism, synaptic inflammation, impaired clearance of amyloid beta-42, LDL receptor impairment [24–27]
CLU	Deposition and metabolism of amyloid beta-42 [28]
ABCA7	Immune mediated and lipid metabolic response alterations [29]
CR1	Aggravated senile plaque formation [30]
CD33	Influences microglia mediated clearance of amyloid beta-42 [31]
MS4A	Dysregulation of intracellular calcium concentration [32]
EPHA1	Axonal guidance changes [33]
SORL1, BIN, CD2AP, PICALM	Alterations in lipid metabolism [34]

8.5 MODELS FOR ALZHEIMER'S DISEASE

Animal models were used in the early phases of research to help understand the pathogenic mechanisms behind Alzheimer's disease. Numerous species, including as Drosophila melanogaster, Caenorhabditis elegans, and Dario rerio, have been used in studies. Nonetheless, because mice might have symptoms that are comparable to those observed in people, they are the main animal used as a model for AD [38]. Alzheimer's disease family mutation-carrying mice exhibit synaptic and cognitive deficits caused by Aβ, but they do not fully reproduce other important clinical manifestations of AD, such as diseases unrelated to neurofibrillary tangles (NFTs) [38, 39].

There is currently a debate concerning the effectiveness of animal models because human experiments with them show a significant failure rate of roughly 99.6% [40]. The impression that neurodegenerative markers are not realistic enough stems from the absence of observable markers such as neuronal death or the formation of neurofibrillary tangles (NFTs). Moreover, it is significant to remember that the morphology and physiology of the mouse and human brains differ anatomically [41, 42]. Because the altered genes cover only a small percentage of cases (FAD) and do not cover the whole spectrum of the diseased condition, animal models show bias in the data they offer [35]. In a similar vein, neurons isolated from AD patients who have passed away have shown elevated quantities of toxic Aβ species and phosphorylated tau. Nevertheless, neither Aβ plaques nor neurofibrillary tangles (NFTs) have been seen in these neurons [43].

Determining the pathogenesis of Alzheimer's disease requires the establishment of an efficient disease model. Over the years, many in vivo and in vitro methods have been used to mimic Alzheimer's disease. Still, none of these methods can precisely reproduce the aberrant features seen in AD patients' brains. Rats are the most common animal model used in in vivo techniques, whereas two- and three-dimensional [2D and 3D) preparations of live cells are used in in vitro techniques (**Figure 8.2**). Thus, the application of stem cell technology in conjunction with developments in 2D and 3D neural differentiation, particularly in relation to human pluripotent stem cells (hPSCs), offers a unique opportunity to overcome this challenge and generate an infinite supply of neurons and human brain cells for in vitro research [44].

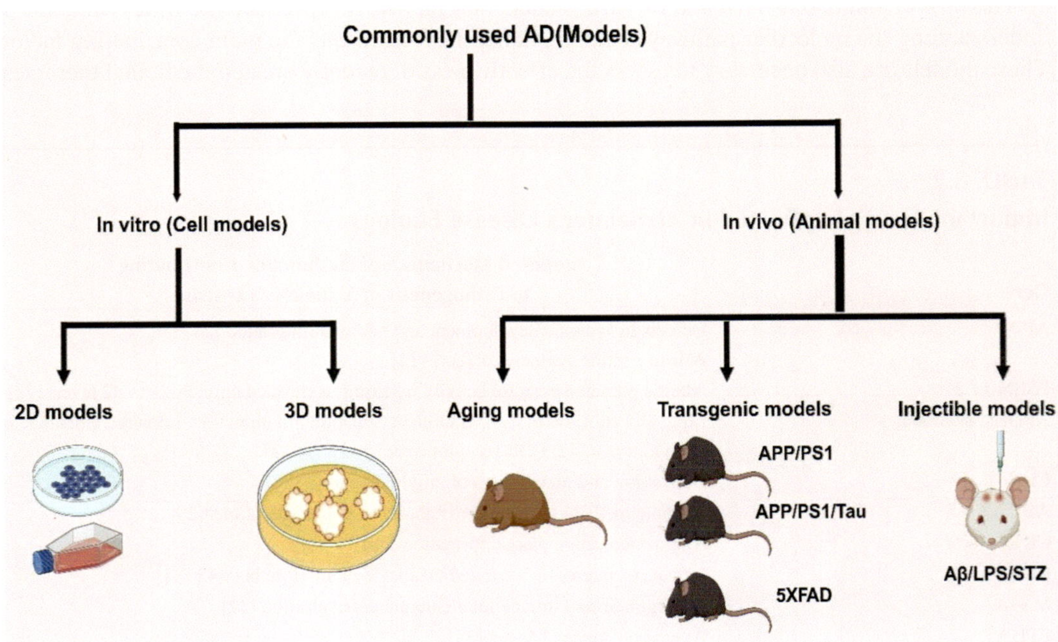

FIGURE 8.2 Most commonly used animal- and cell-based models in Alzheimer's disease.

8.6 AD CELL MODELS

The method by which fibroblast reprogramming-generated induced pluripotent stem cells (iPSCs) differentiates [45]. The idea that these FAD iPSCs can function as a model for FAD is supported by the fact that they developed from neurons with higher A42 levels in patients with FAD. There are intrinsic benefits to 3D cell cultures over 2D cell cultures. First, compared to 2D surroundings, cell-associated markers are more highly expressed and precisely localised in 3D preparations. Synapses in the 3D environment have more complicated morphologies (with 3D neurite extension) and better nerve cell marker localisation, more akin to what is seen in the body. Secondly, 3D cultures have a better ability to replicate the environment of the brain. One promising approach to create a meaningful cellular model of AD is the co-cultivation of neurons and glial cells, which could mimic the complex interplay of brain cells. However, in two-dimensional co-cultures consisting of neurons and glial cells, cell division is limited to the horizontal direction. An unexpected top-to-bottom polarity is produced when the upper layer of neurons grows over the lower layer of glial cells; this can be corrected by 3D co-culturing [46]. Seidel et al. [47] observed an EGFP-fused wild-type tau protein in SH-SY5Y cell lines in both 2D and 3D settings, with the 3D cells containing a noticeably higher concentration of this neuronal protein. Furthermore, cells can interact with one another in a three-dimensional environment. The 3D-plated brain cells are connected by neuronal synapses and function as a bulbous mass with directional intercellular connections. But 2D cultured cells can develop only in two-dimensional monolayers, and neurite interactions and outgrowth mostly take place in the horizontal plane of the substrate. The development of synaptic connections and coordinated interactions between cells is inhibited by this 2D cell spacing. The formation of functional synapses between neurons can therefore be facilitated by a suitable 3D microenvironment, where neurites can develop several millimeters within the 3D bioactive framework. The properties of individual neuronal subpopulations innervating distant targets can be accurately modeled in this context. Neurons produce mature cytoskeletal proteins, have intricate three-dimensional morphologies, and show signs of network connectivity. They also exhibit extensive neurite arborisation in all dimensions [47]. Under 2D-culture circumstances, no strong connections between cells are seen, in contrast to 3D conditions. It follows that the difficulty of A accumulations in AD cells cultured in two dimensions is not surprising [48], as the A produced after cell induction can spread to the culture medium and is removed during medium replacement. It is difficult to initiate subsequent pathogenic cascade processes (e.g. NFTs and tau phosphorylation) when A cannot aggregate [43].

Choi et al. [43] developed human neural progenitor cells (hNPCs) on a Matrigel substrate while overexpressing human APP or APP and PSEN1. In 3D cultures, there are more toxic A deposits and p-tau in hNPCs than in 2D cultures. The combination of the two methods (iPSC-differentiated cells and 3D culture) implies that an in vitro 3D environment can be a legitimate and physiologically accurate model for future research on Alzheimer's disease and evaluating the effectiveness of AD medications [49]. The environment of a 3D hydrogel cell culture can offer a closed system for brain tissue and nerve cells, hence restricting A diffusion and fostering A deposition [50]. Ravi et al. [51] contend that 3D cell-culture settings are superior to 2D circumstances due to their capacity to provide spatial structuring, adhesion, proliferation, signaling, and mechanical cell transduction [52], which 2D cell culture cannot give. Consequently, 3D cell culture is essential for investigating AD pathophysiology [43]. These settings also shed light on the study and development of three-dimensional tissues and organs.

8.7 DEVELOPMENT OF 3D CEREBRAL ORGANOIDS

With its complex vascular and neural connections, the brain is the most sophisticated organ in the human body. It is not feasible to observe pathogenic pathways in vivo within the central nervous system (CNS). The human brain can be studied only by using postmortem or biopsy tissue; it is not possible to observe diseased mechanisms in the central nervous system in vivo. Ever since mouse embryonic stem cells were discovered in 1981 [53], the field of stem-cell technology has developed quickly. Because of their ability to self-organise [54–57], stem cells can grow into complete organoids. Numerous in vitro tissue-culture experiments coupled with the utilisation of human iPSCs (hiPSCs) have established a solid foundation for 3D neural tissue models [58]. In 2007, Takahashi et al. [59] effectively generated hiPSCs by reprogramming the expression of four transcription factors in human somatic cells that have matured and differentiated: Oct3/4, Sox2, Klf4, and c-myc. These grown and mature human somatic cells can now differentiate into 3D neuroepithelial tissues, neural rosettes, and embryoid bodies because of advancements in stem cell technologies. The human brain's genetic makeup may be present in the neuroepithelial tissues, but there is no brain structure or interregional coordination in them [60]. Eiraku et al. [57] 2011 saw the creation of the first 3D neuroepithelial tissue with complete tissue architecture. This brain tissue can self-organise and self-develop to form subregional structures if it is kept in a floating environment. Sato and associates [54] then developed a gel-like extracellular matrix called Matrigel, which has been proven to provide an adequate environment for the self-organisation of several organoids [61–64], including stomach, liver, lung, and kidney organoids. In 2013, Lancaster et al. [65] initially used the techniques outlined in the other studies to distinguish hiPSCs into an organoid representing the entire brain. These hiPSCs gave rise to a neuroectoderm structure that resembled the cerebral cortex's early developmental stage in a human embryo [66]. They created cerebral organoids by modelling human microcephaly using patient-specific iPSCs, and they demonstrated that the microcephaly organoid differs significantly from cerebral organoids made from normal human stem cells. Premature neuronal differentiation [65] is suggested by the patient organoids' low progenitor areas, increased neurons, and decreased radial glial stem cells. This condition could account for the disease phenotype. The findings suggest that human brain organoids can mimic disease and development while having the same genetic makeup as their stem-cell donors.

Dang et al. [67] found that the gene expression pattern in 30-day-old human cerebral organoids is identical to that in eight- to nine-week-old fetal brains using RNA sequencing techniques. There have been published protocols for the creation of organoids that mimic different parts of the human brain, including the cortex [68], basal ganglia [69], hippocampus [70], choroid plexus [71], thalamus [72], retina [56], striatum [73], hypothalamus [74] midbrain [75–77], cerebellum [78, 79], and spinal cord [80, 81]. Furthermore, the development of organoid cultures has been improved by the use of bioreactors. For example, Qian et al. [75] built a small, low-cost, easily operated spinning bioreactor to improve feeding and oxygen diffusion in order to produce human iPSC-derived forebrain, midbrain, and hypothalamus organoids. This bioreactor is a potential advance for the modeling of brain development and disease, as well as for therapeutic screening using organoids. The evolution of cerebral organoids is depicted in **Figure 8.2**. Because the cerebral organoids contain human-specific genotypes and protein expression patterns, they may mimic the developmental processes and anatomical characteristics of early embryonic brains, which makes them highly relevant from a clinical standpoint [82]. First, the folded sulcus and area expansions of the cerebral organoid cortex are similar to those of the cerebral cortex in humans. Both the ventricular zone and the later-formed subventricular zone, which are organised into continuous cortical folds, grow in size and surface area as human cerebral organoids mature. Increased cell proliferation temporarily postponed neuronal differentiation, and an overall rise in neural-progenitor cell population are the causes of this in vitro growth. The cellular and anatomical architecture of the developing human brain is exactly like this type of amplification [83]. Second, Cerebral organoids are rich in synaptic connections and neuronal collaterals. Third, the cellular and structural components of the cerebral organoids and the human brain are identical [84, 85]. All things considered, brain organoids have shown a great deal of promise for creating precise models of CNS diseases like AD and acting as a platform for analysing disease pathways. Later studies on cerebral organoids have used cerebral organoids made from iPSCs drawn from FAD patients [86, 87] and have also produced differentiated SAD cerebral organoids [88] by converting APOE3 to APOE4 in iPSCs taken from SAD patients. AD characteristics, including A peptide, hyperphosphorylated tau protein,

NFTs, synaptic dysfunction, neuronal degeneration, and neuroinflammation, must be present in qualified AD cerebral organoids [89].

8.8 CONCLUSION

As this analysis has shown, 3D models, in particular cerebral organoids (COs), seem to be the most useful models for advancing research on AD. COs have been discovered to recapitulate some of the most important aspects of the AD pathogenesis more effectively than the previous models (2D cells in vitro and in vivo animals). This has made significant progress in the industry possible during the last ten years. A portion of the complexity of the human brain is contained in human COs, a novel model. They are therefore quickly emerging as a very interesting model for the investigation of the cellular and molecular anomalies at the brain's level that characterise AD. These organoids can be created by genetically manipulating control iPS cells from familial or sporadic AD patients, then adding or removing mutations in relevant genes. This may hasten the identification of new therapeutic targets for this illness's possible cure. Furthermore, COs are a fantastic tool for furthering our understanding of the interactions between different phenotypes of brain cells, including microglia, astrocytes, oligodendrocytes, and neurons, in both healthy and pathological conditions. However, increasing the use of these models and improving them is the scientific community's current priority. Despite their present limitations, they are already producing remarkable outcomes that allow for the investigation of possible treatments as well as a deeper knowledge of AD. Brain organoids offer a platform of the future for modelling the causes and mechanisms of Alzheimer's disease development as well as for testing potential treatments.

REFERENCES

1. Prince, M, Bryce, R, Albanese, E, Wimo, A, Ribeiro, W, Ferri, CP. The global prevalence of dementia: A systematic review and metaanalysis. Alzheimers Dement. 2013;9(1):63–75.
2. Lane, CA, Hardy, J, Schott, JM. Alzheimer's disease. Eur J Neurol. 2018;25(1):59–70.
3. Burns, A, Byrne, EJ, Maurer, K. Alzheimer's disease. Lancet. 2002;360(9327):163–5.
4. Amemori, T, Jendelova, P, Ruzicka, J, Urdzikova, LM, Sykova, E. Alzheimer's disease: Mechanism and approach to cell therapy. Int J Mol Sci. 2015;16(11):26417–51.
5. 2020 Alzheimer's disease facts and figures. Alzheimers Dement. 2020.
6. Tuveson, D, Clevers, H. Cancer modeling meets human organoid technology. Science. 2019;364(6444):952–5.
7. Forbes, TA, Howden, SE, Lawlor, K, Phipson, B, Maksimovic, J, Hale, L, et al. Patient-iPSC-derived kidney organoids show functional validation of a ciliopathic renal phenotype and reveal underlying pathogenetic mechanisms. Am J Hum Genet. 2018;102(5):816–31.
8. Carcamo-Orive, I, Hoffman, GE, Cundiff, P, Beckmann, ND, D'Souza, SL, Knowles, JW, et al. Analysis of transcriptional variability in a large human iPSC library reveals genetic and non-genetic determinants of heterogeneity. Cell Stem Cell. 2017;20(4):518–32.e9.
9. Narsinh, KH, Sun, N, Sanchez-Freire, V, Lee, AS, Almeida, P, Hu, S, et al. Single cell transcriptional profiling reveals heterogeneity of human induced pluripotent stem cells. J Clin Invest. 2011;121(3):1217–21.
10. McCracken, KW, Catá, EM, Crawford, CM, Sinagoga, KL, Schumacher, M, Rockich, BE, et al. Modelling human development and disease in pluripotent stem-cell-derived gastric organoids. Nature. 2014;516(7531):400–4.
11. Barage, SH, Sonawane, KD. Amyloid cascade hypothesis: Pathogenesis and therapeutic strategies in Alzheimer's disease. Neuropeptides. 2015;52:1–18.
12. Serrano-Pozo, A, Frosch, MP, Masliah, E, Hyman, BT. Neuropathological alterations in Alzheimer disease. Cold Spring Harb Perspect Med. 2011;1(1):a006189.
13. Selkoe, DJ, Hardy, J. The amyloid hypothesis of Alzheimer's disease at 25 years. EMBO Mol Med. 2016;8(6):595–608.
14. Bernabeu-Zornoza, A, Coronel, R, Palmer, C, Monteagudo, M, Zambrano, A, Liste, I. Physiological and pathological effects of amyloid-β species in neural stem cell biology. Neural Regen Res. 2019;14(12):2035–42.

15. Bernabeu-Zornoza, A, Coronel, R, Palmer, C, López-Alonso, V, Liste, I. Oligomeric and fibrillar species of Aβ42 diversely affect human neural stem cells. Int J Mol Sci. 2021;22(17).
16. Mizwicki, MT, Menegaz, D, Zhang, J, Barrientos-Durán, A, Tse, S, Cashman, JR, et al. Genomic and nongenomic signaling induced by 1α,25(OH)2-vitamin D3 promotes the recovery of amyloid-β phagocytosis by Alzheimer's disease macrophages. J Alzheimers Dis. 2012;29(1):51–62.
17. Masoumi, A, Goldenson, B, Ghirmai, S, Avagyan, H, Zaghi, J, Abel, K, et al. 1alpha,25-dihydroxyvitamin D3 interacts with curcuminoids to stimulate amyloid-beta clearance by macrophages of Alzheimer's disease patients. J Alzheimers Dis. 2009;17(3):703–17.
18. Rygiel, K. Novel strategies for Alzheimer's disease treatment: An overview of anti-amyloid beta monoclonal antibodies. Indian J Pharmacol. 2016;48(6):629–36.
19. Bondi, MW, Edmonds, EC, Salmon, DP. Alzheimer's disease: Past, present, and future. J Int Neuropsychol Soc. 2017;23(9-10):818–31.
20. Chen, YG. Research progress in the pathogenesis of Alzheimer's disease. Chin Med J (Engl). 2018;131(13): 1618–24.
21. Priller, C, Bauer, T, Mitteregger, G, Krebs, B, Kretzschmar, HA, Herms, J. Synapse formation and function is modulated by the amyloid precursor protein. J Neurosci. 2006;26(27):7212–21.
22. Arold, ST, Baur, AS. Dynamic Nef and Nef dynamics: How structure could explain the complex activities of this small HIV protein. Trends Biochem Sci. 2001;26(6):356–63.
23. Schellenberg, GD, Bird, TD, Wijsman, EM, Orr, HT, Anderson, L, Nemens, E, et al. Genetic linkage evidence for a familial Alzheimer's disease locus on chromosome 14. Science. 1992;258(5082):668–71.
24. Mahley, RW, Weisgraber, KH, Huang, Y. Apolipoprotein e: Structure determines function, from atherosclerosis to Alzheimer's disease to AIDS. J Lipid Res. 2009;50(Suppl):S183–8.
25. Herz, J, Chen, Y, Masiulis, I, Zhou, L. Expanding functions of lipoprotein receptors. J Lipid Res. 2009;50(Suppl):S287–92.
26. Zhong, N, Weisgraber, KH. Understanding the association of apolipoprotein E4 with Alzheimer disease: Clues from its structure. J Biol Chem. 2009;284(10):6027–31.
27. Verghese, PB, Castellano, JM, Garai, K, Wang, Y, Jiang, H, Shah, A, et al. ApoE influences amyloid-β (aβ) clearance despite minimal apoE/aβ association in physiological conditions. Proc Natl Acad Sci U S A. 2013;110(19):E1807–16.
28. Nuutinen, T, Suuronen, T, Kauppinen, A, Salminen, A. Clusterin: A forgotten player in Alzheimer's disease. Brain Res Rev. 2009;61(2):89–104.
29. Ramirez, LM, Goukasian, N, Porat, S, Hwang, KS, Eastman, JA, Hurtz, S, et al. Common variants in ABCA7 and MS4A6A are associated with cortical and hippocampal atrophy. Neurobiol Aging. 2016; 39:82–9.
30. Chibnik, LB, Shulman, JM, Leurgans, SE, Schneider, JA, Wilson, RS, Tran, D, et al. CR1 is associated with amyloid plaque burden and age-related cognitive decline. Ann Neurol. 2011;69(3):560–9.
31. Bradshaw, EM, Chibnik, LB, Keenan, BT, Ottoboni, L, Raj, T, Tang, A, et al. CD33 Alzheimer's disease locus: Altered monocyte function and amyloid biology. Nat Neurosci. 2013;16(7):848–50.
32. Karch, CM, Jeng, AT, Nowotny, P, Cady, J, Cruchaga, C, Goate, AM. Expression of novel Alzheimer's disease risk genes in control and Alzheimer's disease brains. PLoS One. 2012;7(11):e50976.
33. Wang, HF, Tan, L, Hao, XK, Jiang, T, Tan, MS, Liu, Y, et al. Effect of EPHA1 genetic variation on cerebrospinal fluid and neuroimaging biomarkers in healthy, mild cognitive impairment and Alzheimer's disease cohorts. J Alzheimers Dis. 2015;44(1):115–23.
34. Naj, AC, Jun, G, Beecham, GW, Wang, LS, Vardarajan, BN, Buros, J, et al. Common variants at MS4A4/MS4A6E, CD2AP, CD33 and EPHA1 are associated with late-onset Alzheimer's disease. Nat Genet. 2011;43(5):436–41.
35. Yang, J, Li, S, He, XB, Cheng, C, Le, W. Induced pluripotent stem cells in Alzheimer's disease: Applications for disease modeling and cell-replacement therapy. Mol Neurodegener. 2016;11(1):39.
36. Kumar, A, Singh, A, Ekavali A review on Alzheimer's disease pathophysiology And its management: An update. Pharmacol Rep. 2015;67(2):195–203.
37. Tincer, G, Mashkaryan, V, Bhattarai, P, Kizil, C. Neural stem/progenitor cells in Alzheimer's disease. Yale J Biol Med. 2016;89(1):23–35.
38. Qin, T, Prins, S, Groeneveld, GJ, Van Westen, G, de Vries, HE, Wong, YC, et al. Utility of animal models to understand human Alzheimer's disease, using the mastermind research approach to avoid unnecessary further sacrifices of animals. Int J Mol Sci. 2020;21(9).
39. Malm, T, Magga, J, Koistinaho, J. Animal models of Alzheimer's disease: Utilization of transgenic Alzheimer's disease models in studies of amyloid Beta clearance. Curr Transl Geriatr Exp Gerontol Rep. 2012;1(1):11–20.

40. Drummond, E, Wisniewski, T. Alzheimer's disease: Experimental models and reality. Acta Neuropathol. 2017;133(2):155–75.
41. Choi, SH, Kim, YH, Quinti, L, Tanzi, RE, Kim, DY. 3D culture models of Alzheimer's disease: A road map to A "cure-in-A-dish. Mol Neurodegener. 2016;11(1):75.
42. Tong, G, Izquierdo, P, Raashid, RA. human induced pluripotent stem cells and the modelling of Alzheimer's disease: The human brain outside the dish. Open Neurol J. 2017;11:27–38.
43. Choi, SH, Kim, YH, Hebisch, M, Sliwinski, C, Lee, S, D'Avanzo, C, et al. A three-dimensional human neural cell culture model of Alzheimer's disease. Nature. 2014;515(7526):274–8.
44. Arber, C, Lovejoy, C, Wray, S. Stem cell models of Alzheimer's disease: Progress and challenges. Alzheimers Res Ther. 2017;9(1):42.
45. Arber, C, Toombs, J, Lovejoy, C, Ryan, NS, Paterson, RW, Willumsen, N, et al. Familial Alzheimer's disease patient-derived neurons reveal distinct mutation-specific effects on amyloid beta. Mol Psychiatry. 2020;25(11):2919–31.
46. Cullen, DK, Wolf, JA, Vernekar, VN, Vukasinovic, J, LaPlaca, MC. Neural tissue engineering and biohybridized microsystems for neurobiological investigation in vitro (part 1). Crit Rev Biomed Eng. 2011;39(3):201–40.
47. Seidel, D, Krinke, D, Jahnke, HG, Hirche, A, Kloß, D, Mack, TG, et al. Induced tauopathy in A novel 3D-culture model mediates neurodegenerative processes: A real-time study on biochips. PLoS One. 2012;7(11):e49150.
48. D'Avanzo, C, Aronson, J, Kim, YH, Choi, SH, Tanzi, RE, Kim, DY. Alzheimer's in 3D culture: Challenges and perspectives. Bioessays. 2015;37(10):1139–48.
49. Zhang, D, Pekkanen-Mattila, M, Shahsavani, M, Falk, A, Teixeira, AI, Herland, A. A 3D Alzheimer's disease culture model and the induction of P21-activated kinase mediated sensing in iPSC derived neurons. Biomaterials. 2014;35(5):1420–8.
50. Liedmann, A, Rolfs, A, Frech, MJ. Cultivation of human neural progenitor cells in a 3-dimensional self-assembling peptide hydrogel. J Vis Exp. 2012(59): e3830.
51. Ravi, M, Paramesh, V, Kaviya, SR, Anuradha, E, Solomon, FD. 3D cell culture systems: Advantages and applications. J Cell Physiol. 2015;230(1):16–26.
52. Baker, BM, Chen, CS. Deconstructing the third dimension: How 3D culture microenvironments alter cellular cues. J Cell Sci. 2012;125(Pt 13):3015–24.
53. Evans, MJ, Kaufman, MH. Establishment in culture of pluripotential cells from mouse embryos. Nature. 1981;292(5819):154–6.
54. Sato, T, Vries, RG, Snippert, HJ, van de Wetering, M, Barker, N, Stange, DE, et al. Single Lgr5 stem cells build crypt-villus structures in vitro without a mesenchymal niche. Nature. 2009;459(7244):262–5.
55. Suga, H, Kadoshima, T, Minaguchi, M, Ohgushi, M, Soen, M, Nakano, T, et al. Self-formation of functional adenohypophysis in three-dimensional culture. Nature. 2011;480(7375):57–62.
56. Nakano, T, Ando, S, Takata, N, Kawada, M, Muguruma, K, Sekiguchi, K, et al. Self-formation of optic cups and storable stratified neural retina from human ESCs. Cell Stem Cell. 2012;10(6):771–85.
57. Eiraku, M, Takata, N, Ishibashi, H, Kawada, M, Sakakura, E, Okuda, S, et al. Self-organizing optic-cup morphogenesis in three-dimensional culture. Nature. 2011;472(7341):51–6.
58. Kelava, I, Lancaster, MA. Dishing out mini-brains: Current progress and future prospects in brain organoid research. Dev Biol. 2016;420(2):199–209.
59. Takahashi, K, Tanabe, K, Ohnuki, M, Narita, M, Ichisaka, T, Tomoda, K, et al. Induction of pluripotent stem cells from adult human fibroblasts by defined factors. Cell. 2007;131(5):861–72.
60. Brennand, KJ, Gage, FH. Modeling psychiatric disorders through reprogramming. Dis Model Mech. 2012;5(1):26–32.
61. Stange, DE, Koo, BK, Huch, M, Sibbel, G, Basak, O, Lyubimova, A, et al. Differentiated Troy+ chief cells act as reserve stem cells to generate all lineages of the stomach epithelium. Cell. 2013;155(2):357–68.
62. Huch, M, Dorrell, C, Boj, SF, van Es, JH, Li, VS, van de Wetering, M, et al. In vitro expansion of single Lgr5+ liver stem cells induced by Wnt-driven regeneration. Nature. 2013;494(7436):247–50.
63. Dye, BR, Hill, DR, Ferguson, MA, Tsai, YH, Nagy, MS, Dyal, R, et al. In vitro generation of human pluripotent stem cell derived lung organoids. Elife. 2015;4.
64. Takasato, M, Er, PX, Chiu, HS, Maier, B, Baillie, GJ, Ferguson, C, et al. Kidney organoids from human iPS cells contain multiple lineages and model human nephrogenesis. Nature. 2015;526(7574):564–8.
65. Lancaster, MA, Renner, M, Martin, CA, Wenzel, D, Bicknell, LS, Hurles, ME, et al. Cerebral organoids model human brain development and microcephaly. Nature. 2013;501(7467):373–9.
66. Lancaster, MA, Knoblich, JA. Generation of cerebral organoids from human pluripotent stem cells. Nat Protoc. 2014;9(10):2329–40.

67. Dang, J, Tiwari, SK, Lichinchi, G, Qin, Y, Patil, VS, Eroshkin, AM, et al. Zika virus depletes neural progenitors in human cerebral organoids through activation of the innate immune receptor TLR3. Cell Stem Cell. 2016;19(2):258–65.
68. Eiraku, M, Watanabe, K, Matsuo-Takasaki, M, Kawada, M, Yonemura, S, Matsumura, M, et al. Self-organized formation of polarized cortical tissues from ESCs and its active manipulation by extrinsic signals. Cell Stem Cell. 2008;3(5):519–32.
69. Birey, F, Andersen, J, Makinson, CD, Islam, S, Wei, W, Huber, N, et al. Assembly of functionally integrated human forebrain spheroids. Nature. 2017;545(7652):54–9.
70. Sakaguchi, H, Kadoshima, T, Soen, M, Narii, N, Ishida, Y, Ohgushi, M, et al. Generation of functional hippocampal neurons from self-organizing human embryonic stem cell-derived dorsomedial telencephalic tissue. Nat Commun. 2015; 6:8896.
71. Pellegrini, L, Bonfio, C, Chadwick, J, Begum, F, Skehel, M, Lancaster, MA. Human CNS barrier-forming organoids with cerebrospinal fluid production. Science. 2020;369(6500).
72. Xiang, Y, Tanaka, Y, Cakir, B, Patterson, B, Kim, KY, Sun, P, et al. hESC-derived thalamic organoids form reciprocal projections when fused with cortical organoids. Cell Stem Cell. 2019;24(3):487–97.e7.
73. Miura, Y, Li, MY, Birey, F, Ikeda, K, Revah, O, Thete, MV, et al. Generation of human striatal organoids and cortico-striatal assembloids from human pluripotent stem cells. Nat Biotechnol. 2020;38(12):1421–30.
74. Huizing, M, Helip-Wooley, A, Westbroek, W, Gunay-Aygun, M, Gahl, WA. Disorders of lysosome-related organelle biogenesis: Clinical and molecular genetics. Annu Rev Genomics Hum Genet. 2008;9:359–86.
75. Qian, X, Nguyen, HN, Song, MM, Hadiono, C, Ogden, SC, Hammack, C, et al. Brain-region-specific organoids using mini-bioreactors for modeling ZIKV exposure. Cell. 2016;165(5):1238–54.
76. Monzel, AS, Smits, LM, Hemmer, K, Hachi, S, Moreno, EL, van Wuellen, T, et al. Derivation of human midbrain-specific organoids from neuroepithelial stem cells. Stem Cell Reports. 2017;8(5):1144–54.
77. Jo, J, Xiao, Y, Sun, AX, Cukuroglu, E, Tran, HD, Göke, J, et al. Midbrain-like organoids from human pluripotent stem cells contain functional dopaminergic and neuromelanin-producing neurons. Cell Stem Cell. 2016;19(2):248–57.
78. Lee, S, Park, C, Han, JW, Kim, JY, Cho, K, Kim, EJ, et al. Direct reprogramming of human dermal fibroblasts into endothelial cells using ER71/ETV2. Circ Res. 2017;120(5):848–61.
79. Muguruma, K, Nishiyama, A, Kawakami, H, Hashimoto, K, Sasai, Y. Self-organization of polarized cerebellar tissue in 3D culture of human pluripotent stem cells. Cell Rep. 2015;10(4):537–50.
80. Shi, Y, Sun, L, Wang, M, Liu, J, Zhong, S, Li, R, et al. Vascularized human cortical organoids (vOrganoids) model cortical development in vivo. PLoS Biol. 2020;18(5):e3000705.
81. Ogura, T, Sakaguchi, H, Miyamoto, S, Takahashi, J. Three-dimensional induction of dorsal, intermediate and ventral spinal cord tissues from human pluripotent stem cells. Development. 2018;145(16).
82. Amin, ND, Paşca, SP. Building models of brain disorders with three-dimensional organoids. Neuron. 2018;100(2):389–405.
83. Li, Y, Muffat, J, Omer, A, Bosch, I, Lancaster, MA, Sur, M, et al. Induction of expansion and folding in human cerebral organoids. Cell Stem Cell. 2017;20(3):385–96.e3.
84. Bagley, JA, Reumann, D, Bian, S, Lévi-Strauss, J, Knoblich, JA. Fused cerebral organoids model interactions between brain regions. Nat Methods. 2017;14(7):743–51.
85. Luo, C, Lancaster, MA, Castanon, R, Nery, JR, Knoblich, JA, Ecker, JR. Cerebral organoids recapitulate epigenomic signatures of the human fetal brain. Cell Rep. 2016;17(12):3369–84.
86. Gonzalez, C, Armijo, E, Bravo-Alegria, J, Becerra-Calixto, A, Mays, CE, Soto, C. Modeling amyloid beta and tau pathology in human cerebral organoids. Mol Psychiatry. 2018;23(12):2363–74.
87. Raja, WK, Mungenast, AE, Lin, YT, Ko, T, Abdurrob, F, Seo, J, et al. Self-organizing 3D human neural tissue derived from induced pluripotent stem cells recapitulate Alzheimer's disease phenotypes. PLoS One. 2016;11(9):e0161969.
88. Lin, YT, Seo, J, Gao, F, Feldman, HM, Wen, HL, Penney, J, et al. APOE4 causes widespread molecular and cellular alterations associated with Alzheimer's disease phenotypes in human iPSC-derived brain cell types. Neuron. 2018;98(6):1141–54.e7.
89. Braak, H, Braak, E. Neuropathological stageing of Alzheimer-related changes. Acta Neuropathol. 1991;82(4):239–59.

9 Importance of Genetic Model in Huntington's Disease

Mehak Bajaj and Devhuti Palan
School of Innovation in Biodesign, Translational Health Science and Technology Institute, Faridabad, India

Mansi Gandhi
School of Innovation in Biodesign, Translational Health Science and Technology Institute, Faridabad, India
and
Department of Chemistry, CHRIST (Deemed to be University), Bengaluru, India

9.1 INTRODUCTION

Huntington's disease (HD) is one of the rare, autosomal dominant disorders that is inherent in nature. It is classified as a neurological disease, and its symptoms usually manifest as psychiatric and cognitive disturbances; involuntary gait movements of proximal and distal muscles; unusual body posture; motor task deficit; dementia; and change in behavior, cognition, judgement, and emotion; anhedonia; antisocial conduct; along with changes in personality [1]. The patients also experience slurred speech and difficulty while eating and swallowing. It was first identified by George Huntington in adults who showed symptoms of chorea, in East Hampton, Long Island, in 1872, and the disease is named after him. He published an article on 'On Chorea'. [2]. The disease shows selective degeneration of medium spiny neurons of the striatum in the early stage, which proceeds to cortical neurons as the disease stage progresses. At the cellular level, protein aggregates are visible in the cytoplasm and the nucleus. Symptoms of the disease usually start showing during mid-life years, and stay progressively till death, which is around 15 to 20 years after onset of the disease. It is an autosomal dominant disorder, which means that only one parent could have the altered gene, and the diseased allele will be transferred to the offspring, thus causing a particular disorder. HD is caused by an abnormal number of cytosine-adenine-guanine (CAG) repeats between 36 and 180 units in length in the coding region of the IT-15 gene. The CAG repeats code for a stretch of polyglutamine (polyQ) in exon 1 of the human huntingtin gene, which is a 350-kDa protein without any known function as of now [1]. There are various animal models that showcase how HD and its neurological abnormalities progress or manifest, and the different biological pathways that are affected during the progression. A mind map for working of HD has been portrayed in **Figure 9.1**. This chapter discusses such genetic models in detail.

9.2 EPIDEMIOLOGY AND ETIOLOGY OF HUNTINGTON'S DISEASE

9.2.1 Epidemiology of Huntington's Disease

The disease was first identified in 1872, in Long Island, New York, by George Huntington. Initial identification was based on the symptoms, as no genetic testing was available. Families with patients with Huntington's were undertaken for individual studies since the disorder was first identified. There are records of extensive data on families in the New England region of

DOI: 10.1201/9781032620442-9

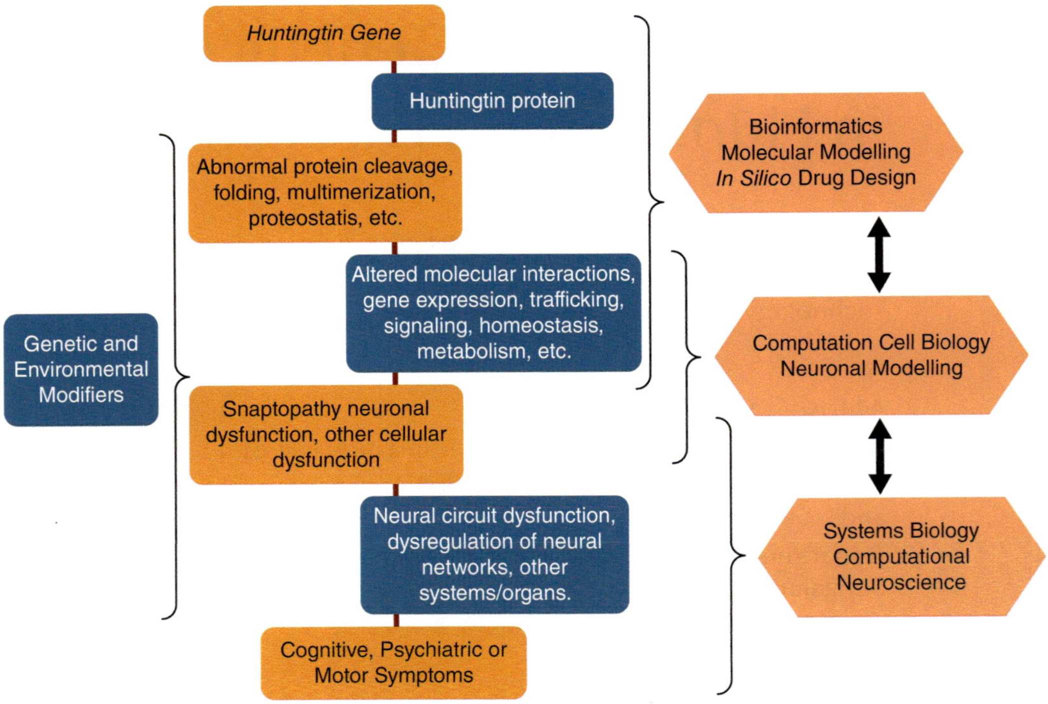

FIGURE 9.1 Mind map of working of huntingtin gene.

the United States [3]. Julia Bell presented more systematic studies for detailed genetic analysis in Britain, and these studies are still validated. The disease is most common and prevalent in Europe and various countries like USA and Australia, as they are of European origin. The occurrence of the disease is lowest in places like Africa, Finland, Japan, and China [4]. The disease manifests in 2.7 people per 100,000 people all over the world. In Europe, the prevalence is high as 10 out of every 100,000 [5]. The prevalence of HD in Asia is lower in comparison, ranging from 0.46 to 0.16 [4]. **Figure 9.2** shows the comparison of cellular pathways of a healthy human versus diseased.

Egypt has shown a high prevalence rate of the disease, with 21 people out of every 100,000 being diagnosed with HD; still this population has been overlooked by researchers [4, 5]. In India, the prevalence rate would be similar to that of European populations because of the similarity in the haplotype [6]. HD in the United States is observed to be prevalent, with a range from 4.1–8.4 per 100,000 people. Some of the isolated populations of western European origin have exhibited an unusually high prevalence of HD, which may be a result of a founder effect. People living in the Lake Maracaibo region in Venezuela have an incidence of 700 per 100,000 people; the people from the island of Mauritius off the South African coast show that 46 per 100,000 people have HD; and in Tasmania, around 17.4 per 100,000 people are diagnosed.

9.2.2 Etiology of the Disease

Huntington's disease is characterised as an autosomal dominant disorder that manifests due to the elongation of cytosine, adenine, and guanine (CAG) repeats, especially on the p arm of chromosome 4p16.3 in the HTT gene, which gives instructions for the huntingtin protein that is crucial for synaptic function as well as the post-embryonic period. HTT gene is considered important because it is believed to protect against the toxic strain of HTT mutation and have anti-apoptotic functions. After various research have been carried out, it has been observed that the mutation can cause both loss and gain of function.

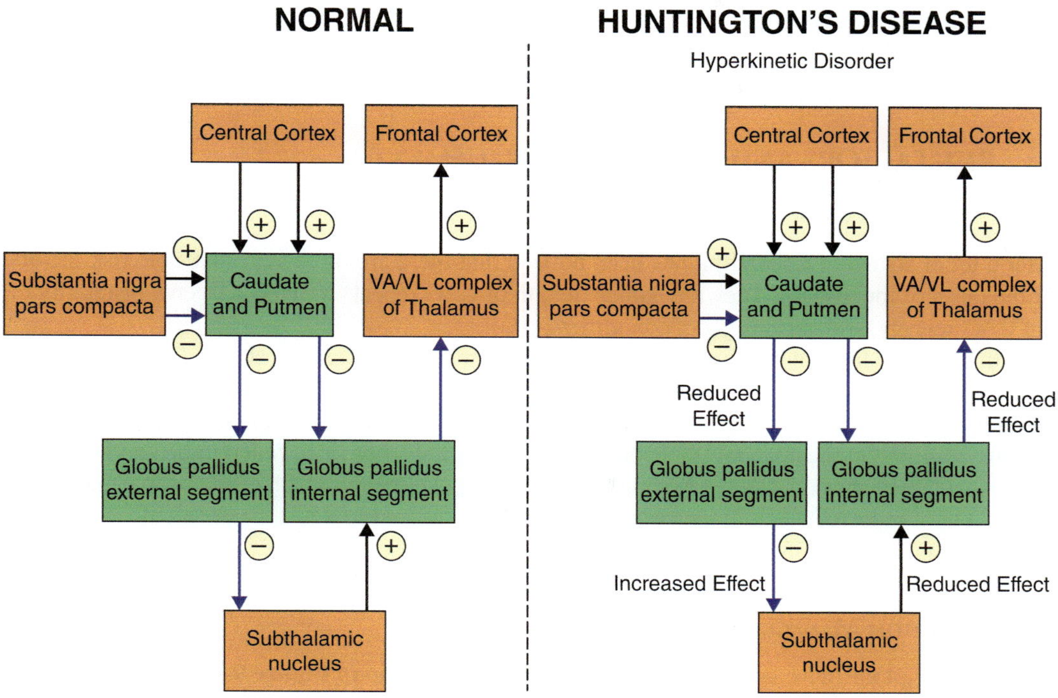

FIGURE 9.2 Comparison of cellular pathways in a healthy human being versus a person suffering from Huntington's disease.

People who are diagnosed with HD are found to have the CAG repeats in the range of 36 to 55 along with the HTT allele. There are some cases of juvenile or early onset of the disease where this number becomes more than 60. **Figure 9.3** shows the causes of various inherent neurological disorders that are commonly observed. Interestingly, the people who have CAG repeats between 27 and 35 don't suffer from the disease, but they are definitely prone to repeat instability. Inverse relation

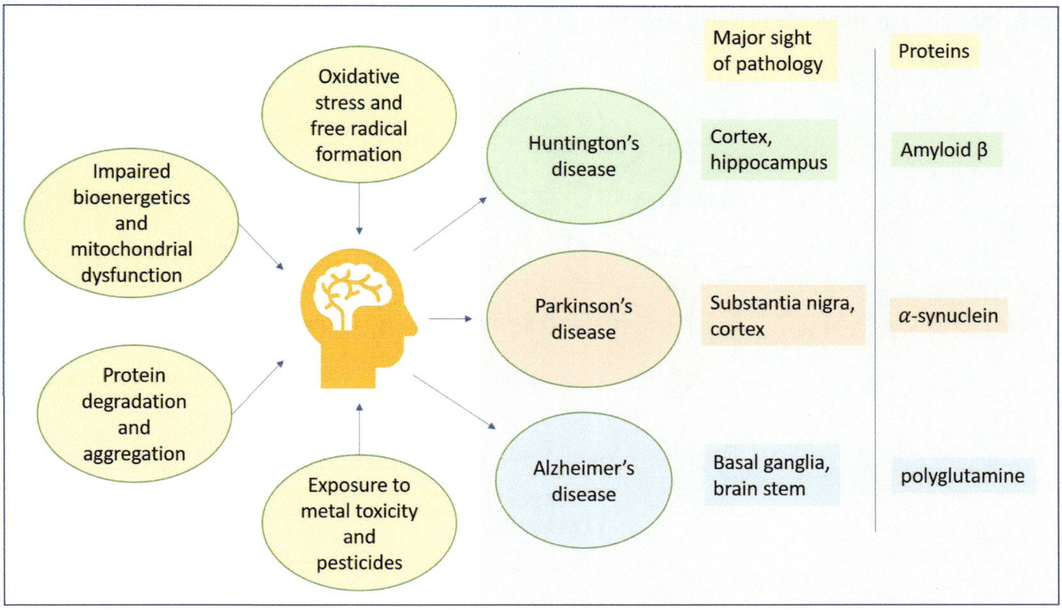

FIGURE 9.3 Causes of various inherent neurological disorders.

is observed between the length of repeats and the age of onset. The phenomenon of anticipation is usually seen in the paternal pattern of inheritance, which is visible due to the instable method of repetition of CAG during spermatogenesis [7].

9.3 RISK FACTORS AND SYMPTOM MANAGEMENT

9.3.1 Risk Factors in HD

After a lot of stringent research, it has been noticed that there are three very plausible risk factors for the onset of the disease. They are namely length of CAG repeats in the HTT gene, instability of that CAG repeats, and various genetic modifiers.

A lot of factors like environmental, demographic, genetic and past medical and clinical risks have been undertaken as parameters for the progression of HD. Also, there is evidence in literature that a higher number of CAG repeats will cause shortening of the lifespan and increase in the urgency of being admitted to the hospital [8]. People with Huntington's were also found to be depressed and anxious, and they showed signs of suicidal ideation [8, 9]. At least one-fifth of the people diagnosed with HD have reported suicidal ideation, which could be related to a history of suicide ideation in pre-manifested and manifested patients of HD [10].

9.3.2 Symptom Management of HD

The basic symptoms have been discussed in form of a mind map in **Figure 9.4**, listing various literature articles [11–17].

9.3.3 Movement Disorder

Patients with HD face loss of coordination and difficulty walking due to deprivation of motor control. Losing balance is also visible as a symptom. Initially, the symptom manifests as dropping items and being unable to maintain balance, which might lead to stumbling. There is no requirement or any kind of routine treatment for this; rather the patient should be made aware of the progression of the symptom along with management. Treatments can suppress the involuntary movements but not clearly stop them. Treatments that target neurotransmitters, like dopamine, glutamate, etc., have been studied, but they are not that effective [12, 17, 18]. Current treatment includes focusing on

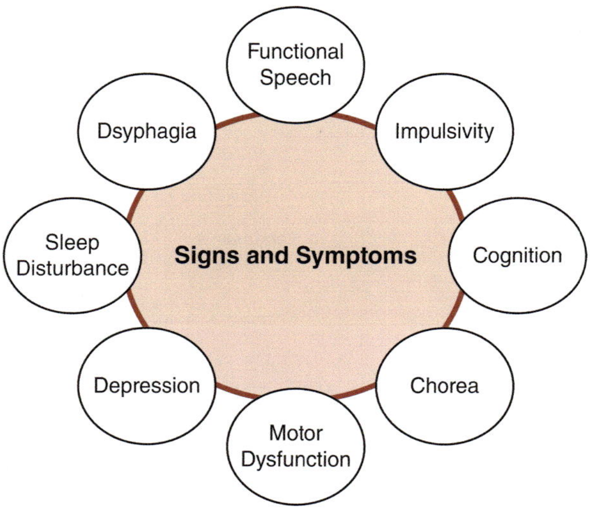

FIGURE 9.4 Signs and symptoms of Huntington's disease.

dopamine depletion with the help of atypical antipsychotics or with tetrabenazine [19]. It is necessary to remember that antipsychotics might cause side effects and apathy, which could lead to worsening of pre-existing movement issues.

Tetrabenazine, when taken in treatment dosage, causes depletion in dopamine, serotonin, and norepinephrine. But it has shown to be a good treatment method for chorea. Even then, its side effects cannot be sidelined as it causes feelings of anxiety, sedation, confusion, and sometimes even depression when co-prescribing. It is not recommended for patients with depression or a tendency/history of suicidal ideation [20].

9.3.4 Mood and Anxiety Disorders

Depression has been one of the common symptoms found in patients with HD, with a very high prevalence of 33%–76%. Recommended treatment includes standard management of depression. But there should be a conscious effort to manage increased suicidal tendency in HD.

9.3.5 Irritability and Psychosis

Medication might prove helpful, but strategies to manage behavior and prevent irritability or aggression outbursts would prove to be a better method or treatment and risk management. Mood stabilisers can be prescribed in the events of aggression or impulsivity.

Non-pharmacological treatment can include speech and language therapy, occupational therapy, and other types of non-pharmacological therapies along with physiotherapy that could help with irregular gait [11, 21].

9.4 GENETIC MODELS AND THEIR IMPORTANCE IN HUNTINGTON'S DISEASE

Many different non-human primates have been utilised to generate transgenic monkey models to express disease genes or exogenous foreign genes. The HD model of monkeys was created by injecting the fertilised oocytes with lentivirus, which expressed the mutated version of the HTT gene. But, this ended up showing more severe versions than that of mouse models. There were cases of early postnatal deaths in the monkey models. Despite this drawback, the modified monkeys were able to show symptoms of HD like chorea, dystonia, and seizures. This result has not been replicated by other animal models, including mice. This observation goes on to show that larger animals are highly sensitive to the toxic mutation of HTT gene. On the basis of this theory, a transgenic pig model was created through somatic cell nuclear transfer. The pig model also supported the observation of early postnatal death. Even though large animals, especially non-human primates would be a better study model to understand the effect of HD in humans as they would be more closely related to humans, their high cost and long duration of breeding pose a difficult challenge [22–24]. A tabulated detailed comparison has been done for the different mouse models for Huntington's disease, especially for transgenic and targeted *Hdh* models in mice as shown in **Tables 9.1** and **9.2**, respectively.

9.4.1 Akt Is Altered in an Animal Model of Huntington's Disease and in Patients

It has been observed that the insulin-like growth factor I (IGF-1)/Akt pathway is a very critical pathway for HD, as is phosphorylate, the causal protein polyQ-huntingtin, and removes its pernicious properties. Hence, it can be inferred that there might be some dysregulation of this pathway to show disease progression in HD. The study used a rat model with HD to scrutinise the presence of Akt in the brain and in tissues of the peripheral region. The model showed altered Akt during neural dysfunction, which when compared to a meager amount of lymphoblasts and lymphocytes in humans suffering from Huntington's disease, showed similar modifications in Akt. This confirmed that there is a dysregulation of Akt [35]. The final observation showed that Akt gets cleaved by

TABLE 9.1
Mouse Model for Huntington's Disease (Transgenic *Hdh*)

Transgenic *Hdh* Models for Huntington's Disease

S. No.	Promotor/Gene Size/ CAG Reports	Symptoms and Phenotypes	Neuropathology	Articles Cited
1.	CMV promoter Full- length gene 89Q, 48Q, 16Q	89Q and 48Q onset 4 months Circling hyperactivity, end-stage hypoactivity, and urinary incontinence	Fewer inclusions throughout brain 20% cell loss in striatum of some animals	[25]
2.	HD YAC transgenic Full-length gene with either 72Q, 46Q, or 18Q	72Q onset by 3 months Hyperactive and circling	Inclusions in striatum Cell loss in striatum	[26]
3.	Mouse prion protein promoter First 171 amino acids 82Q, 44Q, 18Q	82Q onset 5 months Tremors and abnormal gait, hypokinesis, weight loss, early death	Inclusions in striatum, cortex, hippocampus, amygdala Diffuse nuclear accumulation of Huntingtin Neuronal degeneration in striatum Neuronal degeneration in striatum	[27]
4.	Rat neuron specific enolase promoter 3-kb N-terminal fragment 100Q, 48Q, 18Q	100Q onset 3–4 months Hyperactivity to end-stage hypoactivity	Inclusions in Q100 (few in Q46) Systrophic neurites observed 20% cell loss ~8 some animals Brain atrophy in some animals	[28]
5.	CMV promoter Exons 1-3, 89Q	Exons 1–3:89Q onset 4 months Prolonged hyperactivity	Fewer inclusions throughout the brain	[18]
6.	HD promoter; exon 1 R6/2 line: 144Q (expanded to170– 190 Q) R6/1 line 115 Q	Onset 2 months (R6/1 onset 5 months) Tremors and abnormal gait, learning deficit, hypokinesis, diabetes	Intranuclear and neuropil aggregates throughout brain, fewer dentritic spines. Frontal cortex, dorsal striatum and cerebellar Purkinje cells at late stage Overall brain atrophy	[29]
7.	Tet-off (camKIIα-tTA) promoter Exon1 94Q	94Q onset in 50% by 2.5 months Late onset tremor and abnormal gait	Inclusions in striatum, septum, cortex, hippocampus, and reactive astrocytes Brain atrophy and progressive striatal atrophy	[30]

caspase-3, which renders it inactive. This can be used to infer that Akt is an important pathway that is responsible to regulate the toxicity caused by huntingtin protein [36].

9.4.2 Neurological Abnormalities in a Knock-In Mouse Model of Huntington's Disease

To study the effects of HD on a patient, a mouse model was prepared by targeting genes to replace the short CAG repeat with the longer one, to cause the disease. It was found that the mice who had alleles of around 150 CAG repeats units in length showed effects of late-onset HD, with behavioral changes similar to those of HD patients. Since there were longer repeats, it was a more severe phenotype. With this experiment, it could be concluded that this mouse model could be used in experiments with regards to study the molecular nature of neural dysfunction in HD [23, 24, 37–39].

TABLE 9.2
Mouse Model for Huntington's Disease (Targeted *Hdh*)

Targeted *Hdh* Models for Huntington's Disease				
S. No.	**Promotor/Gene Size/ CAG Reports**	**Symptoms and Phenotypes**	**Neuropathology**	**Articles Cited**
1.	Hdh promoter knock-in 80Q, 72Q	Early onset aggressive behavior	Late inclusions, hippocampus. LTP impaired, repeat instability in striatum	[31]
2.	Hdh promoter 150Q	60 weeks heterozygote knock-in homozygote Clasping gait deficit 25 weeks	Glial fibrillary acid protein positive cells	[13]
3.	Endogenous mouse HD gene promoter (*Hdh* prom) knock-in 111Q, 92Q, 50Q	No movement disorder	CAG-and age-dependent huntingtin nuclear re-localisation at 1.5 months, inclusion >6 months	[32]
4.	Hdh promoter knock-in 94Q, 71Q	NMDA sensitivity No movement disorder	Striatal cell swelling in response to NMDA Smaller striatal cells	[33]
5.	Hdh promoter 77Q knock-in	No symptoms	Glial fibrillary acid protein positive cells	[34]
6.	neo and *Hdh* promoter, knock-in 111Q, 20Q	$Hdhn^{Q20/nQ111}$ have onset ~2 months nQ20/nQ111 show disorders similar to Exon1-Q144 mouse	Age-dependent huntingtin nuclear re-localisation and inclusions *Preliminary result*: brain atrophy	[32]

9.4.3 Transgenic Model of Huntington's Disease in a Non-Human Primate

Non-human primates are very important as an animal model to study various disorders so that therapeutic strategies can be developed. But even then, there is little work done or reported while creating transgenic primate models of human diseases that are non-human. There are various rodent models to study HD, but they are not the perfect comparison while studying the changes caused by HD as they aren't able to mimic the behavioral and brain changes in a satisfactory manner. The higher primates are quite similar neurologically and genetically; hence they can offer a comparable model. There is a study that shows the progress in developing a transgenic model with a rhesus monkey that had polyglutamine-expanded HTT, thus expressed HD. The model had the basic characteristics of HD, like nuclear inclusions in the brain, along with physical characteristics of chorea and dystonia. This model shows there is a huge possibility to develop non-human models for various other genetic disorders, which might open new doors towards therapeutic strategies for those disorders [14, 18, 40].

9.4.4 Early Development of Aberrant Synaptic Plasticity in a Mouse Model of HD

Genetic testing for prediction of HD revealed cognitive deficits in an early stage, even where there is no or little evidence for cell death, which implies that impaired cognition is caused from a synaptic or cellular deficit like an aberrant synaptic plasticity. Depression that lasts for a long duration is known to be a developmental but regulated form of plasticity. This declines with age. In the mouse model, the controls behaved in a similar pattern in the expressed long-term depression (LTD) and then lost the ability with age. Unlike controls, the mouse model's synapses later get the ability to again support LTD. This happens because of nuclear localisation of the mutant gene, but this is seen to happen a few months before the nuclear aggregation begins [25, 41, 42].

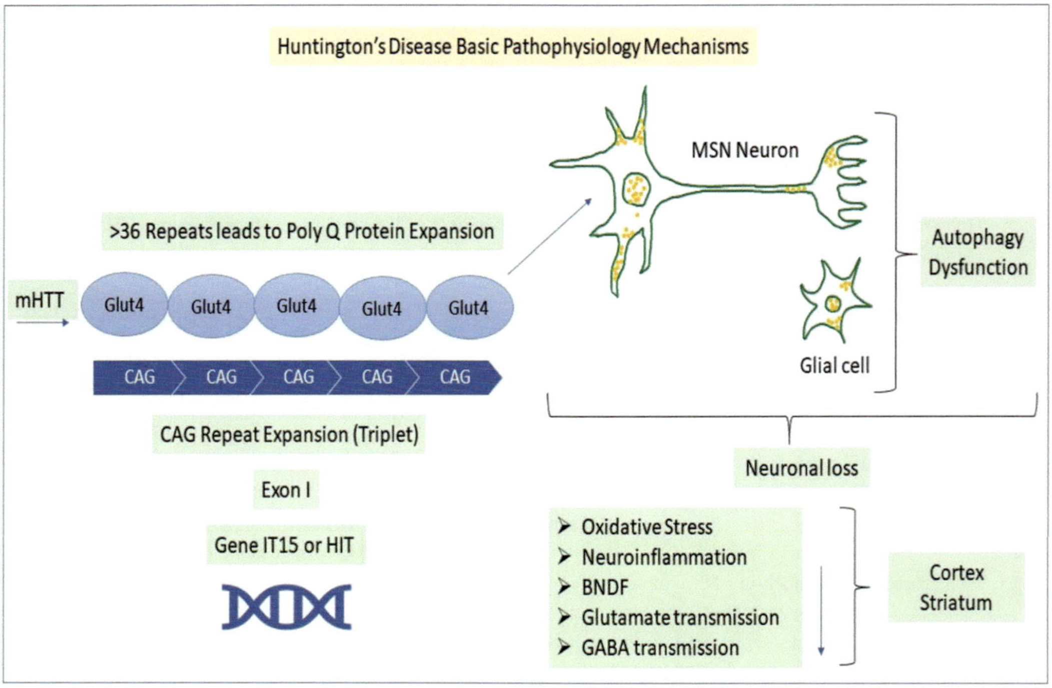

FIGURE 9.5 Pathophysiology of Huntington's disease.

9.4.5 Decreased Expression of Striatal Signaling Genes in a Mouse Model of HD

There is various research that lets us understand the gene expression changes that are caused by a polyglutamine repeat expansion in the HTT gene. One study focused on using oligonucleotide DNA arrays to profile around 6000 striatal mRNAs in the R6/2 strain of mice, which is a transgenic mouse model of HD. The observation mentioned diminishing levels of mRNAs, calcium and retinoid pathways of signaling [16, 43–45].

9.4.6 Kynurenine Pathway Modulates Neurodegeneration in a Drosophila Model of HD

Huntington's disease shows a pathophysiology wherein tryptophan degradation due to kynurenine pathway (KP) and neuroactive metabolites is seen [46]. The research mentions a Drosophila melanogaster model that is transgenic and shows signs of HD to check the potential of KP manipulation in a therapeutic manner. **Figure 9.5** illustrates the pathophysiology of the disease.

This study shows clear evidence that inhibiting KMO and TDO protects against neurodegenerative disease in this model, which implies it may have some relevant therapeutics when it comes to HD [41, 46, 47].

9.5 CONCLUSION AND DISCUSSION

Various research and studies on divergent genetic models of mice exhibiting HD have given the researchers many options and processes for testing at preclinical level of numerous plans of action for therapeutic procedures. Mouse models are preferred for studying various diseases as they have shorter life spans, so data can be accumulated by testing different compounds before the disease manifests. The mouse model created by Yamamoto and the team, which had mutant expression of huntingtin that is mediated by a promoter that is under regulation by tetracycline, contributes towards a hope in the treatment of HD [17]. There have been studies that show that HD might be reversible, even after symptoms manifest, as one mouse model that expressed the mutant transgene of HD on exon 1 started to show degrading motor dysfunction [48]. It was observed that if transgene is switched off in the mice that showed symptoms, then there was a mitigation of motor dysfunction. This suggested that the mutant gene needed to be expressed continuously to maintain the disease condition [49].

ACKNOWLEDGMENT

Mansi Gandhi would like to acknowledge Hebrew University of Jerusalem for completion of her postdoc research.

DECLARATIONS

Conflict of interest: The authors declare no conflict of interest.
Funding: This research received no external funding.
Authors' contribution: *Mansi Gandhi*: Conceptualisation, methodology, writing, project administration, validation, investigation, supervision, writing. *Mehak Bajaj*: Conceptualisation, methodology, writing, project administration, validation, investigation, supervision and writing. *Devhuti Palan*: Validation, writing.

All authors have read and agreed to the published version of the manuscript.

REFERENCES

1. E. Colin, E. Régulier, V. Perrin, A. Dürr, A. Brice, P. Aebischer, N. Déglon, S. Humbert, and F. Saudou, Eur. J. Neurosci. **21**, 1478 (2005).
2. C. Kay, M.R. Hayden, and B.R. Leavitt, Handb. Clin. Neurol. **144**, 31 (2017).
3. P.S. Harper, Hum. Genet. **89**, 365 (1992).
4. T. Pringsheim, K. Wiltshire, L. Day, J. Dykeman, T. Steeves, and N. Jette, Mov. Disord. **27**, 1083 (2012).
5. L. Luxner, 1 (2019).
6. S.C. Warby, H. Visscher, J.A. Collins, C.N. Doty, C. Carter, S.L. Butland, A.R. Hayden, I. Kanazawa, C.J. Ross, and M.R. Hayden, Eur. J. Hum. Genet. **19**, 561 (2011).
7. MDVU, WeMove 1 (2024).
8. T.K. Chao, J. Hu, and T. Pringsheim, Neurotoxicology. **61**, 79 (2017).
9. A. Mcgarry, M.P. Mcdermott, K. Kieburtz, W.L.A. Fung, E. Mccusker, J. Peng, E.A. De Blieck, and M. Cudkowicz, Neurology. **92**, E1643 (2019).
10. P. Honrath, I. Dogan, O. Wudarczyk, K.S. Görlich, M. Votinov, C.J. Werner, B. Schumann, R.T. Overbeck, J.B. Schulz, B.G. Landwehrmeyer, R.E. Gur, U. Habel, and K. Reetz, J. Neurol. **265**, 2548 (2018).
11. C. Patrick, and S. Ritchie, Prescriber. **31**, 23 (2020).
12. G. Bergonzoni, J. Döring, and M. Biagioli, Front. Cell. Neurosci. **15**, 1 (2021).
13. C.H. Lin, S. Tallaksen-Greene, W.M. Chien, J.A. Cearley, W.S. Jackson, A.B. Crouse, S. Ren, X.J. Li, R.L. Albin, and P.J. Detloff, Hum. Mol. Genet. **10**, 137 (2001).
14. S. Yang, P. Cheng, H. Banta, K. Piotrowska, J. Yang, E.C.H. Cheng, B. Snyder, K. Larkin, J. Liu, J. Orkin, Z. Fang, Y. Smith, J. Bachevalier, S.M. Zola, S. Li, X. Li, and A.W.S. Chan, **453**, 921 (2009).
15. A.J. Milnerwood, D.M. Cummings, G.M. Dallérac, J.Y. Brown, S.C. Vatsavayai, M.C. Hirst, P. Rezaie, and K.P.S.J. Murphy, Hum. Mol. Genet. **15**, 1690 (2006).
16. R. Luthi-Carter, A. Strand, N.L. Peters, S.M. Solano, Z.R. Hollingsworth, A.S. Menon, A.S. Frey, B.S. Spektor, E.B. Penney, G. Schilling, C.A. Ross, D.R. Borchelt, S.J. Tapscott, A.B. Young, J.H.J. Cha, and J.M. Olson, Hum. Mol. Genet. **9**, 1259 (2000).
17. J. Muchowski, R. Schwarcz, C.P. Kyriacou, and F. Giorgini, **21**, 961 (2014).
18. D.C. Rubinsztein, Trends Genet. **18**, 202 (2002).
19. F.O. Walker, Lancet. **369**, 218 (2007).
20. J.S. Paulsen, J. Neuropsychiatr. **17**, 496 (2005).

21. B. Bilney, M.E. Morris, and A. Perry, Neurorehabil. Neural Repair. **17**, 12 (2003).
22. S. Yan, S. Li, and X.J. Li, Cell Regen. **8**, 9 (2019).
23. Z.R, Crook, and D. Housman, Neuron. **69**, 423 (2011).
24. J.P. Vonsattel, R.H. Myers, T.J. Stevens, R.J. Ferrante, E.D. Bird, and E.P. Richardson, J. Neuropathol. Exp. Neurol. **44**, 559 (1985).
25. P.H. Reddy, M. Williams, V. Charles, L. Garrett, L. Pike-Buchanan, W.O. Whetsell, G. Miller, and D.A. Tagle, Nat. Genet. **20**, 198 (1998).
26. J.G. Hodgson, N. Agopyan, C.A. Gutekunst, B.R. Leavitt, F. Lepiane, R. Singaraja, D.J. Smith, N. Bissada, K. McCutcheon, J. Nasir, L. Jamot, L. Xiao-Jiang, M.E. Stevens, E. Rosemond, J.C. Roder, A.G. Phillips, E.M. Rubin, S.M. Hersch, and M.R. Hayden, Neuron. **23**, 181 (1999).
27. G. Schilling, M.W. Becher, A.H. Sharp, H.A. Jinnah, K. Duan, J.A. Kotzuk, H.H. Slunt, T. Ratovitski, J.K. Cooper, N.A. Jenkins, N.G. Copeland, D.L. Price, C.A. Ross, and D.R. Borchelt, Hum. Mol. Genet. **8**, 397 (1999).
28. G.A. Laforet, E. Sapp, K. Chase, C. McIntyre, F.M. Boyce, M. Campbell, B.A. Cadigan, L. Warzecki, D.A. Tagle, P.H. Reddy, C. Cepeda, C.R. Calvert, E.S. Jokel, G.J. Klapstein, M.A. Ariano, M.S. Levine, M. DiFiglia, and N. Aronin, J. Neurosci. **21**, 9112 (2001).
29. L. Mangiarini, K. Sathasivam, M. Seller, B. Cozens, A. Harper, C. Hetherington, M. Lawton, Y. Trottier, H. Lehrach, S.W. Davies, and G.P. Bates, Cell. **87**, 493 (1996).
30. A. Yamamoto, R. Hen, and J. Lucas, Cell. **101**, 57 (2000).
31. P.F. Shelbourne, N. Killeen, R.F. Hevner, H.M. Johnston, L. Tecott, M. Lewandoski, M. Ennis, L. Ramirez, Z. Li, C. Iannicola, D.R. Littman, and dR.M. Myers, Hum. Mol. Genet. **8**, 763 (1999).
32. J.K. White, W. Auerbach, M.P. Duyao, J.-P. Vonsattel, J.F. Gusella, A.L. Joyner, and M.E. Macdonald, Nat. Genet. **17**, 404 (1997).
33. M.S. Levine, G.J. Klapstein, A. Koppel, E. Gruen, C. Cepeda, M.E. Vargas, E.S. Jokel, E.M. Carpenter, H. Zanjani, R.S. Hurst, A. Efstratiadis, S. Zeitlin, and M.F. Chesselet, J. Neurosci. Res. **58**, 515 (1999).
34. H. Ishiguro, K. Yamada, H. Sawada, K. Nishii, N. Ichino, M. Sawada, Y. Kurosawa, N. Matsushita, K. Kobayashi, J. Goto, H. Hashida, N. Masuda, I. Kanazawa, and T. Nagatsu, J. Neurosci. Res. **65**, 289 (2001).
35. A. Bellacosa, T.O. Chan, N.N. Ahmed, K. Datta, S. Malstrom, D. Stokoe, F. McCormick, J. Feng, and P. Tsichlis, Oncogene. **17**, 313 (1998).
36. L.P. De Almeida, C.A. Ross, D. Zala, P. Aebischer, and N. Déglon, J. Neurosci. **22**, 3473 (2002).
37. B.L. Zaret, J. Nucl. Cardiol. **8**, 427 (2001).
38. J. Gafni, E. Hermel, J.E. Young, C.L. Wellington, M.R. Hayden, and L.M. Ellerby, J. Biol. Chem. **279**, 20211 (2004).
39. X.-J. Li, S. Li, J.H. Meador-Woodruff, A. Mansour, D.K. Grandy, S.P. Damask, O. Civelli, S.J. Watson, C. Romero-grimaldi, B. Moreno-lo, I. Tsutsui-Kimura, H. Takiue, K. Yoshida, M. Xu, R. Yano, H. Ohta, H. Nishida, Y. Bouchekioua, H. Okano, M. Uchigashima, M. Watanabe, N. Takata, M.R. Drew, H. Sano, M. Mimura, K.F. Tanaka, I. Tsutsui-Kimura, A. Kono, A. Yamanaka, K. Kobayashi, M. Watanabe, M. Mimura, K.F. Tanaka, M.R. Drew, M. Mimura, K.F. Tanaka, Y. Zhuang, P. Xu, C. Mao, L. Wang, B. Krumm, X.E. Zhou, S. Huang, H. Liu, X. Cheng, X.P. Huang, D.D. Shen, T. Xu, Y.F. Liu, Y. Wang, J. Guo, Y. Jiang, H. Jiang, K. Melcher, B.L. Roth, Y. Zhang, C. Zhang, and H.E. Xu, Behavioral Neurobiology of Huntington's Disease and Parkinson's Disease (2008).
40. E. Siemers, Arch. Neurol. **58**, 308 (2001).
41. M.E. MacDonald, C.M. Ambrose, M.P. Duyao, R.H. Myers, C. Lin, L. Srinidhi, G. Barnes, S.A. Taylor, M. James, N. Groot, H. MacFarlane, B. Jenkins, M.A. Anderson, N.S. Wexler, J.F. Gusella, G.P. Bates, S. Baxendale, H. Hummerich, S. Kirby, M. North, S. Youngman, R. Mott, G. Zehetner, Z. Sedlacek, A. Poustka, A.M. Frischauf, H. Lehrach, A.J. Buckler, D. Church, L. Doucette-Stamm, M.C. O'Donovan, L. Riba-Ramirez, M. Shah, V.P. Stanton, S.A. Strobel, K.M. Draths, J.L. Wales, P. Dervan, D.E. Housman, M. Altherr, R. Shiang, L. Thompson, T. Fielder, J.J. Wasmuth, D. Tagle, J. Valdes, L. Elmer, M. Allard, L. Castilla, M. Swaroop, K. Blanchard, F.S. Collins, R. Snell, T. Holloway, K. Gillespie, N. Datson, D. Shaw, and P.S. Harper, Cell. **72**, 971 (1993).
42. T. Foroud, E. Siemers, D. Kleindorfer, D.J. Bill, M.E. Hodes, J.A. Norton, P.M. Conneally, and J.C. Christian, Ann. Neurol. **37**, 657 (1995).
43. J.H.J. Cha, C.M. Kosinski, J.A. Kerner, A. Alsdorf, L. Mangiarini, S.W. Davies, J.B. Penney, G.P. Bates, and A.B. Young, Proc. Natl. Acad. Sci. U. S. A. **95**, 6480 (1998).
44. S.J. Augood, R.L.M. Faull, D.R. Love, and P.C. Emson, Neuroscience. **72**, 1023 (1996).
45. J. Bao, A.H. Sharp, M.V. Wagster, M. Becher, G. Schilling, C.A. Ross, V.L. Dawson, and T.M. Dawson, Proc. Natl. Acad. Sci. U. S. A. **93**, 5037 (1996).

46. X. Zheng, A. Zhang, M. Binnie, K. McGuire, S.P. Webster, J. Hughes, S.E.M. Howie, and D.J. Mole, Exp. Mol. Med. **51** (2019).
47. R. Carpenedo, A. Pittaluga, A. Cozzi, S. Attucci, A. Galli, M. Raiteri, and F. Moroni, Eur. J. Neurosci. **13**, 2141 (2001).
48. T. Palpagama, A.R. Mills, M.W. Ferguson, P. Vikas Ankeal, C. Turner, L. Tippett, B. van der Werf, H.J. Waldvogel, R.L.M. Faull, and A. Kwakowsky, Ann. Neurol. **94**, 895 (2023).
49. J.W. Shin, K.H. Kim, M.J. Chao, R.S. Atwal, T. Gillis, M.E. MacDonald, J.F. Gusella, and J.M. Lee, Hum. Mol. Genet. **25**, 4566 (2016).

10 Natural Sources of Drugs for Neurodegenerative Diseases

Chandrabhan Prajapati, Sachchida Nand Rai, Shashank Shekhar Mishra, and Santosh Kumar Singh
Centre of Experimental Medicine and Surgery, Institute of Medical Sciences, Banaras Hindu University, Varanasi, India

Anurag Kumar Singh
Department of Pharmaceutical Engineering & Technology, Indian Institute of Technology (BHU) Varanasi, India

10.1 INTRODUCTION

Neurons are the fundamental units that comprise the structure and function of the nervous system, functioning as signal transmitters that facilitate uninterrupted communication and control throughout the whole body. The nervous system consists of an intricate neuronal network that supervises both voluntary activities and involuntary processes, comprising two essential divisions: The central nervous system (CNS) comprises the spinal cord, along with the brain, which serves as the main regulatory unit, and the peripheral nervous system (PNS) encompasses neurons and their components beyond the CNS, such as sensory as well as motor neurons, establishing a network of nerves for transmitting signals to and from the CNS [1]. Neurodegenerative disorders (NDDs) are generally chronic and progressive circumstances that trigger nerve cells to deteriorate. These disorders lead to problems with cognition, impaired movement, or senses due to the damage and loss of neurons. Neurodegenerative disorders are caused by various factors comprising genetic as well as environmental factors [2–4] (**Figure 10.1**).

Certain hallmarks characterising neurodegenerative diseases (NDDs) include abnormalities in cytoskeleton proteins, defects in nucleic acids (DNA/RNA), aggregation of pathological proteins, and alterations in energy metabolism etc. [5]. The processes involved in neurodegenerative disorders encompass collective defects in various hallmarks of NDDs. Therefore, multiple-target treatments are required [6–9]. Some examples of NDDs include Parkinson's disease (PD), Huntington's disease' (HD), Alzheimer's disease (AD), amyotrophic lateral sclerosis (ALS), as well as multiple system atrophy (MSA), and frontotemporal dementia (FTD) [10, 11]. With aging, there appears to be a decreased expression of heat shock proteins (HSPs), which are necessary for protein folding and refolding. Consequently, HSPs become potential targets for safeguarding against age-related neurodegeneration through lifestyle modifications, medications, gene therapy, etc. [12–15]

Recent studies reveal that certain plant-derived compounds having antioxidant as well as anti-inflammatory properties might be useful to treat neurodegenerative disorders [16–20] (**Figure 10.2**). There are several drawbacks to employing plant compounds to treat neurodegenerative diseases. There is a lack of rigorous research and clinical trials, differences in composition, reduced bioavailability, delayed response, ecological impact, complex regulations, and less effectiveness for the progression of neurodegenerative disease progression [21–23]. However, natural bioactive compounds have the potential to offer novel treatment options for individuals with neurodegenerative diseases (NDDs) [24]. These alternatives to conventional therapies are cost-effective and would assist millions of individuals worldwide affected by age-associated neurodegenerative diseases [25, 26]. This

DOI: 10.1201/9781032620442-10

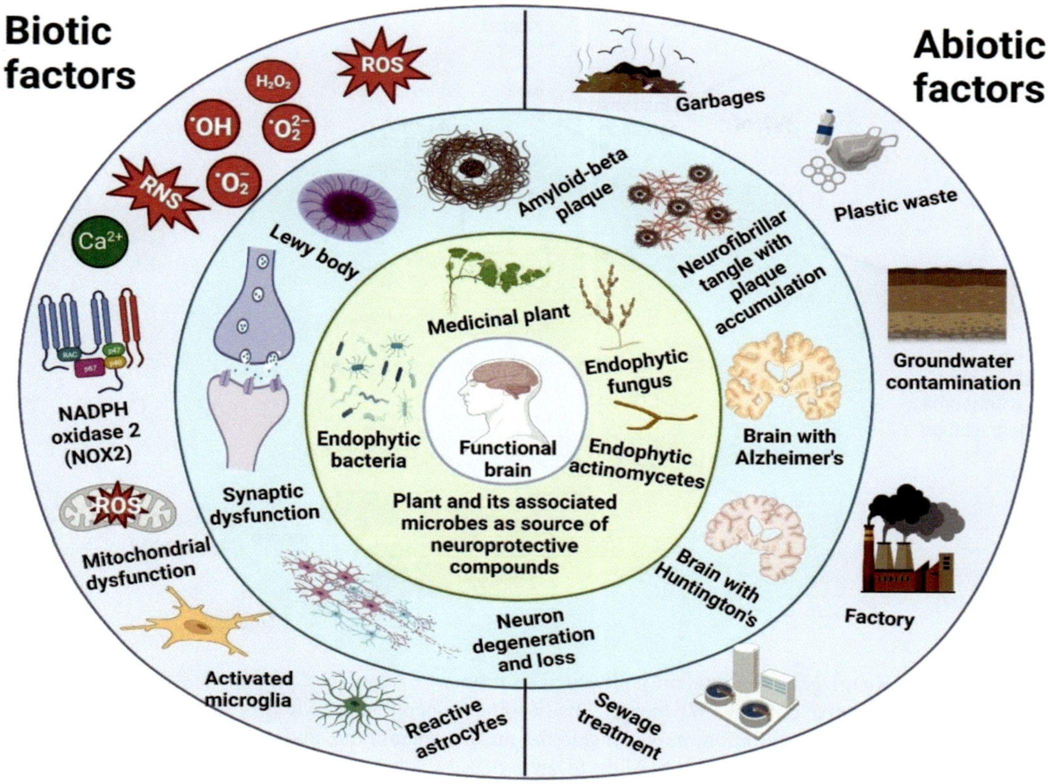

FIGURE 10.1 Causes of neurodegenerative disorders and the potential of plant and associated microbes derived compounds for treatment of neurodegenerative diseases. Various biotic and abiotic factors are responsible for causing neurodegenerative diseases. Production of ROS is part of a common mechanism for action found in metals and other environmental risk factors associated with neurodegenerative diseases. An oxidative stress occurs when there is imbalance between harmful reactive oxygen species (ROS) and our body's ability to defend against them with antioxidants, resulting in mitochondrial dysfunction, impairment in DNA repair system, neuroinflammation, etc. Neurodegenerative disorders can be treated with a variety of bioactive substances produced by medicinal plants as well as endophytes. The mechanism by which these bioactive exert neuroprotective effect include antioxidant activity, anti-inflammatory properties, enhancement of growth factors, etc.

chapter will focus on the natural origins of bioactive chemicals and their function in the treatment of neurodegenerative disorders, emphasising the cellular and molecular level hallmarks.

10.2 BIOACTIVE COMPOUNDS FOR ALZHEIMER'S DISEASE (AD) THERAPY

AD is a complicated neurodegenerative condition marked by several significant characteristics: amyloid plaques, neurofibrillary tangles, oxidative stress, neuroinflammation, imbalance in neurotransmitters, and neuronal loss. These hallmarks ultimately lead to the gradual cognitive impairment as well as memory loss seen in individuals with AD [27–29]. Recent findings have revealed that the cholinergic and glutamatergic systems are the main target sites for the currently available drugs used in AD treatments, and these drugs might assist to partially control the symptoms [30, 31]. Various studies highlighted the effective use of plant-derived bioactives that demonstrate antioxidant activity as well as anti-inflammatory activity towards AD [32–35].

The latest investigation suggests that herbal products and nutraceuticals can serve as emerging sources to uncover new pathways for Alzheimer's treatment [36]. The crude extract from freshly

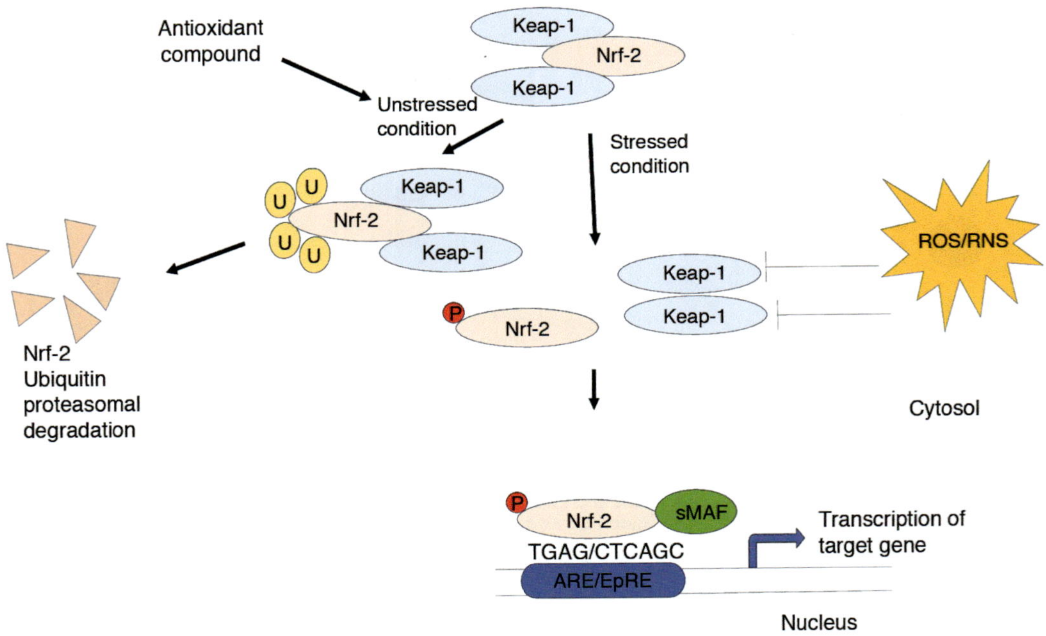

FIGURE 10.2 Keapl/Nrf2 antioxidant pathway. The proposed Keapl/Nrf2 process of antioxidants protects cells from oxidative damage. Whenever cells become susceptible to oxidative damage, Nrf2 becomes detached from the Keapl inhibition, moves inside the nuclear membrane, and stimulates the expression of genes associated with antioxidants by binding to antioxidant-related elements (AREs). This results in the production of antioxidant enzymes, which sustain biochemical equilibrium while protecting cells from oxidative harm.

harvested *Bacopa floribunda* leaves exhibited significant antioxidant effect towards DPPH and NO radicals, having IC_{50} values of 14.90±0.8 and 15.92±1.3 µg/ml, respectively. It also exhibited AChE and BChE inhibition having IC_{50} values of 123.71±3.4 and 84.83±3.1 µg/ml, respectively [37]. Another study showed that water-soluble extract of *Centella asiatica* exhibited significant neuroprotection effect against neurotoxicity triggered by exogenously supplemented along with endogenously produced β-amyloids in the cells of SH-SY5Y as well as MC-65 (human neuroblastoma) cells, respectively. The open-field and Morris water maze tests conducted on Tg2576 mice showed improvements in behavioral abnormalities and enhanced learning ability [38]. The bioactive compounds β-sitosterol and daucosterol derived from the *Salvia syriaca* L., displayed effective inhibition activities of AChE exhibiting IC_{50} values of 24.1 µg/mL as well as 34.3 µg/mL, respectively. The molecular docking of β-sitosterol and daucosterol also exhibited strong hydrogen bonding with the residues of amino acids in the active site of the acetylcholinesterase enzyme. Furthermore, both compounds demonstrated significant binding affinity, with $\Delta G_{Binding}$ values of –11.31 kcal/mol as well as –5.22 kcal/mol, respectively [39].

Another study revealed that acetone extract of the marine algae *Padina gymnospora* exhibited an anti-aggregation effect against beta-amyloid peptide ($A\beta_{25-35}$) along with significant AChE and BuChE inhibition activity. *Padina gymnospora* crude extract contains the bioactive component α-bisabolol, which inhibits AChE (87%) along with BuChE (66%) with a concentration equal to 50 µg/mL [40]. Another similar study revealed that the natural bioactive compound curcumin had potent neuroprotective effect via reducing LDH, and MDA production against $A\beta_{(25-35)}$ triggered oxidative damage to cells of PC12. Curcumin may upregulate NMDAR (N-methyl-D-aspartate receptor), a key protein for learning, along with the associated NR2A subunit, which is essential for memory as well as cognition, thereby enhancing brain function [41].

Another study found that the methanolic extract from the fruit of the *Piper nigrum* L. plant demonstrated significant antioxidant potential analysed by specific activities of SOD (superoxide dismutase), CAT (catalase) as well as GSH (glutathione peroxidase) in the rat hippocampal tissue. It also reduced beta-amyloid$_{(1-42)}$ triggered neuronal oxidative damage by lowering glutathione, MDA (malondialdehyde), and level of carbonyl protein in rat hippocampal tissue. Furthermore, the *Piper nigrum* L. fruit extract was found to significantly enhance memory processes in the rat analysed through Y-maze along with radial arm-maze tests [42].

Recent studies have centered on tau hyperphosphorylation, which could potentially prevent apoptosis by regulating the β-catenin mechanism in Alzheimer's neurodegeneration [43–46]. A recent study revealed that phlorizin, a chalcone class of compound derived from apple plants, exhibited significant reducing effect of oxidative damage triggered by β-amyloid $_{(1-40)}$ in the brain of rats. In the *in vivo* Morris water maze test, phlorizin also indicated improved learning as well as memory functioning. The molecular mechanism showed mRNA expressions of the neuroinflammation markers, comprising CD11b, GFAP, IL-1, and IL-1β as well as TNF-α remained considerably higher in contrast with sham group in β-amyloid $_{(1-40)}$ treated rats. The phlorizin also exhibited improve β-amyloid $_{(1-40)}$ neuronal dysfunction may via inhibition activity of GSK-3β (glycogen synthase kinase-3β) and promoted tau hyper-phosphorylation [47].

10.3 BIOACTIVE COMPOUNDS FOR PARKINSON'S THERAPY

Parkinson's illness is the second leading neurological disorder that results in symptoms such as tremors, stiffness, delayed movement, and unstable posture. These symptoms arise from the destruction of dopamine-producing neurons throughout the brain [48]. PD appears to be marked by a degeneration of neurons that release dopamine within the substantia nigra, the appearance of Lewy bodies containing misfolded alpha-synuclein protein, and sensations of movement such as muscular stiffness, allowing tremors, as well as bradykinesia. Additionally, it is also characterised by non-motor dysfunction that includes impaired cognition, autonomous nervous system (ANS) imbalance, and anxiety, etc. [49–52]

A recent study showed that a cell therapy approach involving the application of GLP-1R (glucagon-like peptide-1 receptor) activator in combination with bioactive compounds emerged as an approach to combat the symptoms of PD [53]. Now researchers have targeted tyrosine hydroxylase enzyme, which is a key enzyme for dopamine synthesis, to treat Parkinson's disease [54, 55].

Another study revealed that regulated autophagy machinery plays a key function in controlling PD [56–58]. The neurons surrounding glial cells, especially microglia and astrocytes, may take up the resulting debris from the loss of dopaminergic neurons, triggering an immune response in the brain [59]. The regulation of transmission of misfolded alpha-synuclein protein is an emerging approach in Parkinson's therapy [60]. The multifactorial nature of Parkinson's disease progression offers several potential therapeutic targets for slowing or even halting neurodegeneration. Recently, researchers have been actively focusing on enhancing the clearance of abnormal proteins, blocking their transmission, improving mitochondrial function, and targeting biomarkers in neuroinflammation to combat the symptoms of PD [61–63]. The plant *Mucuna pruriens* and its potent bioactive compound, ursolic acid (**Figure 10.3**) exhibited neuroprotection against rotenone-induced toxicity by controlling GSK-3β, α-Synuclein, and Ca^{2+} mediated signaling pathways using a Swiss albino male mouse model. The extract of *Mucuna pruriens* and Ursolic acid demonstrated the upregulation of Ca^{2+} ions via GSK-3β signaling. This process assisted to ameliorate mitochondrial dysfunction-mediated apoptosis along with downregulation of α-Synuclein using rotenone-intoxicated Swiss albino male mice [64]. The bioactive compound curcumin, derived from the plant turmeric, demonstrated a reduction in the formation of the A53T α-synuclein protein by SH-SY5Y cells. This impact was accomplished via downregulating the mTOR/p70S6K signaling process [65].

Another similar study revealed that β-Asarone (**Figure 10.3**) isolated from the plant *Acorus tatarinowii Schott* exhibited a protective effect towards 6-OHDA-mediated neurotoxicity. The

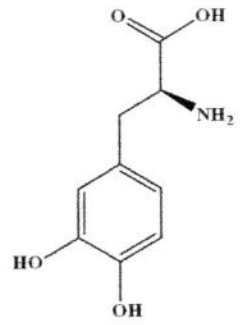

1. L- DOPA

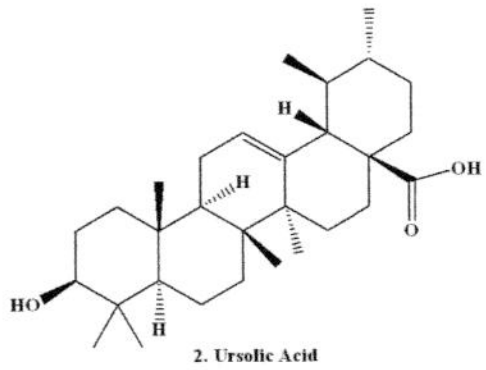

2. Ursolic Acid

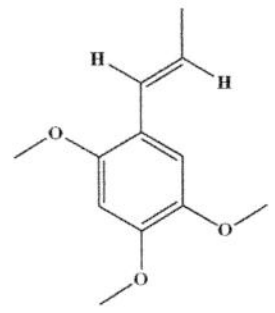

3. Alpha-asarone

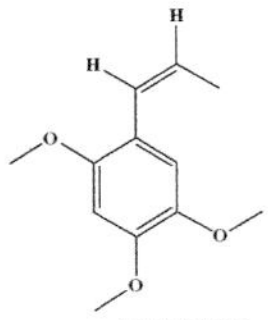

4. Beta-asarone

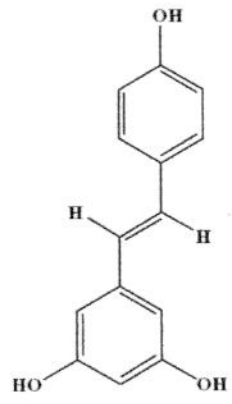

5. Resveratrol

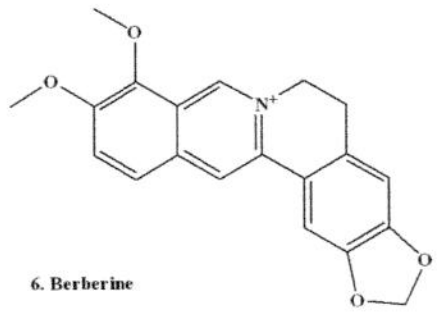

6. Berberine

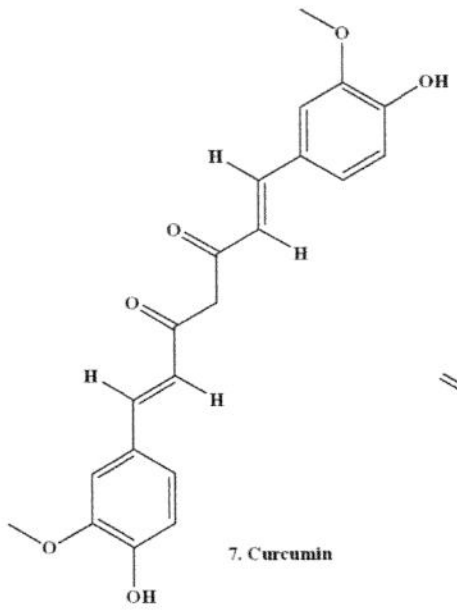

7. Curcumin

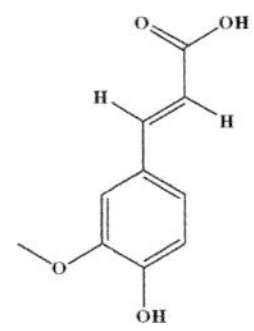

8. Ferulic acid

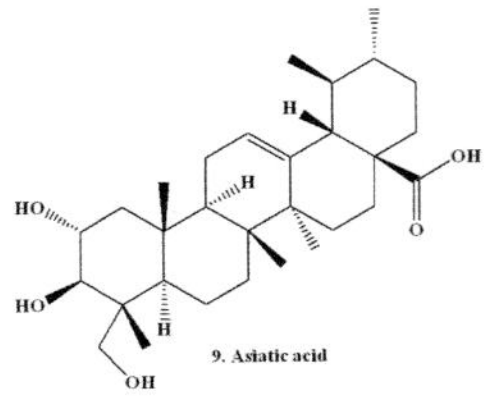

9. Asiatic acid

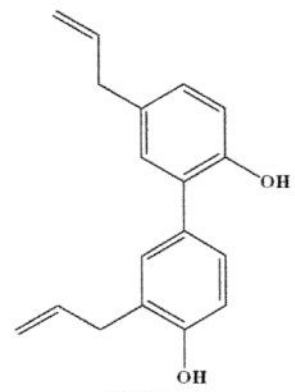

10. Honokiol

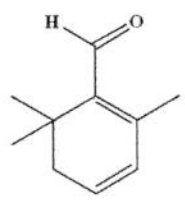

11. Safranal

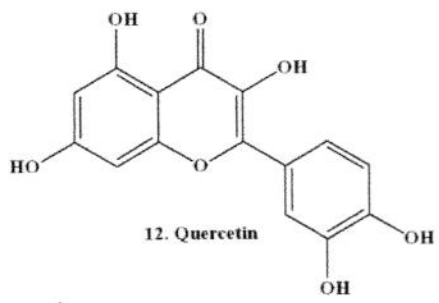

12. Quercetin

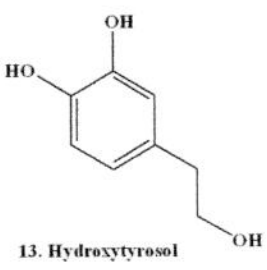

13. Hydroxytyrosol

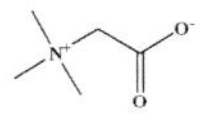

14. Betaine

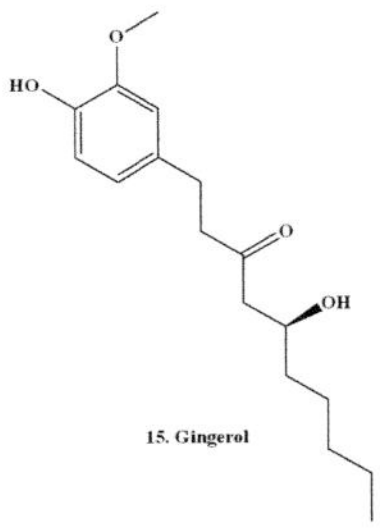

15. Gingerol

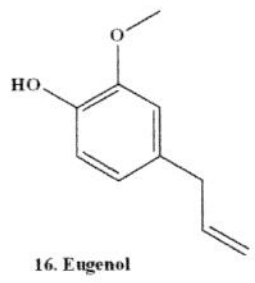

16. Eugenol

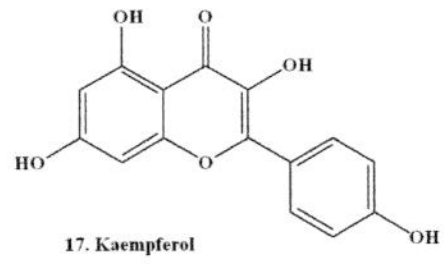

17. Kaempferol

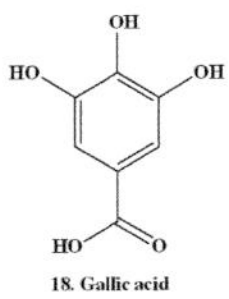

18. Gallic acid

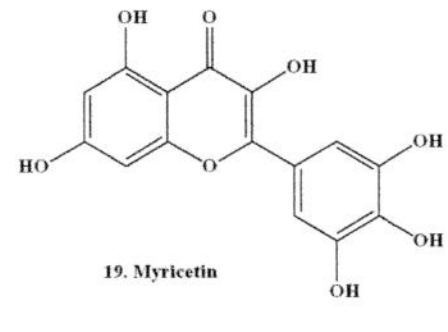

19. Myricetin

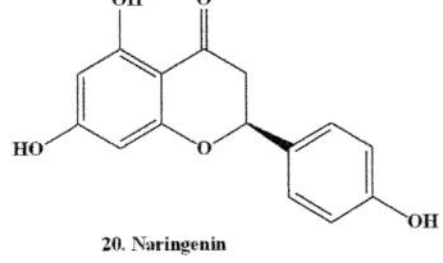

20. Naringenin

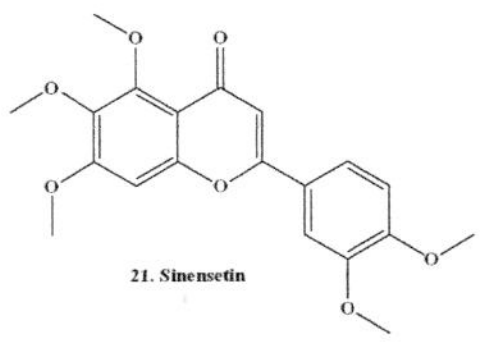

21. Sinensetin

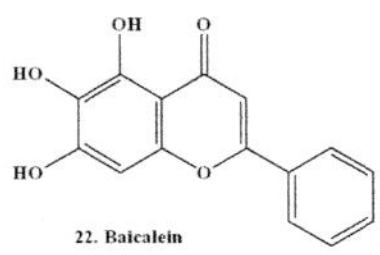

22. Baicalein

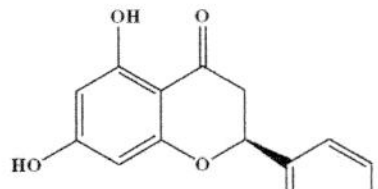

23. Pinocembrin

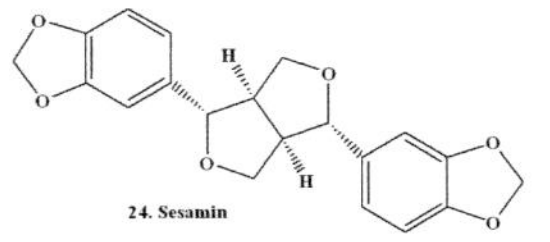

24. Sesamin

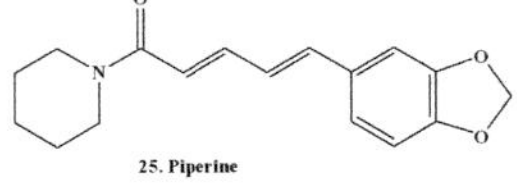

25. Piperine

FIGURE 10.3 The chemical structure of neuroprotective compounds derived from plants and their associated microbes. In this figure, the chemical structure of total 25 neuroprotective metabolites derived from plants and their associated microbes are as follows: L-DOPA **(1)**, ursolic acid **(2)**, alpha-asarone **(3)**, beta-asarone **(4)**, resveratrol **(5)**, berberine **(6)**, curcumin **(7)**, ferulic acid **(8)**, asiatic acid **(9)**, honokiol **(10)**, safranal **(11)**, quercetin **(12)**, hydroxytyrosol **(13)**, betaine **(14)**, gingerol **(15)**, eugenol **(16)**, kaempferol **(17)**, gallic acid **(18)**, myricetin **(19)**, naringenin **(20)**, sinensetin **(21)**, baicalein **(22)**, pinocembrin **(23)**, sesamin **(24)**, and piperine **(25)**.

protection effect was accomplished through the downregulation of GRP78 (a chaperone protein), p-PERK (a protein kinase), CHOP (a transcription factor), and Beclin-1 (involved in autophagy), and upregulation the Bcl-2 protein in 6-OHDA-mediated parkinsonian rat [66]. Another study demonstrated that the methanolic extract from the *Lavandula stoechas* L. plant had neuroprotective action towards 6-OHDA-induced oxidatively damaged PC-12 cells. This protection was accomplished by the regulation of proteins involved in apoptosis such as decreased ratio of phospho SAPK/JNK to SAPK/JNK protein, and upregulation of the p44/42 MAPK (ERK1/2) protein in PC-12 cells [67]. The bioactive compound hesperidin found in citrus fruits exhibited protective effects through preserving mitochondrial membrane potential and alleviates ROS production against 6-OHDA-mediated oxidative damage in SHSY-5Y cells. It is found that protection was accomplished by downregulating the gene expression of *lrrk2*, *gsk3β*, *casp9*, as well as *polg* using a zebrafish model. The hesperidin application also exhibited the improvement the locomotion behavior in the 6-OHDA-induced zebrafish model [68].

Another similar study demonstrated that ethanolic extract of *Tinospora cordifolia* exhibited a neuroprotective effect against 6-OHDA-mediated neurotoxicity in the triatum region of the rat's brain. The treatment with ethanolic extract of *Tinospora cordifolia* also demonstrated the protection of dopamine-producing neurons of substantia nigra against 6-OHDA-mediated oxidative stress. This protection was accomplished via amelioration of mitochondrial function and reducing the ratio of iron asymmetry of substantia nigra in the rat's brain. Additionally, it led to improved locomotor activity [16].

10.4 BIOACTIVE COMPOUNDS FOR HUNTINGTON'S THERAPY

Huntington's disease refers to a hereditary neurodegenerative ailment triggered by a specific gene mutation, which causes gradual neuronal degradation as well as death [69]. It develops by an autosomal-dominant variation at *Htt* gene (first exon of the *IT15* gene), which leads to a stretched and unstable CAG triplet repeat formation, which leads to formation of a mutant polyglutamine strand (poly Q) at the N-terminus of mHtt protein [70]. Huntington's disease has a profound impact on an individual's well-being, giving rise to a varied spectrum of symptoms, including chorea (involuntary, erratic movements), dystonia (characterised by sustained muscle contractions leading to unusual postures), motor ataxia (resulting in coordination and balance problems), cognitive impairment (manifesting as difficulties in thinking, reasoning, and memory), as well as depression and anxiety [71].

Now researcher focus is on epigenetic regulation for treatment of Huntington's disease [72–75]. Advanced molecular techniques such as antisense oligonucleotides, small interfering RNAs, small-molecule splicing modulators target RNA, along with methods of gene modification like zinc finger proteins (ZFP), transcription activator-like effectors, and CRISPR/cas9 are currently used for effective management of HD [76]. A study also showed that oligosaccharides can decrease the polyglutamine aggregates and extend the survival time of Huntington's patients [77]. A recent study demonstrated that ethanolic and ethyl acetate extract of the plant *Ficus religiosa* showed a protective effect towards 3-nitropropionic acid (3 NP)-mediated neurotoxicity using adult male Wistar rat model. Animals treated with *Ficus religiosa* extract showed positive changes in behavior, including improved mobility, muscle coordination, and memory function. There were also beneficial biochemical changes, such as reduced lipid peroxidation and increased expressions of catalase and superoxide dismutase, as well as glutathione enzymes. Additionally, the neurochemical acetylcholinesterase enzyme was shown to be reduced using a 3-NP-stimulated Huntington's animal study [78]. The root crude extract of *Withania somnifera* showed the neuroprotection effect against 3-NP-mediated oxidatively damaged of striatum as well as cortex region of the rat's brain. Rats administered with crude extract at respective dosages of 100 and 200 mg/kgbwt for up to two weeks demonstrated dose-related improvements in behavioral changes produced by 3-NP, including hypolocomotion, muscular incoordination, and competence. The crude extract's protective activity was also demonstrated through biochemical improvements. These included decreased lipid peroxidation, nitrite, and lactate dehydrogenase enzyme levels, along with increased expressions of enzymes like superoxide dismutase and catalase. Furthermore, mitochondrial function enhancement was observed in the rat brain's striatum and cortical areas [79].

Chrysin, also referred to as 5,7-dihydroxyflavone, is a flavonoid that is found in the propolis plant. The bioactive compound chrysin also exhibited neuroprotective effect towards 3-NP-mediated neurotoxicity in Wistar male rats. This neuroprotective action was also accomplished by amelioration of mitochondrial dysfunction, reduced oxidative stress markers, along with boosted antioxidant defenses in striatal region of the rat brain. It prevented apoptosis via upregulation of anti-apoptotic gene *Bcl-2* and the downregulation of pro-apoptotic genes *Bax* and *Bad* using the 3-NP-stimulated condition. Chrysin also improve locomotory behavior [80].

A recent study revealed that the bioactive compound licofelone competitively inhibited COX-1, COX-2, as well as 5-LOX isoenzymes, providing protection against neurotoxicity stimulated by 3-NP using a Huntington's animal model. This study is significant for exploring the roles of the COX and lipoxygenase (LOX) pathways in Huntington's disease [81].

10.5 BIOACTIVE COMPOUNDS FOR AMYOTROPHIC LATERAL SCLEROSIS THERAPY

ALS, commonly referred to as Lou Gehrig's disease, exhibits neurodegenerative symptoms that primarily affect the motor cortex, brain stem, and pinal cord, along with leading to progressive muscle weakness, speech difficulties, and eventual paralysis [82]. The specific etiology of ALS appears uncertain. Some recent studies demonstrated that changes in environmental factors, oxidative damage of mitochondrial functioning, and neurotransmitter dysfunction could potentially assist in the development of ALS [83–85]. Studies conducted recently have investigated the possible neuroprotective impact of bioactive substances originating from plants, such as epigallocatechin gallate, ginkgolides, resveratrol, protopanaxatriol, cassoside and genistein, etc., against ALS. These compounds might regulate through different mechanisms, including the upregulated expression of the Bcl-2 gene, activation of ERK1/2 as well as Akt-dependent signaling processes, along with upregulation of the Nrf2-HO antioxidant pathway, etc. [85–87]

A recent study revealed that riluzole reduces glutamate release, masitinib targets tyrosine kinases, and edaravone acts as an antioxidant. These treatments can slow ALS progression and manage symptoms [88]. Currently, researchers are exploring treatment approaches for ALS that focus on the role of non-neuronal glial cells, neuroinflammation, oxidative damages, mitochondrial dysfunction as well as glutamate excitotoxicity as crucial factors [89]. In ALS, phytochemicals appear to operate through interconnected mechanisms that involve antioxidants, anti-inflammatories, heat shock proteins (molecular chaperones), and molecular signaling pathways that support cell survival and adapt to stress [90–94]. Ongoing research aims to uncover the precise mechanisms and specific factors that could lead to effective treatment strategies for ALS.

10.6 BIOACTIVE COMPOUNDS FOR MULTIPLE SYSTEM ATROPHY THERAPY

Multiple system atrophy (MSA) refers to a rare neurological disorder that affects both the central and autonomic nervous systems, resulting in symptoms related to dysfunction of the autonomic system, parkinsonism, cerebellar impairments, issues with swallowing, and sleep disturbances [95]. The actual cause of MSA remains uncertain, and it may be influenced by environmental or genetic factors [96–98]. MSA is marked by factors like phosphoprotein-25α (p25α), which promote the aggregation of α-synuclein, decrease the expression of GDNF (glial cell line-derived neurotrophic factor), disturbances in autophagy, mitochondrial dysfunction, and neuroinflammation. Ongoing research aims to unravel the precise mechanisms involved in treating MSA [99, 100].

10.7 BIOACTIVE COMPOUNDS FOR FRONTOTEMPORAL DEMENTIA THERAPY

Frontotemporal dementia (FTD) appears as an uncommon neurodegenerative condition that primarily impacts the frontal as well as temporal regions of the brain, leading to a range of symptoms including behavioral changes, language problems, and alterations in personality [101, 102]. FTD represents a neurodegenerative disorder with autosomal dominant characteristics linked to a genetic trait on the pericentromeric region of chromosome number 3 [103]. FTD is marked by neuronal degeneration and loss along with gliosis, as well as microvacuolation occurring in lamina I to III parts of the frontal as well as temporal brain regions [104, 105]. The actual cause of FTD is also unclear to date. A recent study demonstrated that dysfunction of the ubiquitin-based protein degradation system might be a possible cause of FTD disease [106]. Plant-based drugs hold promise as dementia treatments, targeting various risk factors like inadequate cerebral blood flow, toxic chemical exposure, mitochondrial malfunction, as well as oxidative harm. However, conclusive evidence regarding the therapeutic effectiveness of plant-based drugs utilised in major clinical studies is currently ongoing [107].

10.8 ENDOPHYTE-DERIVED BIOACTIVE COMPOUNDS IN NDDs TREATMENTS

Endophytes are microorganisms living inside plants, showing a mutual relationship and providing benefits like stress tolerance and pathogen protection without causing harm [108]. Endophytes, encompassing fungi, bacteria, as well as actinomycetes, play a crucial role in the production of bioactive compounds [109, 110]. Horizontal gene transfer is a mechanism by which endophytes can generate substances that mimic those found in their host plants. This process aids endophytes in adapting and thriving within their host's tissues, thus contributing to the complex interactions between microorganisms and plants in natural ecosystems [111, 112]. It was recently found that endophyte-derived compounds can upregulate the transcription factor NF-E2-related Factor-2 (NRF2), a crucial target for combating neurodegenerative diseases [113] (**Figure 10.2**).

Another similar study demonstrated that a bioactive compound, corynesidone A, derived from endophytes, exhibited significant inhibitory activity against enzymes such as cholinesterases, and beta-secretase 1 (BACE1), as well as monoamine oxidase (MAO), all of which are crucial target sites among various neurodegenerative disorders [114]. A recent study revealed that the fungal endophyte *Neosartorya fischeri* JS0553, was isolated from the root of the *Glehnia littoralis* plant and produced the bioactive compound fischerin. Fischerin exhibited significant neuroprotective properties by preventing glutamate dependent neurotoxicity of HT22 cells. This protection was achieved by reducing reactive oxygen species (ROS), blocking calcium (Ca^{2+}) ion influx, and inhibiting the phosphorylation of mitogen-activated protein kinases, including c-Jun N-terminal kinase, and extracellular signal-regulated kinase, as well as p38 protein [115].

Another similar study demonstrated that *Nigrospora oryzae* (GL15) an endophytic fungus survive in the host plant *Tinospora cordifolia*, exhibited neuroprotective effects through inhibition of the acetylcholinesterase enzyme. Spectroscopic analyses, including UV, FTIR spectra, HPLC, ^{1}H

NMR, and ESI-MS, identified quercetin (**Figure 10.3**) as the bioactive compound present in the extract, believed to contribute to its neuroprotective effects. This extract demonstrated a capacity to mitigate dementia-like symptoms in a scopolamine (SCO)-induced study, including improvements in memory of spatial recognition as well as latency periods [116]. In a recent study, it was found that the fungal endophyte *Penicillium chrysogenum* No. 005, dwelling in the host plant *Cistanche deserticola* Y. C. Ma, generated an alkaloid known as chrysogenamide A, which exhibited neuroprotective properties against oxidative damage in SH-SY5Y cells [117].

10.9 CONCLUSION AND FUTURE PROSPECTS

Neurodegenerative disorders (NDs) are among the most common diseases that particularly create adverse impacts on the brain along with nerves. NDs involve the gradual breakdown and eventual death of nerve cells. Recently, there continues to be a lot of study into the therapeutic potential of plant-derived biologically active components to combat a number of NDs, including Alzheimer's, Huntington's, and Parkinson's [118]. The important role of plant metabolites is supported by the fact that more than 80 of the 371 pharmacological monographs published in the 9th edition of the International Pharmacopoeia have been devoted to plant-derived bioactives [119].

There are number of mechanisms by which these plant-based bioactives may work – through inhibition of acetylcholinesterase, aggregation of anti-amyloid, antioxidant effect, modification of monoamines, etc., [120–122]. In preclinical trials plant-based bioactives have been shown to have neuroprotective qualities against neurodegenerative disorders, but proving these properties in human clinical trials has been difficult. The difficulties and restrictions that are related to natural and their isolated natural products consist of their lower bioavailability, low water solubility, quick metabolism, and potential to cross the blood-brain barrier [123, 124]. For example, curcumin, a polyphenol, has been found effective in combating the symptoms of Alzheimer's. However, its effectiveness has been hindered by low absorption along with restricted bioavailability [125].

To overcome these issues, administering phytochemicals as bioconjugates employing nanotechnology as well as nanocarrier-based approaches could assist in improving therapeutic effects and effectiveness. The most frequent types of bioconjugates include nanoparticles of polymer, solid-lipid nanoparticles, crystal nanoparticles, liposomes, nanogels, micelles, and dendrimer complexes [126, 127]. Apart from medicinal plants, fungal endophyte-derived bioactives have been found effective to combat the symptoms of NDs. Endophytes have unique characteristics to produce plant-derived chemicals and their effective derivative form as well. Chemical components obtained from the identified endophytes have been used in treatment of neurodegenerative diseases. Bioactives synthesized by fungal endophytes are vital for improving growth factors, strengthening antioxidant defenses, reducing neuroinflammation and inhibiting apoptotic pathways [122, 128, 129].

To conclude, natural sources can act as an important source for management of various neurodegenerative disorders. The advancement in the development of novel medications and therapies that are efficient therapeutics against various neurodegenerative disorder would be enhanced overall by research in plant-based bioactives along with fungal endophytes.

ACKNOWLEDGMENTS

SKS lab would like to acknowledge funding from the Institution of Eminence Incentive Grant Scheme, Trans Disciplinary Research Scheme, Department of Science and technology, SEED grant, New Delhi, and funding from the Institute of Medical Sciences intramural grant, Banaras Hindu University, Varanasi, India. CP & SNR would like to acknowledge CSIR (09/0013(12277)/2021-EMR-I) and ICMR (3/1/3/196/Neuro/2021-NCD-I) for providing fellowship assistance respectively. I also would like to acknowledge the utilisation of BioRender software in Figure 10.1 of this chapter.

DECLARATION

Conflict of Interest: Authors declare there is no conflict of interest.

REFERENCES

1. M. Taussig, *The nervous system*. Routledge, 2012.
2. V. N. Uversky, "α-synuclein misfolding and neurodegenerative diseases," *Curr. Protein Pept. Sci*, vol. 9, no. 5, pp. 507–540, 2008.
3. K. J. Thompson, S. Shoham, and J. R. Connor, "Iron and neurodegenerative disorders," *Brain Res. Bull*, vol. 55, no. 2, pp. 155–164, 2001.
4. L. Migliore, and F. Coppedè, "Genetics, environmental factors and the emerging role of epigenetics in neurodegenerative diseases," *Mutat. Res. Mol. Mech. Mutagen*, vol. 667, no. 1–2, pp. 82–97, 2009.
5. D. M. Wilson, M. R. Cookson, L. Van Den Bosch, H. Zetterberg, D. M. Holtzman, and I. Dewachter, "Hallmarks of neurodegenerative diseases," *Cell*, vol. 186, no. 4, pp. 693–714, 2023.
6. G. Zeng, Z. Zhang, L. Lu, D. Xiao, S. Zong, and J. He, "Protective effects of ginger root extract on alzheimer disease-induced behavioral dysfunction in rats," *Rejuvenation Res*, vol. 16, no. 2, pp. 124–133, 2013.
7. E. Huh *et al.*, "Ginger fermented with schizosaccharomyces pombe alleviates memory impairment via protecting hippocampal neuronal cells in amyloid beta 1–42 plaque injected mice," *Food Funct.*, vol. 9, no. 1, pp. 171–178, 2018.
8. J. Olloquequi *et al.*, "Protein tyrosine phosphatase 1B (PTP1B) as a potential therapeutic target for neurological disorders," *Biomed. Pharmacother.*, vol. 155, p. 113709, 2022.
9. W. Zahra *et al.*, "Neuroprotection of rotenone-induced parkinsonism by ursolic acid in PD mouse model," *CNS Neurol. Disord. Targets (Formerly Curr. Drug Targets-CNS Neurol. Disord.*, vol. 19, no. 7, pp. 527–540, 2020.
10. J. Cummings, "The role of neuropsychiatric symptoms in research diagnostic criteria for neurodegenerative diseases," *Am. J. Geriatr. Psychiatry*, vol. 29, no. 4, pp. 375–383, 2021.
11. P. Wan, W. Su, and Y. Zhuo, "The role of long noncoding RNAs in neurodegenerative diseases," *Mol. Neurobiol*, vol. 54, pp. 2012–2021, 2017.
12. E. Grünblatt *et al.*, "Gene expression profiling of parkinsonian substantia nigra pars compacta; Alterations in ubiquitin-proteasome, heat shock protein, iron and oxidative stress regulated proteins, cell adhesion/cellular matrix and vesicle trafficking genes," *J. Neural Transm.*, vol. 111, pp. 1543–1573, 2004.
13. V. Calabrese *et al.*, "Increased expression of heat shock proteins in rat brain during aging: Relationship with mitochondrial function and glutathione redox state," *Mech. Ageing Dev.*, vol. 125, no. 4, pp. 325–335, 2004.
14. O. Hansson, J. Nylandsted, R. F. Castilho, M. Leist, M. Jäättelä, and P. Brundin, "Overexpression of heat shock protein 70 in R6/2 Huntington's disease mice has only modest effects on disease progression," *Brain Res*, vol. 970, no. 1–2, pp. 47–57, 2003.
15. H. H. Kampinga, and S. Bergink, "Heat shock proteins as potential targets for protective strategies in neurodegeneration," *Lancet Neurol*, vol. 15, no. 7, pp. 748–759, 2016.
16. J. Kosaraju, S. Chinni, P. D. Roy, E. Kannan, A. S. Antony, and M. N. S. Kumar, "Neuroprotective effect of tinospora cordifolia ethanol extract on 6-hydroxy dopamine induced parkinsonism," *Indian J. Pharmacol*, vol. 46, no. 2, p. 176, 2014.
17. K. Faust, S. Gehrke, Y. Yang, L. Yang, M. F. Beal, and B. Lu, "Neuroprotective effects of compounds with antioxidant and anti-inflammatory properties in a drosophila model of Parkinson's disease," *BMC Neurosci*, vol. 10, pp. 1–17, 2009.
18. J. Pérez-Hernández, V. J. Zaldívar-Machorro, D. Villanueva-Porras, E. Vega-Ávila, and A. Chavarría, "A potential alternative against neurodegenerative diseases: Phytodrugs," *Oxid. Med. Cell. Longev.*, vol. 2016, 2016.
19. A. Pandey, S. Bani, P. Dutt, N. K. Satti, K. A. Suri, and G. N. Qazi, "Multifunctional neuroprotective effect of withanone, a compound from *Withania somnifera* roots in alleviating cognitive dysfunction," *Cytokine*, vol. 102, pp. 211–221, 2018.
20. S. Fazel Nabavi, N. Braidy, S. Habtemariam, A. Sureda, A. Manayi, and S. Mohammad Nabavi, "Neuroprotective effects of fisetin in Alzheimer's and Parkinson's diseases: From chemistry to medicine," *Curr. Top. Med. Chem*, vol. 16, no. 17, pp. 1910–1915, 2016.
21. G. Ratheesh *et al.*, "Role of medicinal plants in neurodegenerative diseases," *Biomanufacturing Rev.*, vol. 2, pp. 1–16, 2017.

22. S. Hagl, A. Kocher, C. Schiborr, N. Kolesova, J. Frank, and G. P. Eckert, "Curcumin micelles improve mitochondrial function in neuronal PC12 cells and brains of NMRI mice–Impact on bioavailability," *Neurochem. Int*, vol. 89, pp. 234–242, 2015.
23. M. C. Cittadini *et al.*, "Effects of bioavailable phenolic compounds from Ilex paraguariensis on the brain of mice with lung adenocarcinoma," *Phyther. Res.*, vol. 33, no. 4, pp. 1142–1149, 2019.
24. P. H. Reddy *et al.*, "Protective effects of a natural product, curcumin, against amyloid β induced mitochondrial and synaptic toxicities in Alzheimer's disease," *J. Investig. Med.*, vol. 64, no. 8, pp. 1220–1234, 2016.
25. F. Pohl, and P. Kong Thoo Lin, "The potential use of plant natural products and plant extracts with antioxidant properties for the prevention/treatment of neurodegenerative diseases: In vitro, In vivo and clinical trials," *Molecules*, vol. 23, no. 12, p. 3283, 2018.
26. N. Haleagrahara, and K. Ponnusamy, "Neuroprotective effect of Centella asiatica extract (CAE) on experimentally induced parkinsonism in aged Sprague-Dawley rats," *J. Toxicol. Sci*, vol. 35, no. 1, pp. 41–47, 2010.
27. A. S. Abdul Manap *et al.*, "Synergistic effects of curcumin and piperine as potent acetylcholine and amyloidogenic inhibitors with significant neuroprotective activity in SH-SY5Y cells via computational molecular modeling and in vitro assay," *Front. Aging Neurosci.*, vol. 11, p. 206, 2019.
28. T. Retinasamy, M. F. Shaikh, Y. Kumari, S. A. Zainal Abidin, and I. Othman, "*Orthosiphon stamineus* standardized extract reverses streptozotocin-induced Alzheimer's disease-like condition in a rat model," *Biomedicines*, vol. 8, no. 5, p. 104, 2020.
29. E. Rojas-Gutierrez *et al.*, "Alzheimer's disease and metabolic syndrome: A link from oxidative stress and inflammation to neurodegeneration," *Synapse*, vol. 71, no. 10, p. e21990, 2017.
30. D. K. Ingram *et al.*, "Rodent models of memory dysfunction in Alzheimer's disease and normal aging: Moving beyond the cholinergic hypothesis," *Life Sci.*, vol. 55, no. 25–26, pp. 2037–2049, 1994.
31. E. L. Schaeffer, and W. F. Gattaz, "Cholinergic and glutamatergic alterations beginning at the early stages of Alzheimer disease: Participation of the phospholipase a 2 enzyme," *Psychopharmacology (Berl)*, vol. 198, pp. 1–27, 2008.
32. A. C. Allison, R. Cacabelos, V. R. M. Lombardi, X. A. Álvarez, and C. Vigo, "Celastrol, a potent antioxidant and anti-inflammatory drug, as a possible treatment for Alzheimer's disease," *Prog. Neuro-Psychopharmacology Biol. Psychiatry*, vol. 25, no. 7, pp. 1341–1357, 2001.
33. G. García-Morales *et al.*, "Anti-inflammatory, antioxidant and anti-acetylcholinesterase activities of bouvardia ternifolia: Potential implications in Alzheimer's disease," *Arch. Pharm. Res.*, vol. 38, pp. 1369–1379, 2015.
34. Y. Zhang, and M. Zhang, "Neuroprotective effects of morinda officinalis how.: Anti-inflammatory and antioxidant roles in Alzheimer's disease," *Front. Aging Neurosci*, vol. 14, p. 963041, 2022.
35. S. Liu *et al.*, "Resveratrol oligomers from *Paeonia suffruticosa* protect mice against cognitive dysfunction by regulating cholinergic, antioxidant and anti-inflammatory pathways," *J. Ethnopharmacol.*, vol. 260, p. 112983, 2020.
36. C. Calfio, A. Gonzalez, S. K. Singh, L. E. Rojo, and R. B. Maccioni, "The emerging role of nutraceuticals and phytochemicals in the prevention and treatment of Alzheimer's disease," *J. Alzheimer's Dis*, vol. 77, no. 1, pp. 33–51, 2020.
37. F. O. Adetuyi, E. S. Akintimehin, and K. O. Karigidi, "Comparative analysis of freshly harvested and stored bacopa floribunda leaves: HPLC phenolic fingerprinting, antioxidant and cholinergic enzyme inhibition properties," *Adv. Tradit. Med*, vol. 23, no. 1, pp. 261–271, 2023.
38. A. Soumyanath *et al.*, "Centella asiatica extract improves behavioral deficits in a mouse model of Alzheimer's disease: Investigation of a possible mechanism of action," *Int. J. Alzheimer's Dis.*, vol. 2012, 2012.
39. M. B. Bahadori, L. Dinparast, H. Valizadeh, M. M. Farimani, and S. N. Ebrahimi, "Bioactive constituents from roots of *Salvia syriaca* L.: Acetylcholinesterase inhibitory activity and molecular docking studies," *South African J. Bot*, vol. 106, pp. 1–4, 2016.
40. B. Shanmuganathan, D. Sheeja Malar, S. Sathya, and K. Pandima Devi, "Antiaggregation potential of *Padina gymnospora* against the toxic Alzheimer's beta-amyloid peptide 25–35 and cholinesterase inhibitory property of its bioactive compounds," *PLoS One*, vol. 10, no. 11, p. e0141708, 2015.
41. W. Qian, H. Li, N. Pan, and C. Zhang, "Curcumin treatment is associated with increased expression of the n-methyl-d-aspartate receptor (NMDAR) subunit, NR2A, in a rat PC12 cell line model of Alzheimer's disease treated with the acetyl amyloid-β peptide, aβ (25–35," *Med. Sci. Monit. Int. Med. J. Exp. Clin. Res*, vol. 24, p. 2693, 2018.
42. L. Hritcu, J. A. Noumedem, O. Cioanca, M. Hancianu, V. Kuete, and M. Mihasan, "Methanolic extract of *Piper nigrum* fruits improves memory impairment by decreasing brain oxidative stress in amyloid beta (1–42) rat model of Alzheimer's disease," *Cell. Mol. Neurobiol*, vol. 34, pp. 437–449, 2014.

43. F. Zhou, S. Chen, J. Xiong, Y. Li, and L. Qu, "Luteolin reduces zinc-induced tau phosphorylation at Ser262/356 in an ROS-dependent manner in SH-SY5Y cells," *Biol. Trace Elem. Res*, vol. 149, pp. 273–279, 2012.
44. D. Frost *et al.*, "β-carboline compounds, including harmine, inhibit DYRK1A and tau phosphorylation at multiple Alzheimer's disease-related sites," *PLoS One*, vol. 6, no. 5, p. e19264, 2011.
45. M. J. Jembrek, N. Slade, P. R. Hof, and G. Šimić, "The interactions of p53 with tau and aß as potential therapeutic targets for Alzheimer's disease," *Prog. Neurobiol*, vol. 168, pp. 104–127, 2018.
46. K. R. Brunden, J. Q. Trojanowski, and V. M.-Y. Lee, "Advances in tau-focused drug discovery for Alzheimer's disease and related tauopathies," *Nat. Rev. Drug Discov*, vol. 8, no. 10, pp. 783–793, 2009.
47. Y. Tian *et al.*, "Phlorizin administration ameliorates cognitive deficits by reducing oxidative stress, tau hyper-phosphorylation, and neuroinflammation in a rat model of Alzheimer's disease," *J. Food Biochem.*, vol. 42, no. 6, p. e12644, 2018.
48. A. Dey, and J. N. De, "Possible anti-Parkinson's disease therapeutics from nature: A review," *Stud. Nat. Prod. Chem*, vol. 44, pp. 447–520, 2015.
49. Y. Bang, Y. Kwon, M. Kim, S. H. Moon, K. Jung, and H. J. Choi, "Ursolic acid enhances autophagic clearance and ameliorates motor and non-motor symptoms in Parkinson's disease mice model," *Acta Pharmacol. Sin*, vol. 44, no. 4, pp. 752–765, 2023.
50. Q. Sun *et al.*, "Effect of a leucine-rich repeat kinase 2 variant on motor and non-motor symptoms in Chinese Parkinson's disease patients," *Aging Dis.*, vol. 7, no. 3, p. 230, 2016.
51. D. S. Prediger, "Effects of caffeine in Parkinson's disease: From neuroprotection to the management of motor and non-motor symptoms," *J. Alzheimer's Dis*, vol. 20, no. s1, pp. S205–S220, 2010.
52. T. Long *et al.*, "Ferulic acid exerts neuroprotective effects via autophagy induction in c. elegans and cellular models of Parkinson's disease," *Oxid. Med. Cell. Longev.*, vol. 2022, 2022.
53. O. F. Elabi, J. S. Davies, and E. L. Lane, "L-dopa-dependent effects of GLP-1R agonists on the survival of dopaminergic cells transplanted into a rat model of Parkinson disease," *Int. J. Mol. Sci*, vol. 22, no. 22, p. 12346, 2021.
54. W.-D. Rausch, F. Wang, and K. Radad, "From the tyrosine hydroxylase hypothesis of Parkinson's disease to modern strategies: A short historical overview," *J. Neural Transm*, vol. 129, no. 5–6, pp. 487–495, 2022.
55. S. O. Adeosun *et al.*, "Allopregnanolone reinstates tyrosine hydroxylase immunoreactive neurons and motor performance in an MPTP-lesioned mouse model of Parkinson's disease," *PLoS One*, vol. 7, no. 11, p. e50040, 2012.
56. B. Spencer *et al.*, "Beclin 1 gene transfer activates autophagy and ameliorates the neurodegenerative pathology in α-synuclein models of Parkinson's and lewy body diseases," *J. Neurosci.*, vol. 29, no. 43, pp. 13578–13588, 2009.
57. X. Hou, J. O. Watzlawik, F. C. Fiesel, and W. Springer, "Autophagy in Parkinson's disease," *J. Mol. Biol*, vol. 432, no. 8, pp. 2651–2672, 2020.
58. G. S. Virdi *et al.*, "Protein aggregation and calcium dysregulation are hallmarks of familial Parkinson's disease in midbrain dopaminergic neurons," *Npj Park. Dis.*, vol. 8, no. 1, p. 162, 2022.
59. S. Phani, J. D. Loike, and S. Przedborski, "Neurodegeneration and inflammation in Parkinson's disease," *Parkinsonism Relat. Disord*, vol. 18, pp. S207–S209, 2012.
60. A. Elkouzi, V. Vedam-Mai, R. S. Eisinger, and M. S. Okun, "Emerging therapies in Parkinson disease—repurposed drugs and new approaches," *Nat. Rev. Neurol*, vol. 15, no. 4, pp. 204–223, 2019.
61. J. Pan *et al.*, "Curcumin inhibition of JNKs prevents dopaminergic neuronal loss in a mouse model of Parkinson's disease through suppressing mitochondria dysfunction," *Transl. Neurodegener.*, vol. 1, no. 1, pp. 1–9, 2012.
62. S. Abrahams, H. C. Miller, C. Lombard, F. H. van der Westhuizen, and S. Bardien, "Curcumin pre-treatment may protect against mitochondrial damage in LRRK2-mutant Parkinson's disease and healthy control fibroblasts," *Biochem. Biophys. Reports*, vol. 27, p. 101035, 2021.
63. A. Picca *et al.*, "Mitochondrial dysfunction, protein misfolding and neuroinflammation in Parkinson's disease: Roads to biomarker discovery," *Biomolecules*, vol. 11, no. 10, p. 1508, 2021.
64. W. Zahra *et al.*, "Anti-parkinsonian effect of *Mucuna pruriens* and ursolic acid on GSK3β/Calcium signaling in neuroprotection against rotenone-induced parkinsonism," *Phytomedicine Plus*, vol. 2, no. 4, p. 100343, 2022.
65. T.-F. Jiang *et al.*, "Curcumin ameliorates the neurodegenerative pathology in A53T α-synuclein cell model of Parkinson's disease through the downregulation of mTOR/p70S6K signaling and the recovery of macroautophagy," *J. Neuroimmune Pharmacol.*, vol. 8, pp. 356–369, 2013.

66. B. Ning, Q. Zhang, N. Wang, M. Deng, and Y. Fang, "β-asarone regulates ER stress and autophagy via inhibition of the PERK/CHOP/Bcl-2/Beclin-1 pathway in 6-OHDA-induced parkinsonian rats," *Neurochem. Res*, vol. 44, pp. 1159–1166, 2019.
67. Z. Tayarani-Najaran, E. Hadipour, S. M. S. Mousavi, S. A. Emami, L. Mohtashami, and B. Javadi, "Protective effects of *Lavandula stoechas* L. Methanol extract against 6-OHDA-induced apoptosis in PC12 cells," *J. Ethnopharmacol*, vol. 273, p. 114023, 2021.
68. S. Kesh, R. R. Kannan, K. Sivaji, and A. Balakrishnan, "Hesperidin downregulates kinases lrrk2 and gsk3β in a 6-OHDA induced Parkinson's disease model," *Neurosci. Lett*, vol. 740, p. 135426, 2021.
69. P. Kumar, H. Kalonia, and A. Kumar, "Huntington's disease: Pathogenesis to animal models," *Pharmacol. Reports*, vol. 62, no. 1, pp. 1–14, 2010.
70. P. T. Lum, M. Sekar, S. H. Gan, S. R. Bonam, and M. F. Shaikh, "Protective effect of natural products against Huntington's disease: An overview of scientific evidence And understanding their mechanism of action," *ACS Chem. Neurosci*, vol. 12, no. 3, pp. 391–418, 2021.
71. R. C. S. R. Danduga, S. R. Dunderpate, P. K. Kola, L. Grace, R. V. B. Tadigiri, and V. K. Kanakaraju, "Neuroprotective activity of tetramethylpyrazine against 3-nitropropionic acid induced Huntington's disease-like symptoms in rats," *Biomed. Pharmacother*, vol. 105, pp. 1254–1268, 2018.
72. M. Mukhi, and J. Jai, "Evaluating the potency of three plant compounds as HDAC inhibitors for the treatment of Huntington's disease: An in silico study," *Int. J. Herb. Med*, vol. 8, no. 5, pp. 10–13, 2020.
73. E. A. Thomas *et al.*, "The HDAC inhibitor 4b ameliorates the disease phenotype and transcriptional abnormalities in Huntington's disease transgenic mice," *Proc. Natl. Acad. Sci.*, vol. 105, no. 40, pp. 15564–15569, 2008.
74. E. Hockly *et al.*, "Suberoylanilide hydroxamic acid, a histone deacetylase inhibitor, ameliorates motor deficits in a mouse model of Huntington's disease," *Proc. Natl. Acad. Sci.*, vol. 100, no. 4, pp. 2041–2046, 2003.
75. H. Jia, C. D. Morris, R. M. Williams, J. F. Loring, and E. A. Thomas, "HDAC inhibition imparts beneficial transgenerational effects in Huntington's disease mice via altered DNA and histone methylation," *Proc. Natl. Acad. Sci*, vol. 112, no. 1, pp. E56–E64, 2015.
76. K. M. Shannon, "Recent advances in the treatment of Huntington's disease: Targeting DNA and RNA," *CNS Drugs*, vol. 34, no. 3, pp. 219–228, 2020.
77. M. Tanaka *et al.*, "Trehalose alleviates polyglutamine-mediated pathology in a mouse model of Huntington disease," *Nat. Med.*, vol. 10, no. 2, pp. 148–154, 2004.
78. J. O. Bhangale, N. S. Acharya, and S. R. Acharya, "Protective effect of Ficus religiosa (l.) Against 3-nitropropionic acid induced Huntington disease," *Orient. Pharm. Exp. Med*, vol. 16, pp. 165–174, 2016.
79. P. Kumar, and A. Kumar, "Possible neuroprotective effect of *Withania somnifera* root extract against 3-nitropropionic acid-induced behavioral, biochemical, and mitochondrial dysfunction in an animal model of Huntington's disease," *J. Med. Food*, vol. 12, no. 3, pp. 591–600, 2009.
80. S. Thangarajan, S. Ramachandran, and P. Krishnamurthy, "Chrysin exerts neuroprotective effects against 3-nitropropionic acid induced behavioral despair—Mitochondrial dysfunction and striatal apoptosis via upregulating bcl-2 Gene and downregulating Bax—Bad Genes in male Wistar rats," *Biomed. Pharmacother*, vol. 84, pp. 514–525, 2016.
81. P. Kumar, H. Kalonia, and A. Kumar, "Expression of concern: Role of LOX/COX pathways in 3-nitropropionic acid-induced Huntington's disease-like symptoms in rats: Protective effect of licofelone," *Br. J. Pharmacol*, vol. 164, no. 2b, pp. 644–654, 2011.
82. L. C. Wijesekera, and P. Nigel Leigh, "Amyotrophic lateral sclerosis," *Orphanet J. Rare Dis*, vol. 4, pp. 1–22, 2009.
83. P. R. Heath, and P. J. Shaw, "Update on the glutamatergic neurotransmitter system and the role of excitotoxicity in amyotrophic lateral sclerosis," *Muscle Nerve Off. J. Am. Assoc. Electrodiagn. Med*, vol. 26, no. 4, pp. 438–458, 2002.
84. F. Trojsi, M. R. Monsurrò, and G. Tedeschi, "Exposure to environmental toxicants and pathogenesis of amyotrophic lateral sclerosis: State of the art and research perspectives," *Int. J. Mol. Sci*, vol. 14, no. 8, pp. 15286–15311, 2013.
85. R. H. Mir *et al.*, "Plant-derived natural compounds for the treatment of amyotrophic lateral sclerosis: An update," *Curr. Neuropharmacol.*, vol. 20, no. 1, p. 179, 2022.
86. C. Zhang *et al.*, "Neuroprotection by urate on the mutant hSOD1-related cellular and drosophila models of amyotrophic lateral sclerosis: Implication for GSH synthesis via activating Akt/GSK3β/Nrf2/GCLC pathways," *Brain Res. Bull.*, vol. 146, pp. 287–301, 2019.
87. J. M. Silva *et al.*, "Secondary metabolites with antioxidant activities for the putative treatment of amyotrophic lateral sclerosis (ALS):'Experimental evidences,'" *Oxid. Med. Cell. Longev.*, vol. 2020, 2020.

88. L. M. G. de Oliveira *et al.*, "Impact of plant-derived compounds on amyotrophic lateral sclerosis," *Neurotox. Res.*, vol. 41, no. 3, pp. 288–309, 2023.
89. E. F. Goodall, and K. E. Morrison, "Amyotrophic lateral sclerosis (motor neuron disease): Proposed mechanisms and pathways to treatment," *Expert Rev. Mol. Med*, vol. 8, no. 11, pp. 1–22, 2006.
90. D. Kieran, B. Kalmar, J. R. T. Dick, J. Riddoch-Contreras, G. Burnstock, and L. Greensmith, "Treatment with arimoclomol, a coinducer of heat shock proteins, delays disease progression in ALS mice," *Nat. Med*, vol. 10, no. 4, pp. 402–405, 2004.
91. M. Kiaei *et al.*, "Integrative role of cPLA2 with COX-2 and the effect of non-steriodal anti-inflammatory drugs in a transgenic mouse model of amyotrophic lateral sclerosis," *J. Neurochem.*, vol. 93, no. 2, pp. 403–411, 2005.
92. A. Neymotin *et al.*, "Neuroprotective effect of Nrf2/ARE activators, CDDO ethylamide and CDDO trifluoroethylamide, in a mouse model of amyotrophic lateral sclerosis," *Free Radic. Biol. Med.*, vol. 51, no. 1, pp. 88–96, 2011.
93. V. Novak, B. Rogelj, and V. Župunski, "Therapeutic potential of polyphenols in amyotrophic lateral sclerosis and frontotemporal dementia," *Antioxidants*, vol. 10, no. 8, p. 1328, 2021.
94. K. Pilipović, R. Jurišić Grubešić, P. Dolenec, N. Kučić, L. Juretić, and J. Mršić-Pelčić, "Plant-based antioxidants for prevention and treatment of neurodegenerative diseases: Phytotherapeutic potential of *Laurus nobilis*, *Aronia melanocarpa*, and celastrol," *Antioxidants*, vol. 12, no. 3, p. 746, 2023.
95. F. Krismer, and G. K. Wenning, "Multiple system atrophy: Insights into a rare and debilitating movement disorder," *Nat. Rev. Neurol*, vol. 13, no. 4, pp. 232–243, 2017.
96. P. A. Hanna, J. Jankovic, and J. B. Kirkpatrick, "Multiple system atrophy: The putative causative role of environmental toxins," *Arch. Neurol*, vol. 56, no. 1, pp. 90–94, 1999.
97. V. Planté-Bordeneuve, O. Bandmann, G. Wenning, N. P. Quinn, S. E. Daniel, and A. E. Harding, "CYP2D6-debrisoquine hydroxylase gene polymorphism in multiple system atrophy," *Mov. Disord. Off. J. Mov. Disord. Soc*, vol. 10, no. 3, pp. 277–278, 1995.
98. C. L. Kragh *et al.*, "α-synuclein aggregation and ser-129 phosphorylation-dependent cell death in oligodendroglial cells," *J. Biol. Chem.*, vol. 284, no. 15, pp. 10211–10222, 2009.
99. L. Ciolli, F. Krismer, F. Nicoletti, and G. K. Wenning, "An update on the cerebellar subtype of multiple system atrophy," *Cerebellum & Ataxias*, vol. 1, no. 1, pp. 1–12, 2014.
100. F. K. Nakamoto *et al.*, "The pathogenesis linked to coenzyme Q10 insufficiency in iPSC-derived neurons from patients with multiple-system atrophy," *Sci. Rep.*, vol. 8, no. 1, p. 14215, 2018.
101. G. M. McKhann, M. S. Albert, M. Grossman, B. Miller, D. Dickson, and J. Q. Trojanowski, "Clinical and pathological diagnosis of frontotemporal dementia: Report of the work group on frontotemporal dementia and Pick's disease," *Arch. Neurol*, vol. 58, no. 11, pp. 1803–1809, 2001.
102. I. V Viskontas, K. L. Possin, and B. L. Miller, "Symptoms of frontotemporal dementia provide insights into orbitofrontal cortex function and social behavior," *Ann. N. Y. Acad. Sci*, vol. 1121, no. 1, pp. 528–545, 2007.
103. S. Gydesen *et al.*, "Chromosome 3 linked frontotemporal dementia (FTD-3)," *Neurology*, vol. 59, no. 10, pp. 1585–1594, 2002.
104. M. Broe, J. Kril, and G. M. Halliday, "Astrocytic degeneration relates to the severity of disease in frontotemporal dementia," *Brain*, vol. 127, no. 10, pp. 2214–2220, 2004.
105. O. Mukherjee *et al.*, "HDDD2 is a familial frontotemporal lobar degeneration with ubiquitin-positive, tau-negative inclusions caused by a missense mutation in the signal peptide of progranulin," *Ann. Neurol. Off. J. Am. Neurol. Assoc. Child Neurol. Soc.*, vol. 60, no. 3, pp. 314–322, 2006.
106. I. E. Holm, E. Englund, I. R. A. Mackenzie, P. Johannsen, and A. M. Isaacs, "A reassessment of the neuropathology of frontotemporal dementia linked to chromosome 3," *J. Neuropathol. Exp. Neurol*, vol. 66, no. 10, pp. 884–891, 2007.
107. D Tewari *et al.*, "Ethnopharmacological approaches for dementia therapy and significance of natural products and herbal drugs," *Front. Aging Neurosci.*, vol. 10, p. 3, 2018.
108. R. Lata, S. Chowdhury, S. K. Gond, and J. F. White Jr, "Induction of abiotic stress tolerance in plants by endophytic microbes," *Lett. Appl. Microbiol*, vol. 66, no. 4, pp. 268–276, 2018.
109. M. Singh, A. Kumar, R. Singh, and K. D. Pandey, "Endophytic bacteria: A new source of bioactive compounds," *3 Biotech*, vol. 7, pp. 1–14, 2017.
110. A. Ranjan *et al.*, "Characterization and evaluation of mycosterol secreted from endophytic strain of *Gymnema sylvestre* for inhibition of α-glucosidase activity," *Sci. Rep.*, vol. 9, no. 1, pp. 1–13, 2019.
111. G. A. Strobel, W. M. Hess, E. Ford, R. S. Sidhu, and X. Yang, "Taxol from fungal endophytes and the issue of biodiversity," *J. Ind. Microbiol*, vol. 17, pp. 417–423, 1996.
112. H. Wang *et al.*, "Horizontal gene transfer of Fhb7 from fungus underlies fusarium head blight resistance in wheat," *Science (80-.)*, vol. 368, no. 6493, p. eaba5435, 2020.

113. D. P. W. Wong *et al.*, "Regulation of the NRF2 transcription factor by andrographolide and organic extracts from plant endophytes," *PLoS One*, vol. 13, no. 10, p. e0204853, 2018.
114. A. O. Falade, K. E. Adewole, A. A. Ishola, G. A. Gyebi, and N. R. Olajide, "Computational studies on the cholinesterase, beta-secretase 1 (BACE1) and monoamine oxidase (MAO) inhibitory activities of endophytes-derived compounds: Towards discovery of novel neurotherapeutics," *J. Biomol. Struct. Dyn*, vol. 41, no. 6, pp. 2540–2554, 2023.
115. S. Bang *et al.*, "Neuroprotective secondary metabolite produced by an endophytic fungus, *Neosartorya fischeri* JS0553, isolated from *Glehnia littoralis*," *J. Agric. Food Chem.*, vol. 67, no. 7, pp. 1831–1838, 2019.
116. R. Vig, F. Bhadra, S. K. Gupta, K. Sairam, and M. Vasundhara, "Neuroprotective effects of quercetin produced by an endophytic fungus *Nigrospora oryzae* isolated from *Tinospora cordifolia*," *J. Appl. Microbiol*, vol. 132, no. 1, pp. 365–380, 2022.
117. Z. Lin, J. Wen, T. Zhu, Y. Fang, Q. Gu, and W. Zhu, "Chrysogenamide a from an endophytic fungus associated with *Cistanche deserticola* and its neuroprotective effect on SH-SY5Y cells," *J. Antibiot. (Tokyo)*, vol. 61, no. 2, pp. 81–85, 2008.
118. M. H. Rahman *et al.*, "Therapeutic potential of natural products in treating neurodegenerative disorders and their future prospects and challenges," *Molecules*, vol. 26, no. 17, p. 5327, 2021.
119. N. Choudhary *et al.*, "Plant based food bioactives: A boon or bane for neurological disorders," *Crit. Rev. Food Sci. Nutr.*, vol. 64, no. 11, pp. 3279–3325, 2024.
120. L. L. Nwidu, E. Elmorsy, J. S. Aprioku, I. Siminialayi, and W. G. Carter, "In vitro anti-cholinesterase and antioxidant activity of extracts of *Moringa oleifera* plants from rivers state, Niger delta, Nigeria," *Medicines*, vol. 5, no. 3, p. 71, 2018.
121. R. Luthra, and A. Roy, "Role of medicinal plants against neurodegenerative diseases," *Curr. Pharm. Biotechnol*, vol. 23, no. 1, pp. 123–139, 2022.
122. J. Xu *et al.*, "Cytotoxic and neuroprotective activities of constituents from *Alternaria alternate*, a fungal endophyte of *Psidium littorale*," *Bioorg. Chem.*, vol. 90, p. 103046, 2019.
123. A. Dubey, N. S. Ghosh, N. Agnihotri, A. Kumar, M. Pandey, and S. Nishad, "Herbs derived bioactive compounds and their potential for the treatment of neurological disorders," *Clin. Schizophr. Relat. Psychoses*, vol. 16, no. 2, 2022.
124. Y. Liu, and N. Feng, "Nanocarriers for the delivery of active ingredients and fractions extracted from natural products used in traditional Chinese medicine (TCM)," *Adv. Colloid Interface Sci*, vol. 221, pp. 60–76, 2015.
125. M. Tang, and C. Taghibiglou, "The mechanisms of action of curcumin in Alzheimer's disease," *J. Alzheimer's Dis*, vol. 58, no. 4, pp. 1003–1016, 2017.
126. M. I. Teixeira, C. M. Lopes, M. H. Amaral, and P. C. Costa, "Current insights on lipid nanocarrier-assisted drug delivery in the treatment of neurodegenerative diseases," *Eur. J. Pharm. Biopharm*, vol. 149, pp. 192–217, 2020.
127. A. Chowdhury, S. Kunjiappan, T. Panneerselvam, B. Somasundaram, and C. Bhattacharjee, "Nanotechnology and nanocarrier-based approaches on treatment of degenerative diseases," *Int. Nano Lett*, vol. 7, pp. 91–122, 2017.
128. P. Manikandan *et al.*, "Neuroprotective effect of endophytic fungal antioxidant polyphenols on cerebral ischemic stroke-induced albino rats; Memory impairments, brain damage, and upregulation of metabolic proteins," *J. King Saud Univ.*, vol. 35, no. 1, p. 102433, 2023.
129. P. Semwal *et al.*, "The neuroprotective potential of endophytic fungi and proposed molecular mechanism: A current update," *Evidence-Based Complement. Altern. Med.*, vol. 2022, 2022.

11 Limitations of Modeling in Different Neurodegenerative Diseases

Chandra Prakash, Shyam Sunder Rabidas, and Deepak Sharma
Neurobiology Laboratory, School of Life Sciences, Jawaharlal Nehru University, New Delhi, India

11.1 INTRODUCTION

The term 'neurodegenerative diseases' refers to a group of chronic pathologies that harm the nervous system, particularly the brain. These disorders are mostly known to affect the elderly population and increase the number of disabled persons, thus putting an extra load on their families, and consequentially affecting their socioeconomic status. Additionally, neurodegenerative diseases are a serious public health concern as these are the leading cause of mortality among the elderly population. The pathophysiology of these diseases is highly diverse, as they cause memory and cognitive impairments and limit the ability to move, speak, and breathe [1]. Although there is a desperate need for effective therapeutic strategies for neurodegenerative diseases, there are currently none available in modern medicine. The only way to gain knowledge about the potential therapeutic choices is to thoroughly examine the origins and mechanisms of these disorders.

Neurodegenerative disorders can either be inherited or arise sporadically. The most common cause of neurodegenerative diseases seems to be defects in the structure and function of specific proteins. The major neurodegenerative diseases emerging due to abnormal protein aggregation are Parkinson's disease (PD), Alzheimer's disease (AD), and amyotrophic lateral sclerosis (ALS). All these diseases exhibit similar mechanistic and pathological changes, such as impaired neuronal development, neuronal death, glia activation and production and spreading of aggregates of misfolded proteins [2, 3]. Additionally, it has been discovered that several neuropathologic traits, including synaptic dysfunction, oxidative stress, neuroinflammation, and neuronal loss, can also play a role in the development of these disorders. Possibly, all these processes may influence each other and amplify them and are responsible for causing neurodegeneration in these diseases.

To investigate the pathogenesis of neurodegenerative diseases and evaluate their possible therapies, animal models are essential. Recently, various preclinical investigations have demonstrated the therapeutic efficacy of potential drug candidates. However, they did not work in human trials. The first reason for this may be the rapidly growing body of knowledge regarding neurodegenerative diseases, making outdated in vivo models insufficient. The second reason for the failure of medications and therapies in clinical trials may be their effectiveness only in the development and testing of cell cultures. One of the main issues for modeling these diseases is the lack of a single model that includes all mechanisms and processes involved in neurodegeneration. Thus, the limitations of animal models for common neurodegenerative diseases like AD, PD, and ALS are covered in this chapter.

DOI: 10.1201/9781032620442-11

11.2 ANIMAL MODELS OF AD AND THEIR LIMITATIONS

11.2.1 Pathophysiology of AD

AD is a neurodegenerative disease affecting the cortex and hippocampus regions of the brain, and it results in the formation of Aβ plaques and neurofibrillary tangles (NFTs) [4]. The major signs and symptoms of AD are retardation, learning disabilities, memory loss, and intellectual and cognitive impairment. The onset of these symptoms during AD is influenced by numerous risk factors like older age, gender, genetic makeup, and family history. The main genetic causes of autosomal dominant early-onset AD include mutations in the amyloid precursor protein (APP), presenilin 1 (PSEN1), and presenilin 2 (PSEN2); and APOE4 allele is an important risk factor for late-onset AD. Moreover, at least 25 more loci and unknown coding mutations have been linked to disease risk, and many of these genetic connections imply that the immune system is crucial to AD. Over the years various animal models have been developed to assess the highly complex mechanism of AD as well as its therapeutic approaches [5, 6].

11.2.2 Age-related Animal Models of AD

11.2.2.1 Natural Aging Model

One of the risk factors that influence the likelihood of developing AD is aging. Hence, naturally aging animals are utilised as AD models for research purposes. Though these animals exhibit age-related symptoms like cognitive deficits, morphologic alterations, and cholinergic hypofunction, they do not develop Aβ or tau pathologies; thus, such models are less helpful for research. Moreover, high maintenance costs and long experimental cycles severely limit the preclinical utilisation of such models [7].

11.2.2.2 SAM Model

An important animal model of aging-related AD is the senescence-accelerated mouse (SAM), which was created via phenotypic selection using a genetic pool of AKR/J. The model consists of nine sub-strains of SAM-prone mice that exhibit several aging symptoms like memory impairment, thinning of hairs and osteoporosis, deposition of hyperphosphorylated tau, and neurotransmitter dysfunction [7]. SAM models are not commonly utilised as they are known to have high mortality and a poor rate of drug absorption. Furthermore, the pathogenic changes in the SAM model are also inconsistent with AD-related pathophysiology.

11.2.3 Chemical-induced Animal Models of AD

11.2.3.1 STZ-Induced Model

Sporadic Alzheimer's disease (sAD) is a complex neurodegenerative disease caused by several factors such as metabolic, genetic, epigenetic, and environmental changes. Disruption of energy and glucose metabolism have been noticed in the early stage of the disease's progression. The streptozotocin (STZ)-induced model is an extensively used animal model of sAD and is developed by the administration of intracerebroventricular (ICV) STZ to generate insulin-signaling dysfunction. The administration of STZ results in AD-like symptoms as well as cerebral glucose and energy metabolism impairment and decreased insulin resistance [8]. These pathological changes often match the characteristics of sAD patients. However, they have some limitations that make them unsuitable for preclinical research. First, despite the hyperphosphorylation of tau protein, STZ-induced AD mice do not create obvious NFTs. Second, because STZ is carcinogenic to humans, prolonged or repeated exposure to it could be harmful to researchers' health. Third, the short half-life of the STZ solution makes it unstable, which occasionally leads to model failure. Moreover, female rats are less susceptible to the pathologic changes produced by STZ injection than male rats.

11.2.3.2 D-gal-Induced Model

Under normal physiological conditions, a reducing sugar D-galactose (D-gal) metabolises in the body; but in excess it can be converted to aldose and hydroperoxide, which produces reactive oxygen species (ROS). The enhanced ROS level causes oxidative stress, which accelerates the aging process [9]. Chronic intraperitoneal (i.p.) D-gal injections damage cholinergic neurons in mice, impair their ability to learn and remember things, and cause mitochondrial dysfunction, oxidative stress, activation of microglia and astrocytic cells, and learning and memory deficits [10, 11]. The tau or Aβ pathologies do not develop in the D-gal-induced AD model. Additionally, male mice are more susceptible to D-gal than female mice, thus generating oxidative stress and other pathogenic changes.

11.2.3.3 Amyloid-Induced Model

An alternative to STZ is the ICV, or intrahippocampal, administration of Aβ-42, Aβ-40, and Aβ-25-30. These amyloids have been classified based on the number of amino acids they contain and possess varying degrees of pathogenesis to AD, with Aβ-42 being the most pathogenic [12, 13]. Furthermore, the type of Aβ injected and the duration between Aβ administration and behavioral testing can also be critical and responsible for the severity of cognitive impairment. However, the model has not been reported to generate NFT, hence implying that Aβ infusion does not develop key neuropathologic features of AD.

11.2.3.4 $AlCl_3$-Induced Model

Chronic aluminium exposure can lead to a variety of toxicities. Aluminium chloride ($AlCl_3$) injected (4 mg/kg or 40 mg/kg, i.p.) for almost 40 days can be used to develop AD in rats or mice [14, 15]. In experimental mice, Al treatment causes tauopathy either by suppressing dephosphorylation of tau or increasing its aggregation and improving the activity of the two major tau kinases, GSK-3 and CDK5. Al exposure also influences Aβ aggregation and causes damage to endothelial cells and neurons [16, 17]. In vivo experimental animals treated with Al exhibit neuropathological and neurobehavioral changes; nevertheless, its poor bioavailability restricts its usefulness. Thus, $AlCl_3$ is preferred over Al to develop animal models of AD; however, the bioavailability of $AlCl_3$ is also limited as an aqueous solution of $AlCl_3$ can partially hydrolyze to volatile hydrogen chloride. Furthermore, no Aβ plaques or NFTs are produced in the Al-induced AD model.

11.2.3.5 Scopolamine-Induced Model

Acetylcholine, a crucial neurotransmitter associated with memory processing is known to improve the synaptic connections. Thus, an animal model of AD can be created by blocking cholinergic neurotransmission. Scopolamine is an anticholinergic drug that acts as a powerful inhibitor of muscarinic receptors [18]. It has been reported that scopolamine acts by increasing acetylcholine esterase activity, thus accelerating the breakdown of acetylcholine. The administration of scopolamine to rodents produces AD-associated pathogenic features, such as decreased acetylcholine levels, loss of cholinergic neurons, oxidative damage, and inflammation [19]. Therefore, an experimental approach to investigate the ways by which the cholinergic system affects cognition and evaluate preclinical efficacy of cholinomimetics provided by the scopolamine-induced animal model. However, this model's main flaw is that it lacks typical pathologic features of AD and the course of the disease.

11.2.4 Transgenic Models of AD

11.2.4.1 Aβ Pathology-Based Transgenic Models

Transgenic animal models of AD can promote the generation of Aβ aggregates and mimic amyloid deposition in senile plaques and, in certain circumstances, cerebrovascular amyloids. The APP, PSEN1, and PSEN2 mutations of AD in humans behave similarly in the mice models [20, 21]. APP mutations either produce more total Aβ or, more frequently, produce Aβ 42, which is prone to aggregation [22]. PSEN1/2 mutations change mouse APP endogenous processing but do not cause amyloid accumulation. The amyloid phenotype is also accelerated by the co-expression of mutants of PSEN1/2 with APP by enhancing Aβ-42, or Aβ-43 synthesis [23]. The majority of APP rodent models show rather weak correlation between behavioral changes and accumulation of Aβ aggregates [24]. Knock-in models and other mice models that encourage Aβ generation and accumulation without APP overexpression either show no abnormal behavior at all or show behaviors that are more subtly exhibited and more in line with the onset of amyloid pathology [25].

The Aβ pathology-based transgenic animal models do not seem to be true AD models as they do not produce AD-induced cognitive and behavioral dysfunction. The tau pathology, severe neurodegeneration, and aberrant neurotransmitter levels that go hand in hand with the symptomatic phases of actual AD are conspicuously absent in these models, even though they may closely resemble several elements of preclinical asymptomatic AD.

11.2.4.2 Tau Pathology-Based Transgenic Models

The overexpression or mutations of the MAPT gene that induce frontotemporal dementia (FTD) are responsible for the development of substantial neuronal tauopathy in transgenic animals [26]. These animals display remarkable neurodegenerative changes in the brain in contrast to the Aβ models. However, the applicability of these models is questionable for tauopathy in case of AD. The inclusions in mice often look more similar to pick bodies than to the usual NFTs linked to AD. Furthermore, not all variants of FTD-MAPT are associated with tau dysfunction, even if FTD-linked P301L or P301S MAPT mutations may indicate tau dysfunction, though these mutations are the basis for the two most commonly utilised tau transgenic animals [27]. While tau-based mice models exhibit significant tau pathology, there is a deficiency in Aβ plaques and amyloid pathology.

11.3 ANIMAL MODELS OF PD AND THEIR LIMITATIONS

11.3.1 Pathophysiology of PD

PD is a neurological disorder characterised by the gradual degeneration of dopaminergic neurons (DA) as well as the distribution of α-synuclein (a misfolded protein) in Lewy bodies (LBs) and neurites [28, 29]. Degeneration of DA neurons in PD patients exhibit symptoms like rigidity, bradykinesia, resting tremor, postural deformities, and instabilities [30, 31]. PD can be classified as sporadic PD (sPD) and familial PD (fPD). The majority of cases with sPD have late onset [32] and are well known to be influenced by many genetic and environmental factors [33]. Moreover, chronic exposure to various toxins like pesticides, metals, and solvents can also lead to the development of sPD pathology [34]. About 10% of PD patients have familial PD, which is brought on by genetic mutations [35].

Mutations to α-synuclein and leucine repeat kinase 2 (LRRK2) cause autosomal dominant PD, while mutations to genes like parkin, PTEN-induced kinase 1 (PINK1), and DJ-1 are associated with autosomal recessive PD [36, 37]. Moreover, mutations to genes like ATP13A2, DNAJC1, SYNJ1, PLA2G6, and FBX07 are also associated with PD-like disorders. The mutational changes to genes like eiF4G1, CHCHD2 and VPS35 can develop a rare form of PD with varying degrees of penetration [38, 39]. The analysis of the distribution pattern of α-synuclein reveals that sPD affects the entire nervous system. The animal models of this disease are chemical-based, α-synuclein pathology-based and transgenic models.

11.3.2 Chemical-Induced Animal Models of PD

11.3.2.1 Reserpine-Induced Model

The first PD animal model was created by injecting reserpine intraperitoneally, which inhibits monoamine transporters, resulting in the depletion of monoamines such as dopamine, 5-hydroxytryptamine, and norepinephrine without affecting the neurons [40, 41]. The main application of this model is studying the therapeutic potential of L-dopa in PD. The major limitation of the model is that pathophysiological changes of PD do not persist for a long time and only appear for a limited period. It is mostly useful for assessing the efficacy of striatum dopamine replacements and dopamine agonists.

11.3.2.2 6-OHDA-Induced Model

The structural analogue of an important neurotransmitter DA is called 6-hydroxydopamine (6-OHDA). This compound can bind with dopamine transporters and allow them to enter the cytosol, where they generate intracellular ROS by auto-oxidisation. Usually, 6-OHDA is administered directly into the striatum or SNpc of animals to mimic PD because it cannot easily cross the blood-brain barrier [42, 43]. It is possible to inject 6-OHDA unilaterally; however, doing so bilaterally results in adipsia and aphagia, as well as death. Similar to others, the 6-OHDA-induced model can cause acute neurotoxicity without producing progressive and age-dependent features of PD [44]. Moreover, this model does not produce LBs or α-synuclein-positive aggregates.

11.3.2.3 MPTP-Induced Model

1-methyl-4-phenyl-1,2,3,6-tetrahydropyridine (MPTP) is an organic molecule that undergoes radial conversion to 1-methyl-4-phenylpyridinium inside the body. It can penetrate the DA neurons and interrupt the mitochondrial respiratory complex I, causing neurotoxic effects [45]. The MPTP model is commonly used to explore the effect of drugs on the DA neurons [46]. Continuous exposure to low-level of MPTP induces the production of LBs-like inclusion in SNpc neurons, while it does not affect the formation of α-synuclein-positive aggregates [47]. As low dose of MPTP is unable to induce the production of LBs, there are conflicting reports regarding the production of inclusions that resemble LBs in MPTP models. However, only a few MPTP models indicated the presence or absence of inclusions that resembled LBs.

11.3.2.4 Rotenone-Induced Model

Rotenone is a popular pesticide known for functional inhibition of mitochondrial complex I. It is lipophilic, rapidly penetrates the blood-brain barrier, and exhibits PD-like symptoms, along with the formation of LBs, and α-synuclein aggregation [48]. In the brain cells of experimental animals, chronic exposure to low doses of rotenone causes the impairment of mitochondrial electron transport chain. A study showed that rotenone-induced PD in rats demonstrates remarkable changes in the nuclear transcription factor 2 and mitochondrial respiratory activity in the SNpc [49]. Rotenone exposure also causes substantial loss of TH-positive neurons, while surviving TH-positive neurons are highly immunoreactive to α-synuclein [50]. Even though rotenone replicates several PD characteristics, its high inter-individual variability continues to be a major drawback. This could be the reason why the availability of research is limited on the rotenone model of PD, as a larger sample size would be needed to account for variability.

11.3.2.5 Paraquat-Induced Model

Another chemical compound widely used to develop the PD model in rodents is paraquat, which is a herbicide that affects photosynthesis in plants. The compound shows a structural resemblance to MPP^+ and generates ROS, thus causing oxidative stress in the brain [51]. It has been noticed that young mice (less than eight weeks old) are more vulnerable to paraquat concerning adults once. Irrespective of mice age and doses, paraquat has been reported to impair motor functions and degenerate DA neurons [52, 53]. Additionally, it can upregulate and aggregate α-synuclein, forming LBs-like aggregates in certain DA neurons of the SNpc region.

11.3.3 α-synuclein Pathology-Based Transgenic Model

Mutational changes to genes encoding for α-synuclein result in autosomal dominant PD with common pathological features showing LBs. The α-synuclein pathology-based transgenic models can be developed by the point mutations and genomic multiplications of α-synuclein gene. The α-synuclein lesions observed in transgenic mice are the consequences of overexpression of mutant as well as wild-type α-synuclein genes [54, 55]. These α-synuclein transgenic mice also show severe nondopaminergic impairments such as anxiety, gastrointestinal dysfunction, and motor dysfunction [56, 57]. Moreover, the transgenic mice exhibit age-dependent neuropathology along with cognitive and locomotor dysfunctions. However, the relationship between these symptoms and the α-synuclein lesions is unclear.

The α-synuclein expression in these models is limited to the SNpc, and without exhibiting the degeneration of DA neurons in the nigrostriatal pathway. Another drawback is that in neurons that are normally unaffected in PD, certain wide promoters express themselves at exceptionally high levels, resulting in different patterns of cell death from those observed in PD patients. Furthermore, these mouse models replicate the pathological symptoms and anomalies reported in early PD, such as autonomic dysfunction and olfactory deficits.

11.3.4 Other Transgenic Models of PD

11.3.4.1 LRRK2 Transgenic Model

The leucine-rich repeat kinase 2 (LRRK2) is the product of the PARK8 gene. Based on its domain structure and sequence similarity with known kinases, LRRK2 has been connected to protein translation, axonal growth, and brain aging as well as kinase activity [58]. Mutational changes to LRRK2 are some of the common genetic features of PD and may cause disease by boosting the activity of its substrate phosphorylation kinase via active region recombination. Endogenous LRRK2 homologs in rodents express a protein that is structurally and functionally similar to human LRRK2, as well as conserved DA neuroanatomical pathways [59]. As a result, numerous knock-in/knock-out models have been developed, with R1441C knock-in mice among the first to model an LRRK2-based disease [60].

11.3.4.2 PINK1 Transgenic Model

PINK1, a serine/threonine kinase of mitochondrial inner and outer membranes is known to have anti-apoptotic and anti-oxidant properties [61, 62]. The mutational changes to PINK1 gene cause oxidative damage and lead to the development of autosomal recessive premature PD. The animals with genetic errors to PINK1 and parkin generate mitochondrial oxidative stress and accumulate protein misfolds, indicating a critical role of PINK1 and parkin in the survival of DA neurons and defects in mitochondria [63]. In mouse models, PINK1 deficiency causes functional impairment and degeneration of DA neurons. Moreover, PINK1 knock-out rats exhibit pathogenic features of endogenous α-synuclein, including progressive loss of SNpc neurons [64, 65]. This model can be employed to investigate the therapeutic approaches that target mitochondrial dysfunction and minimise α-synuclein containing Lewy bodies in PD.

11.3.4.3 Parkin Transgenic Model

This is the first gene linked to recessive genetic PD that was identified as PARK2, which codes for the parkin protein [66, 67]. The central nervous system expresses a high amount of E3-ubiquitin ligase, a parkin protein that can identify and break down misfolded proteins like α-synuclein. Genetic mutations to PARK2 induce loss of function and produce improper E3-ubiquitin ligase, thus affecting the elimination of defective proteins and causing the production of α-synuclein aggregates [68].

11.3.4.4 DJ-1 Transgenic Model

Autosomal recessive early-onset PD can result from a mutation in PARK7, a critical gene that codes for DJ-1 and is linked to the development of LBs [69]. Since DJ-1 functions as a redox-dependent chaperone that prevents the aggregation of α-synuclein, its interaction has a direct effect on this crucial aspect of PD. Reportedly, the generation of oxidative stress and aggregation of proteins like α-synuclein are aggravated by DJ-1 depletion [70], suggesting that DJ-1 is critical for the development and progression of PD. Furthermore, the binding of DJ-1 with mitochondrial complex I, modulates its activity, thus helping in the regulation of mitochondrial function, oxidative stress, and neuroprotection [71].

The fact that fPD is linked with mutational changes in genes like LRRK2, PARK7, parkin, and PINK1 makes determining the difference between disease-causing and comorbid conditions more difficult. All of these familial forms of PD share the degradation of nigrostriatal DA neurons; however, the underlying mechanisms may differ greatly because LBs are not present in all familial forms of PD. Even though mice models based on these more recent mutations have yielded molecular insights, they do not demonstrate loss of DA neurons in the SNpc.

11.4 ANIMAL MODELS OF ALS AND FTD AND THEIR LIMITATIONS

11.4.1 Pathophysiology of ALS and FTD

ALS, also referred to as Lou Gehrig's disease, is a neurological disorder that affects both upper and lower motor neurons in patients. The main clinical characteristic of ALS is muscular atrophy resulting from the degeneration of upper and spinal motor neurons and their axons [72, 73]. It has been observed that certain ALS patients who exhibit cognitive impairments early in their clinical phase develop behavioral variations to FTD. According to reports, about 15% of total FTD patients show characteristics that are similar to ALS, whereas approximately 50% of ALS patients exhibit abnormalities on neuropsychological tests that suggest FTD [74]. People with FTD-ALS usually have shorter lifespan than people with FTD or ALS alone. About 10% of the total cases of ALS are familial ALS (fALS), while the rest of them are sporadic [75]. It has been reported that more than 50 genes show mutational changes and, hence are known to be linked to fALS [76]. The genes that undergo the most frequent mutations include superoxide dismutase-1 (SOD1), TAR DNA-binding protein (TARDBP), fused in sarcoma (FUS), and chromosome 9 open reading frame 72 (C9orf72) [77]. ALS and FTD-TDP53 cosegregate with mutational changes in the TDP-43 gene. However, most of the familial FTD-TDP cases are associated with mutational changes in the progranulin gene (GRN), highlighting the differential pathogenesis of the disorder [78], while mutations in the FUS gene and, less frequently, the SOD1 gene generate sporadic forms of familial ALS [79].

11.4.2 SOD1-based Transgenic Model of ALS

Several SOD1 transgenic rat models (including G93A, D83, D85, D86, D90, and G37R) have been produced by utilising the mutation-related information obtained over the past 28 years from the ALS patients. These different models exhibit weakening and tremors in the hind limbs as well as locomotor impairments and paralysis followed by early mortality [80]. Moreover, these animals had gliosis, degeneration of motor neurons, and SOD1-positive inclusions in their neuropathology. In these models, a combination of autonomous glial cell dysfunction and different gain-of-function processes acting on neurons causes neurodegeneration to start. The key clinical and pathological symptoms of this form of familial ALS are well recapitulated by the mutant SOD1 animals [81]. Some of these models' significant drawbacks include their inability to replicate the pathophysiology of TDP-43 and the lack of degenerating motor neurons in the cerebral cortex region of the brain.

11.4.3 TDP-43 Pathology-Based Transgenic Model

The first mouse model in the emerging field of TDP-43 proteinopathies has demonstrated the upregulation of wild-type and mutant TDP-43 genes, consequently resulting in TDP-43 inclusions and loss of cortical and motor neurons, hallmarks of FTD-TDP and ALS diseases [82]. Patients with

FTD or TDP-43 mutations have nuclear TDP-43 that redistributes into the cytoplasm and creates cytoplasmic inclusions. Contrary to this, mouse models exhibit modest levels of TDP-43 in the cytoplasm with predominant nuclear localisation and do not duplicate the crucial cytoplasmic mislocalisation [83]. These findings raise questions about the reliability of these mouse models for ALS.

11.4.4 Other Transgenic Models of ALS-FTD

The upregulation of either the wild-type or mutant FUS protein in mice is dangerous and shortens the life span. Conversely, mice with an FUS mutation knock-in exhibit gradual degradation of their motor neurons but not fatal paralysis [84]. Notably, transgenic mice with UBQLN294 or PFN1 mutations have been found to have paralytic effects, offering other models for studies beyond SOD1 transgenic animals. Models expressing higher C9orf72 gene repeats in transgenic mice produce pathological features associated with disease, including phospho-TDP-43, DPRs, and RNA foci accumulation. There have been reports of cognitive deficits, substantial, partially penetrant neurodegeneration in the brain and spinal cord, and various degrees of cortical and/or hippocampal neurodegeneration [85].

11.5 CONCLUSION

This chapter provided a detailed overview of animal models of neurodegenerative diseases and characteristics of spontaneous, chemical-induced and transgenic animal models. Some phenotypic and neuropathological alterations in neurodegenerative diseases, especially AD, PD, ALS, etc., can be recapitulated in animals using genetic and pharmaceutical interventions. Despite the obstacles and constraints, some of the existing models are extremely useful for understanding the pathogenic mechanisms underlying various neurodegenerative diseases and investigating new therapeutic approaches.

REFERENCES

1. Dugger, B. N.; Dickson, D. W. Pathology of Neurodegenerative Diseases. *Cold Spring Harb Perspect Biol*, **2017**, 9 (7), a028035. https://doi.org/10.1101/cshperspect.a028035.
2. Brettschneider, J.; Tredici, K. Del; Lee, V. M.-Y.; Trojanowski, J. Q. Spreading of Pathology in Neurodegenerative Diseases: A Focus on Human Studies. *Nat Rev Neurosci*, **2015**, 16 (2), 109–120. https://doi.org/10.1038/nrn3887.
3. Gitler, A. D.; Dhillon, P.; Shorter, J. Neurodegenerative Disease: Models, Mechanisms, and a New Hope. *Dis Model Mech*, **2017**, 10 (5), 499–502. https://doi.org/10.1242/dmm.030205.
4. De-Paula, V. J.; Radanovic, M.; Diniz, B. S.; Forlenza, O. V. Alzheimer's Disease. *Subcell Biochem*, **2012**, 65, 329–352. https://doi.org/10.1007/978-94-007-5416-4_14.
5. Vitek, M. P.; Araujo, J. A.; Fossel, M.; Greenberg, B. D.; Howell, G. R.; Rizzo, S. J. S.; Seyfried, N. T.; Tenner, A. J.; Territo, P. R.; Windisch, M.; et al. Translational Animal Models for Alzheimer's Disease: An Alzheimer's Association Business Consortium Think Tank. *Alzheimer's & Dementia: Translational Research & Clinical Interventions*, **2020**, 6 (1). https://doi.org/10.1002/trc2.12114.
6. Akhtar, A.; Gupta, S. M.; Dwivedi, S.; Kumar, D.; Shaikh, Mohd. F.; Negi, A. Preclinical Models for Alzheimer's Disease: Past, Present, and Future Approaches. *ACS Omega*, **2022**, 7 (51), 47504–47517. https://doi.org/10.1021/acsomega.2c05609.
7. Van Dam, D.; De Deyn, P. P. Animal Models in the Drug Discovery Pipeline for Alzheimer's Disease. *Br J Pharmacol*, **2011**, 164 (4), 1285–1300. https://doi.org/10.1111/j.1476-5381.2011.01299.x.
8. Ravelli, K. G.; Rosário, B. dos A.; Camarini, R.; Hernandes, M. S.; Britto, L. R. Intracerebroventricular Streptozotocin as a Model of Alzheimer's Disease: Neurochemical and Behavioral Characterization in Mice. *Neurotox Res*, **2017**, 31 (3), 327–333. https://doi.org/10.1007/s12640-016-9684-7.
9. Azman, K. F.; Zakaria, R. D-Galactose-Induced Accelerated Aging Model: An Overview. *Biogerontology*, **2019**, 20 (6), 763–782. https://doi.org/10.1007/s10522-019-09837-y.
10. Haider, S.; Liaquat, L.; Shahzad, S.; Sadir, S.; Madiha, S.; Batool, Z.; Tabassum, S.; Saleem, S.; Naqvi, F.; Perveen, T. A High Dose of Short Term Exogenous D-Galactose Administration in Young Male Rats Produces Symptoms Simulating the Natural Aging Process. *Life Sci*, **2015**, 124, 110–119. https://doi.org/10.1016/j.lfs.2015.01.016.

11. Hao, L.; Huang, H.; Gao, J.; Marshall, C.; Chen, Y.; Xiao, M. The Influence of Gender, Age and Treatment Time on Brain Oxidative Stress and Memory Impairment Induced by d-Galactose in Mice. *Neurosci Lett*, **2014**, 571, 45–49. https://doi.org/10.1016/j.neulet.2014.04.038.
12. Yamada, M.; Chiba, T.; Sasabe, J.; Nawa, M.; Tajima, H.; Niikura, T.; Terashita, K.; Aiso, S.; Kita, Y.; Matsuoka, M.; et al. Implanted Cannula-Mediated Repetitive Administration of Aβ25–35 into the Mouse Cerebral Ventricle Effectively Impairs Spatial Working Memory. *Behavioural Brain Research*, **2005**, 164 (2), 139–146. https://doi.org/10.1016/j.bbr.2005.03.026.
13. Sipos, E.; Kurunczi, A.; Kasza, Á.; Horváth, J.; Felszeghy, K.; Laroche, S.; Toldi, J.; Párducz, Á.; Penke, B.; Penke, Z. β-Amyloid Pathology in the Entorhinal Cortex of Rats Induces Memory Deficits: Implications for Alzheimer's Disease. *Neuroscience*, **2007**, 147 (1), 28–36. https://doi.org/10.1016/j.neuroscience.2007.04.011.
14. Garcia, T.; Esparza, J. L.; Nogués, M. R.; Romeu, M.; Domingo, J. L.; Gómez, M. Oxidative Stress Status and RNA Expression in Hippocampus of an Animal Model of Alzheimer's Disease after Chronic Exposure to Aluminum. *Hippocampus*, **2010**, 218–225. https://doi.org/10.1002/hipo.20612.
15. Zhao, Y.; Dang, M.; Zhang, W.; Lei, Y.; Ramesh, T.; Priya Veeraraghavan, V.; Hou, X. Neuroprotective Effects of Syringic Acid against Aluminium Chloride Induced Oxidative Stress Mediated Neuroinflammation in Rat Model of Alzheimer's Disease. *J Funct Foods*, **2020**, 71, 104009. https://doi.org/10.1016/j.jff.2020.104009.
16. Drago, D.; Folin, M.; Baiguera, S.; Tognon, G.; Ricchelli, F.; Zatta, P. Comparative Effects of Aβ(1-42)-Al Complex from Rat and Human Amyloid on Rat Endothelial Cell Cultures. *Journal of Alzheimer's Disease*, **2007**, 11 (1), 33–44. https://doi.org/10.3233/JAD-2007-11107.
17. Ricchelli, F.; Drago, D.; Filippi, B.; Tognon, G.; Zatta, P. Aluminum-Triggered Structural Modifications and Aggregation of β-Amyloids. *Cellular and Molecular Life Sciences*, **2005**, 62 (15), 1724–1733. https://doi.org/10.1007/s00018-005-5141-0.
18. Klinkenberg, I.; Blokland, A. The Validity of Scopolamine as a Pharmacological Model for Cognitive Impairment: A Review of Animal Behavioral Studies. *Neurosci Biobehav Rev*, **2010**, 34 (8), 1307–1350. https://doi.org/10.1016/j.neubiorev.2010.04.001.
19. Rahimzadegan, M.; Soodi, M. Comparison of Memory Impairment and Oxidative Stress Following Single or Repeated Doses Administration of Scopolamine in Rat Hippocampus. *Basic and Clinical Neuroscience Journal*, **2018**, 9 (1), 5–14. https://doi.org/10.29252/nirp.bcn.9.1.5.
20. Price, D. L.; Tanzi, R. E.; Borchelt, D. R.; Sisodia, S. S. ALZHEIMER'S DISEASE: Genetic Studies and Transgenic Models. *Annu Rev Genet*, **1998**, 32 (1), 461–493. https://doi.org/10.1146/annurev.genet.32.1.461.
21. Sasaguri, H.; Nilsson, P.; Hashimoto, S.; Nagata, K.; Saito, T.; De Strooper, B.; Hardy, J.; Vassar, R.; Winblad, B.; Saido, T. C. APP Mouse Models for Alzheimer's Disease Preclinical Studies. *EMBO J*, **2017**, 36 (17), 2473–2487. https://doi.org/10.15252/embj.201797397.
22. Golde, T. E.; Schneider, L. S.; Koo, E. H. Anti-Aβ Therapeutics in Alzheimer's Disease: The Need for a Paradigm Shift. *Neuron*, **2011**, 69 (2), 203–213. https://doi.org/10.1016/j.neuron.2011.01.002.
23. Haass, C.; Strooper, B. De. The Presenilins in Alzheimer's Disease–Proteolysis Holds the Key. *Science (1979)*, **1999**, 286 (5441), 916–919. https://doi.org/10.1126/science.286.5441.916.
24. Ashe, K. H.; Zahs, K. R. Probing the Biology of Alzheimer's Disease in Mice. *Neuron*, **2010**, 66 (5), 631–645. https://doi.org/10.1016/J.NEURON.2010.04.031.
25. Kim, J.; Chakrabarty, P.; Hanna, A.; March, A.; Dickson, D. W.; Borchelt, D. R.; Golde, T.; Janus, C. Normal Cognition in Transgenic BRI2-Aβ Mice. *Mol Neurodegener*, **2013**, 8 (1). https://doi.org/10.1186/1750-1326-8-15.
26. McGowan, E.; Eriksen, J.; Hutton, M. A Decade of Modeling Alzheimer's Disease in Transgenic Mice. *Trends Genet*, **2006**, 22 (5), 281–289. https://doi.org/10.1016/J.TIG.2006.03.007.
27. Strang, K. H.; Croft, C. L.; Sorrentino, Z. A.; Chakrabarty, P.; Golde, T. E.; Giasson, B. I. Distinct Differences in Prion-like Seeding and Aggregation between Tau Protein Variants Provide Mechanistic Insights into Tauopathies. *J Biol Chem*, **2018**, 293 (7), 2408–2421. https://doi.org/10.1074/JBC.M117.815357.
28. Rizek, P.; Kumar, N.; Jog, M. S. An Update on the Diagnosis and Treatment of Parkinson's Disease. *Can Med Assoc J*, **2016**, 188 (16), 1157–1165. https://doi.org/10.1503/cmaj.151179.
29. Jankovic, J.; Tan, E. K. Parkinson's Disease: Etiopathogenesis and Treatment. *J Neurol Neurosurg Psychiatry*, **2020**, 91 (8), 795–808. https://doi.org/10.1136/jnnp-2019-322338.
30. Jankovic, J. Parkinson's Disease: Clinical Features and Diagnosis. *J Neurol Neurosurg Psychiatry*, **2008**, 79 (4), 368–376. https://doi.org/10.1136/jnnp.2007.131045.
31. Váradi, C. Clinical Features of Parkinson's Disease: The Evolution of Critical Symptoms. *Biology (Basel)*, **2020**, 9 (5), 103. https://doi.org/10.3390/biology9050103.

32. Hamza, T. H.; Zabetian, C. P.; Tenesa, A.; Laederach, A.; Montimurro, J.; Yearout, D.; Kay, D. M.; Doheny, K. F.; Paschall, J.; Pugh, E.; et al. Common Genetic Variation in the HLA Region Is Associated with Late-Onset Sporadic Parkinson's Disease. *Nat Genet*, **2010**, 42 (9), 781–785. https://doi.org/10.1038/ng.642.
33. Singh, N.; Banerjee, B.; Bala, K.; Chhillar, M.; Chhillar, N. Gene-Gene and Gene-Environment Interaction on the Risk of Parkinson's Disease. *Curr Aging Sci*, **2014**, 7 (2), 101–109. https://doi.org/10.2174/1874609807666140805123621.
34. Goldman, S. M. Environmental Toxins and Parkinson's Disease. *Annu Rev Pharmacol Toxicol*, **2014**, 54 (1), 141–164. https://doi.org/10.1146/annurev-pharmtox-011613-135937.
35. Lin, M. K.; Farrer, M. J. Genetics and Genomics of Parkinson's Disease. *Genome Med*, **2014**, 6 (6), 48. https://doi.org/10.1186/gm566.
36. Martin, I.; Dawson, V. L.; Dawson, T. M. Recent Advances in the Genetics of Parkinson's Disease. *Annu Rev Genomics Hum Genet*, **2011**, 12 (1), 301–325. https://doi.org/10.1146/annurev-genom-082410-101440.
37. Funayama, M.; Nishioka, K.; Li, Y.; Hattori, N. Molecular Genetics of Parkinson's Disease: Contributions and Global Trends. *J Hum Genet*, **2023**, 68 (3), 125–130. https://doi.org/10.1038/s10038-022-01058-5.
38. Scott, L.; Dawson, V. L.; Dawson, T. M. Trumping Neurodegeneration: Targeting Common Pathways Regulated by Autosomal Recessive Parkinson's Disease Genes. *Exp Neurol*, **2017**, 298, 191–201. https://doi.org/10.1016/j.expneurol.2017.04.008.
39. Lunati, A.; Lesage, S.; Brice, A. The Genetic Landscape of Parkinson's Disease. *Rev Neurol (Paris)*, **2018**, 174 (9), 628–643. https://doi.org/10.1016/j.neurol.2018.08.004.
40. Santos, J. R.; Cunha, J. A. S.; Dierschnabel, A. L.; Campêlo, C. L. C.; Leão, A. H. F. F.; Silva, A. F.; Engelberth, R. C. G. J.; Izídio, G. S.; Cavalcante, J. S.; Abílio, V. C.; et al. Cognitive, Motor and Tyrosine Hydroxylase Temporal Impairment in a Model of Parkinsonism Induced by Reserpine. *Behavioural Brain Research*, **2013**, 253, 68–77. https://doi.org/10.1016/J.BBR.2013.06.031.
41. Rijntjes, M.; Meyer, P. T. No Free Lunch with Herbal Preparations: Lessons from a Case of Parkinsonism and Depression Due to Herbal Medicine Containing Reserpine. *Front Neurol*, **2019**, 10 (JUN), 464366. https://doi.org/10.3389/FNEUR.2019.00634/BIBTEX.
42. Breese, G. R.; Knapp, D. J.; Criswell, H. E.; Moy, S. S.; Papadeas, S. T.; Blake, B. L. The Neonate-6-Hydroxydopamine-Lesioned Rat: A Model for Clinical Neuroscience and Neurobiological Principles. *Brain Res Rev*, **2005**, 48 (1), 57–73. https://doi.org/10.1016/j.brainresrev.2004.08.004.
43. Becker, B.; Demirbas, M.; Johann, S.; Zendedel, A.; Beyer, C.; Clusmann, H.; Haas, S. J. P.; Wree, A.; Tan, S. K. H.; Kipp, M. Effect of Intrastriatal 6-OHDA Lesions on Extrastriatal Brain Structures in the Mouse. *Mol Neurobiol*, **2018**, 55 (5), 4240–4252. https://doi.org/10.1007/S12035-017-0637-9.
44. Blesa, J.; Phani, S.; Jackson-Lewis, V.; Przedborski, S. Classic and New Animal Models of Parkinson's Disease. *J Biomed Biotechnol*, **2012**, 2012. https://doi.org/10.1155/2012/845618.
45. Meredith, G. E.; Rademacher, D. J. MPTP Mouse Models of Parkinson's Disease: An Update. *J Parkinsons Dis*, **2011**, 1 (1), 19–33. https://doi.org/10.3233/JPD-2011-11023.
46. Betarbet, R.; Sherer, T. B.; Greenamyre, J. T. Animal Models of Parkinson's Disease. *BioEssays*, **2002**, 24 (4), 308–318. https://doi.org/10.1002/bies.10067.
47. Fornai, F.; Schlüter, O. M.; Lenzi, P.; Gesi, M.; Ruffoli, R.; Ferrucci, M.; Lazzeri, G.; Busceti, C. L.; Pontarelli, F.; Battaglia, G.; et al. Parkinson-like Syndrome Induced by Continuous MPTP Infusion: Convergent Roles of the Ubiquitin-Proteasome System and Alpha-Synuclein. *Proc Natl Acad Sci U S A*, **2005**, 102 (9), 3413–3418. https://doi.org/10.1073/PNAS.0409713102.
48. Sherer, T. B.; Kim, J. H.; Betarbet, R.; Greenamyre, J. T. Subcutaneous Rotenone Exposure Causes Highly Selective Dopaminergic Degeneration and α-Synuclein Aggregation. *Exp Neurol*, **2003**, 179 (1), 9–16. https://doi.org/10.1006/exnr.2002.8072.
49. Zhang, L.; Hao, J.; Zheng, Y.; Su, R.; Liao, Y.; Gong, X.; Liu, L.; Wang, X. Fucoidan Protects Dopaminergic Neurons by Enhancing the Mitochondrial Function in a Rotenone-Induced Rat Model of Parkinson's Disease. *Aging Dis*, **2018**, 9 (4), 590–604. https://doi.org/10.14336/AD.2017.0831.
50. Inden, M.; Kitamura, Y.; Abe, M.; Tamaki, A.; Takata, K.; Taniguchi, T. Parkinsonian Rotenone Mouse Model: Reevaluation of Long-Term Administration of Rotenone in C57BL/6 Mice. *Biol Pharm Bull*, **2011**, 34 (1), 92–96. https://doi.org/10.1248/BPB.34.92.
51. Berry, C.; La Vecchia, C.; Nicotera, P. Paraquat and Parkinson's Disease. *Cell Death Differ*, **2010**, 17 (7), 1115–1125. https://doi.org/10.1038/CDD.2009.217.
52. Brooks, A. I.; Chadwick, C. A.; Gelbard, H. A.; Cory-Slechta, D. A.; Federoff, H. J. Paraquat Elicited Neurobehavioral Syndrome Caused by Dopaminergic Neuron Loss. *Brain Res*, **1999**, 823 (1–2), 1–10. https://doi.org/10.1016/S0006-8993(98)01192-5.
53. Thiruchelvam, M.; McCormack, A.; Richfield, E. K.; Baggs, R. B.; Tank, A. W.; Di Monte, D. A.; Cory-Slechta, D. A. Age-Related Irreversible Progressive Nigrostriatal Dopaminergic Neurotoxicity

in the Paraquat and Maneb Model of the Parkinson's Disease Phenotype. *Eur J Neurosci*, **2003**, 18 (3), 589–600. https://doi.org/10.1046/J.1460-9568.2003.02781.X.
54. Hatami, A.; Chesselet, M. F. Transgenic Rodent Models to Study Alpha-Synuclein Pathogenesis, with a Focus on Cognitive Deficits. *Curr Top Behav Neurosci*, **2015**, 22, 303–330. https://doi.org/10.1007/7854_2014_355.
55. Visanji, N. P.; Brotchie, J. M.; Kalia, L. V.; Koprich, J. B.; Tandon, A.; Watts, J. C.; Lang, A. E. α-Synuclein-Based Animal Models of Parkinson's Disease: Challenges and Opportunities in a New Era. *Trends Neurosci*, **2016**, 39 (11), 750–762. https://doi.org/10.1016/J.TINS.2016.09.003.
56. Dawson, T. M.; Ko, H. S.; Dawson, V. L. Genetic Animal Models of Parkinson's Disease. *Neuron*, **2010**, 66 (5), 646–661. https://doi.org/10.1016/J.NEURON.2010.04.034.
57. McDowell, K.; Chesselet, M.-F. Animal Models of the Non-Motor Features of Parkinson's Disease. *Neurobiol Dis*, **2012**, 46 (3), 597–606. https://doi.org/10.1016/j.nbd.2011.12.040.
58. Tsika, E.; Nguyen, A. P. T.; Dusonchet, J.; Colin, P.; Schneider, B. L.; Moore, D. J. Adenoviral-Mediated Expression of G2019S LRRK2 Induces Striatal Pathology in a Kinase-Dependent Manner in a Rat Model of Parkinson's Disease. *Neurobiol Dis*, **2015**, 77, 49–61. https://doi.org/10.1016/j.nbd.2015.02.019.
59. Giaime, E.; Tong, Y.; Wagner, L. K.; Yuan, Y.; Huang, G.; Shen, J. Age-Dependent Dopaminergic Neurodegeneration and Impairment of the Autophagy-Lysosomal Pathway in LRRK-Deficient Mice. *Neuron*, **2017**, 96 (4), 796–807.e6. https://doi.org/10.1016/j.neuron.2017.09.036.
60. Volta, M.; Melrose, H. LRRK2 Mouse Models: Dissecting the Behavior, Striatal Neurochemistry and Neurophysiology of PD Pathogenesis. *Biochem Soc Trans*, **2017**, 45 (1), 113–122. https://doi.org/10.1042/BST20160238.
61. Pickrell, A. M.; Youle, R. J. The Roles of PINK1, Parkin, and Mitochondrial Fidelity in Parkinson's Disease. *Neuron*, **2015**, 85 (2), 257–273. https://doi.org/10.1016/J.NEURON.2014.12.007.
62. Nguyen, T. N.; Padman, B. S.; Lazarou, M. Deciphering the Molecular Signals of PINK1/Parkin Mitophagy. *Trends Cell Biol*, **2016**, 26 (10), 733–744. https://doi.org/10.1016/J.TCB.2016.05.008.
63. Moisoi, N.; Fedele, V.; Edwards, J.; Martins, L. M. Loss of PINK1 Enhances Neurodegeneration in a Mouse Model of Parkinson's Disease Triggered by Mitochondrial Stress. *Neuropharmacology*, **2014**, 77, 350–357. https://doi.org/10.1016/J.NEUROPHARM.2013.10.009.
64. Creed, R. B.; Goldberg, M. S. Analysis of α-Synuclein Pathology in PINK1 Knockout Rat Brains. *Front Neurosci*, **2019**, 12 (JAN). https://doi.org/10.3389/FNINS.2018.01034.
65. Ren, X.; Butterfield, D. A. Fidelity of the PINK1 Knockout Rat to Oxidative Stress and Other Characteristics of Parkinson's Disease. *Free Radic Biol Med*, **2021**, 163, 88–101. https://doi.org/10.1016/J.FREERADBIOMED.2020.12.004.
66. Lücking, C. B.; Dürr, A.; Bonifati, V.; Vaughan, J.; De Michele, G.; Gasser, T.; Harhangi, B. S.; Meco, G.; Denèfle, P.; Wood, N. W.; et al. Association between Early-Onset Parkinson's Disease and Mutations in the Parkin Gene. *N Engl J Med*, **2000**, 342 (21), 1560–1567. https://doi.org/10.1056/NEJM200005253422103.
67. Truban, D.; Hou, X.; Caulfield, T. R.; Fiesel, F. C.; Springer, W. PINK1, Parkin, and Mitochondrial Quality Control: What Can We Learn about Parkinson's Disease Pathobiology? *J Parkinsons Dis*, **2017**, 7 (1), 13–29. https://doi.org/10.3233/JPD-160989.
68. Lu, X. H.; Fleming, S. M.; Meurers, B.; Ackerson, L. C.; Mortazavi, F.; Lo, V.; Hernandez, D.; Sulzer, D.; Jackson, G. R.; Maidment, N. T.; et al. Bacterial Artificial Chromosome Transgenic Mice Expressing a Truncated Mutant Parkin Exhibit Age-Dependent Hypokinetic Motor Deficits, Dopaminergic Neuron Degeneration, and Accumulation of Proteinase K-Resistant Alpha-Synuclein. *J Neurosci*, **2009**, 29 (7), 1962–1976. https://doi.org/10.1523/JNEUROSCI.5351-08.2009.
69. Goldberg, M. S.; Pisani, A.; Haburcak, M.; Vortherms, T. A.; Kitada, T.; Costa, C.; Tong, Y.; Martella, G.; Tscherter, A.; Martins, A.; et al. Nigrostriatal Dopaminergic Deficits and Hypokinesia Caused by Inactivation of the Familial Parkinsonism-Linked Gene DJ-1. *Neuron*, **2005**, 45 (4), 489–496. https://doi.org/10.1016/J.NEURON.2005.01.041.
70. Bonilha, V. L.; Bell, B. A.; Rayborn, M. E.; Yang, X.; Kaul, C.; Grossman, G. H.; Samuels, I. S.; Hollyfield, J. G.; Xie, C.; Cai, H.; et al. Loss of DJ-1 Elicits Retinal Abnormalities, Visual Dysfunction, and Increased Oxidative Stress in Mice. *Exp Eye Res*, **2015**, 139, 22–36. https://doi.org/10.1016/j.exer.2015.07.014.
71. Almikhlafi, M. A.; Stauch, K. L.; Villeneuve, L. M.; Purnell, P. R.; Lamberty, B. G.; Fox, H. S. Deletion of DJ-1 in Rats Affects Protein Abundance and Mitochondrial Function at the Synapse. *Sci Rep*, **2020**, 10 (1), 13719. https://doi.org/10.1038/s41598-020-70486-0.
72. Miller, R.; Mitchell, J.; Lyon, M.; Moore, D. Riluzole for Amyotrophic Lateral Sclerosis (ALS)/Motor Neuron Disease (MND). In *Cochrane Database of Systematic Reviews*; Miller, R., Ed.; John Wiley & Sons, Ltd: Chichester, UK, **2002**. https://doi.org/10.1002/14651858.CD001447.

73. Hardiman, O.; Al-Chalabi, A.; Chio, A.; Corr, E. M.; Logroscino, G.; Robberecht, W.; Shaw, P. J.; Simmons, Z.; van den Berg, L. H. Correction: Amyotrophic Lateral Sclerosis. *Nat Rev Dis Primers*, **2017**, 3 (1), 17085. https://doi.org/10.1038/nrdp.2017.85.
74. Abramzon, Y. A.; Fratta, P.; Traynor, B. J.; Chia, R. The Overlapping Genetics of Amyotrophic Lateral Sclerosis and Frontotemporal Dementia. *Front Neurosci*, **2020**, 14. https://doi.org/10.3389/fnins.2020.00042.
75. Boylan, K. Familial Amyotrophic Lateral Sclerosis. *Neurol Clin*, **2015**, 33 (4), 807–830. https://doi.org/10.1016/j.ncl.2015.07.001.
76. Zou, Z.-Y.; Zhou, Z.-R.; Che, C.-H.; Liu, C.-Y.; He, R.-L.; Huang, H.-P. Genetic Epidemiology of Amyotrophic Lateral Sclerosis: A Systematic Review and Meta-Analysis. *J Neurol Neurosurg Psychiatry*, **2017**, 88 (7), 540–549. https://doi.org/10.1136/jnnp-2016-315018.
77. Shatunov, A.; Al-Chalabi, A. The Genetic Architecture of ALS. *Neurobiol Dis*, **2021**, 147, 105156. https://doi.org/10.1016/j.nbd.2020.105156.
78. Laird, A. S.; Van Hoecke, A.; De Muynck, L.; Timmers, M.; Van Den Bosch, L.; Van Damme, P.; Robberecht, W. Progranulin Is Neurotrophic In Vivo and Protects against a Mutant TDP-43 Induced Axonopathy. *PLoS One*, **2010**, 5 (10), e13368. https://doi.org/10.1371/journal.pone.0013368.
79. Kwiatkowski, T. J.; Bosco, D. A.; LeClerc, A. L.; Tamrazian, E.; Vanderburg, C. R.; Russ, C.; Davis, A.; Gilchrist, J.; Kasarskis, E. J.; Munsat, T.; et al. Mutations in the FUS/TLS Gene on Chromosome 16 Cause Familial Amyotrophic Lateral Sclerosis. *Science (1979)*, **2009**, 323 (5918), 1205–1208. https://doi.org/10.1126/science.1166066.
80. Philips, T.; Rothstein, J. D. Rodent Models of Amyotrophic Lateral Sclerosis. *Curr Protoc Pharmacol*, **2015**, 69 (1), 5.67.1–5.67.21. https://doi.org/10.1002/0471141755.ph0567s69.
81. Ittner, L. M.; Halliday, G. M.; Kril, J. J.; Götz, J.; Hodges, J. R.; Kiernan, M. C. FTD and ALS—Translating Mouse Studies into Clinical Trials. *Nature Reviews Neurology 2015 11:6*, **2015**, 11 (6), 360–366. https://doi.org/10.1038/nrneurol.2015.65.
82. White, M. A.; Kim, E.; Duffy, A.; Adalbert, R.; Phillips, B. U.; Peters, O. M.; Stephenson, J.; Yang, S.; Massenzio, F.; Lin, Z.; et al. TDP-43 Gains Function Due to Perturbed Autoregulation in a Tardbp Knock-in Mouse Model of ALS-FTD. *Nature Neuroscience 2018 21:4*, **2018**, 21 (4), 552–563. https://doi.org/10.1038/s41593-018-0113-5.
83. Walker, A. K.; Spiller, K. J.; Ge, G.; Zheng, A.; Xu, Y.; Zhou, M.; Tripathy, K.; Kwong, L. K.; Trojanowski, J. Q.; Lee, V. M. Y. Functional Recovery in New Mouse Models of ALS/FTLD after Clearance of Pathological Cytoplasmic TDP-43. *Acta Neuropathol*, **2015**, 130 (5), 643–660. https://doi.org/10.1007/S00401-015-1460-X/METRICS.
84. Devoy, A.; Kalmar, B.; Stewart, M.; Park, H.; Burke, B.; Noy, S. J.; Redhead, Y.; Humphrey, J.; Lo, K.; Jaeger, J.; et al. Humanized Mutant FUS Drives Progressive Motor Neuron Degeneration without Aggregation in 'FUSDelta14' Knockin Mice. *Brain*, **2017**, 140 (11), 2797–2805. https://doi.org/10.1093/BRAIN/AWX248.
85. Jiang, J.; Zhu, Q.; Gendron, T. F.; Saberi, S.; McAlonis-Downes, M.; Seelman, A.; Stauffer, J. E.; Jafar-nejad, P.; Drenner, K.; Schulte, D.; et al. Gain of Toxicity from ALS/FTD-Linked Repeat Expansions in C9ORF72 Is Alleviated by Antisense Oligonucleotides Targeting GGGGCC-Containing RNAs. *Neuron*, **2016**, 90 (3), 535–550. https://doi.org/10.1016/J.NEURON.2016.04.006.

12 Significance of 3D Organoid Models in Neurodegenerative Diseases

Ishita Bhardwaj, Atifa Haseeb Ansari, Sippy Singh and Durgesh Singh
S.S. Khanna Girls' Degree College (A Constituent College of University of Allahabad), Prayagraj, India

Raghvendra Pratap
Prof. Rajendra Singh (Rajju Bhaiya) University, Prayagraj, India

12.1 INTRODUCTION

In stem cell research, 3D organoid technology has come to be widely used over the past decade. The term 'organoids' refers to small three-dimensional groups of derivatives of pluripotent cells. Organoids are considered miniaturised and simplified versions of the organ from which in vitro production is carried out but could reconcile the essential features of human organ development. Organoids are classified on the basis of a simple version of the organ derived in vitro but cannot recapitulate the basic characteristics of human organ development. These are an indispensable instrument for biomedical study, and they offer a novel avenue to model human diseases in vitro and valuable diagnosis of diseases in humans through better analysis [1] and can imitate a wide variety of aspects of in vivo organs, they have the capacity to renew themselves and show the capability of self-organisation. They represent different brain parts such as prosencephalon, mesencephalon, cerebellum, cortical and hippocampal regions, and they have been used widely in medical issues and research, with respect to the transformed tumor cells and cancer treatment [2].

12.2 HISTORY OF BRAIN ORGANOIDS

It was in 1907 when H.V. Wilson found that sponge cells that are mechanically separated are capable of reconstructing and automating into a new functional sponge in 3D culture so that the origin of organoids could be traced [3]. Traditionally, the manufacturing of organoids has included fragments from a variety of tissue [1]. With the development of embryonic stem cells (ESCs) and (iPSCs) the meaning of 'organoid' has evolved. Specific characteristics such as regeneration into different type of cells are shared between ESCs and iPSCs. Homospecific models help in addressing in neuronal formations. Lancaster and his colleagues have begun to address this complex issue [4]. It has been reported that the analysis of 3D cell cultures from developing human brains, called cerebral organoids, is done by using iPSCs from healthy person. The structure closely resembles the cytoarchitecture of a human embryo's brain [5]. In addition, organoids are likely to promote the development of new methods in regenerative medicine. On other hand, it is likely that a more advanced setup will be required to analyse more complex interactions involving interaction between neurons and non-neuronal cells [6].

DOI: 10.1201/9781032620442-12

12.3 BRAIN ORGANOIDS AS MINI BRAINS

There is a great diversity of cells in the human brain. This diversity comprises cell types from the neuroectodermal lineage of progenitor's neurons, glial cells, oligodendrocytes, microglia, and vascular cells [7]. In the central nervous system, the major cellular types that are important for normal brain functioning and are involved in neurodegeneration are glial cells and neurons. Neuroepithelial cells, known as neural stem cells, are produced during brain development and have been called neurons, astrocytes, or oligodendrocytes. In various neurodegenerative disorders, both gilia and neurons are found to be affected, for example in Alzheimer's disease (AD), Huntington's disease, and Parkinson's disease (PD) [8]. In vivo, models can provide a complex level that is used to understand brain function in neurodegenerative diseases [9]. For many years 2D culture has been utilised to investigate cell–cell communication and response towards the stimuli, normally as a monolayer on rigid substrates, the 2D cell culture approach is much more rapid and easier to do than the 3D method. It's less expensive and more efficient at manipulating, reproducing, or analysing results [10]. The cultivation of a monolayer may result in morphological variations, which usually affect structure and pathways [11]. The use of animal models to mimic human beings is common, and there are significant failures in translating animal diseases. Animal-based models lack accuracy in accurately modeling all critical features that characterise most human neurological disorders, for example, microcephaly is a developmental disorder of the brain. The limitation of 2D culture is that in comparison with a 3D environment, the monocellular culture can cause alteration in expression of genes [12]. In the field of waste disposal, this pattern will repeat itself, which is different from what happens in vivo [13]. Furthermore, due to their irregular morphology, the drugs that have been tested on 2D cells do not usually offer a promising pharmacotherapeutic potential [14]. Newly sophisticated, human-derived models are promising factors to overcome some of the limits. Human embryonic stem cells (hESC) are like human-derived induced pluripotent stem cells (hiPSCs) in terms of cell proliferation. iPSC has ability to distinguish some DNA methylation due to which iPSC profile is being used in cellular reprogramming process In vitro models are valuable in the recapitulation of diseases and understanding their mechanisms; therefore, one can easily study cell interaction mechanisms whether cellular or molecular [9].

12.4 GENERATION OF 3D BRAIN ORGANOIDS

Methodologies have been developed since 1990 to induce differentiation of neural regions in 3D from PSCs where few of them can be regarded as brain organoids. For example, neural spheres derived from neural progenitor cells (NPCs) [15] can be assembled by artificial methods [16]. Generally, brain organoids can be generated using two types of methods: an unguided method and a guidance method. Unguided approaches use intrinsic signaling and organisational functions of hiPSCs to separate the organoids into tissues imitating brain development. There are often heterogeneous tissues, resembling different areas of the brain, resulting from these cerebral organoids. To develop spheroids, which are specific for a type of tissue, guidance approaches are used on tiny molecules and growth factors. Brain region-specific organoid methods are used to determine progenitor fate at an early stage using patterning factors. Those factors are eliminated at the next phase of differentiation [17]. 3D organoids from hiPSCs are likely to replicate cellular 3D cytoarchitecture patterns in the cerebral cortex, which provides new opportunities for exploring disease pathogenesis within our brains [18]. The 3D human neural model was proposed by Hogberg [19] and is the most widely used technique for the generation of 3D brain tissue [18].

12.5 IMPORTANCE OF MODELING NEURODEGENERATIVE DISORDERS WITH 3D BRAIN ORGANOIDS

Neurodegeneration mostly appears in maturity [20]. Organoids are the derivatives of embryonic stem (ES) cells and induced pluripotent stem (iPS) cells [21]. The fetal characteristics of hPSC-derived neural cells should be considered to recapitulate disease onset [22]. 3D organoids that can be grown in a lab are miniaturised, simplified versions of the human brain. They are receiving significant attention and importance in the field of neurological and medical research because they are key factors that are helpful in giving information with neurological injuries. The significance of 3D brain organoids has some key aspects.

12.5.1 Bridging the Gap

To obtain accurate interpretations of human brain tissue, 3D brain organoids are bridging the gap between traditional cell cultures and animal models. This may result in faster, more translational research that directly relates to the health of humans.

12.5.2 Infectious Disease Modeling

Organoids are ideal for in vitro studies of infectious diseases, especially pathogens that are limited to human beings and depend on special cell types. They state salient characteristics of development of the cortex and also recognised glial cells. It has been notified that Zika-virus infection in organoids leads to cause neural destruction [22, 23].

12.5.3 Unveiling Hereditary Disease

Organoids are widely used in study of genes and hereditary diseases. Antitrypsin deficiency in patients causes a fatal effect on the liver [22, 24, 25]. COMMD1 protein can be used to recover the toxicity that is caused by deposition of copper (Cu).

12.5.4 Personalised Medicine

Patient-specific brain organoids may be produced from neural stem cells in individuals who have neurological disorders. This personalised approach may help to determine treatment strategy and test the effectiveness of medicinal products against the patient's own brain cells, which could lead to more effective treatments. Ideal treatment schemes can be selected because of this strategy. One of the most important examples is the treatment of cystic fibrosis (CF). The mutated CF transmembrane conductance regulator (CFTR) gene causes a deleterious effect on internal organs [26–28]. The modeling of disease and the screening of personalised drugs open a new path to precision medicine [26].

12.5.5 Drug Discovery

There are limits to current medical discoveries in human diseases, e.g. individual diversities for patients with drug trials [26, 29]. Organoids based on induced pluripotent cells may be a potential new platform for drug discovery, due to their clinical characteristics that correspond to those of patients with primary cancer, infection, and development disorders. In 2017, Broutier identified extracellular signal-related kinases (ERK) inhibitors as promising therapy in cancer treatment [26, 30].

12.5.6 Drug Toxicity

There are certain limitations to toxic level and information about medicinal products. During this phase toxicity levels of some medicines are gradually identified in trials or in the early stages of their development [31]. Organoids may be used to screen toxicity assessment and their impact on internal organs [32]. These organoids are cultured in a shared medium and equipped with biosensors, enabling the evaluation of drug impacts and toxic reactions on complete organs [26, 33].

12.5.7 Gene Therapy

By utilising gene repair techniques, a range of illnesses like carcinoma, cardiac arrest, and acquired immunodeficiency syndrome (AIDS) can be addressed through the modification of specific genes within the body's cells. Notably, the achievement of producing and implanting genetically engineered Lgr5+ stem cell-derived organoids into impaired tissues signifies a significant advancement [26, 34, 35]. The CRISPR/Cas9-mediated system founded by Schwank et al. for homology-dependent repair has the potential to result in unintended double-strand breaks, which could be harmful. This capability is particularly beneficial for restoring the functionality impaired by CFTR mutations. With this significant advancement, the prospect of mending genetic mutations in patient-derived organoids (PDOs) becomes increasingly viable, thereby hastening the transition of gene repair to the clinical stage [26, 36, 37].

12.5.8 Transplantation Therapy

Organ transplantation is one of the treatments essential for treating patients with organ failure. Research is providing a way to verify the feasibility of organoid transplantation to make it a reality [26]. For example, Watson et al. transplanted an intestinal organoid derived from hPSCs into a host mouse and found marked expansion and maturation in the transplanted organoid [26, 38]. Despite the relatively immature nature of transplanted organoids compared to host natural organs, their therapeutic use will continue to be an important substitute for transplantation due to heterotypic cell interactions [26]. In addition, several other studies have been performed around organoid transplantation, and these provide an important theoretical basis for exploring transplant therapy [39, 40].

12.5.9 Understanding Brain Evolution

Cerebral organoids are becoming increasingly influential in examining the distinctions in brain development between human and non-human primate (NHP) species [41]. The revelation that cerebral organoids, resembling brain structures, can be cultivated through the differentiation of stem cells presents a unique opportunity for the direct exploration of various facets of human brain development. These cerebral organoids are 3D cellular clusters originating from PSCs, encompassing numerous cell types typically present in embryonic brains, displaying local organisation, and functioning akin to cells observed in their natural, in vivo counterparts [2, 41–44].

12.6 INVESTIGATING NEURAL DEVELOPMENTAL DISORDERS

These 3D organoid models hold promise in comprehending and potentially addressing neural developmental disorders like microcephaly and lissencephaly. They allow researchers to replicate atypical brain development in a laboratory setting and investigate possible therapeutic interventions (**Figure 12.1**). In patients with neurodegenerative disorders (NDDs), personalised therapies have been proposed to be developed using organoids. Such models have great potential in giving information about disease mechanisms and new approaches in NDDs [1]. For example, in Alzheimer's disease, organoid modeling subsequent research has been done and there have been advancements in 3D organoid techniques, leading to formation of organoids using cells obtained from individuals with Alzheimer's disease [45]. Organoids that are derived from Alzheimer's disease patients, accurately replicated the pathological characteristics associated with the disease.

These characteristics include elevated levels of Aβ (amyloid-beta). ALS is a lethal neurodisorder marked by the gradual deterioration of both types of motor neurons located in the central nervous system (CNS) [46, 47]. The objective of organoids in this disease is to screen the pathological characteristics associated with ALS, which encompass the aggregation of proteins like TDP-43 and SOD1, oxidative stress, and neuroinflammation [46].

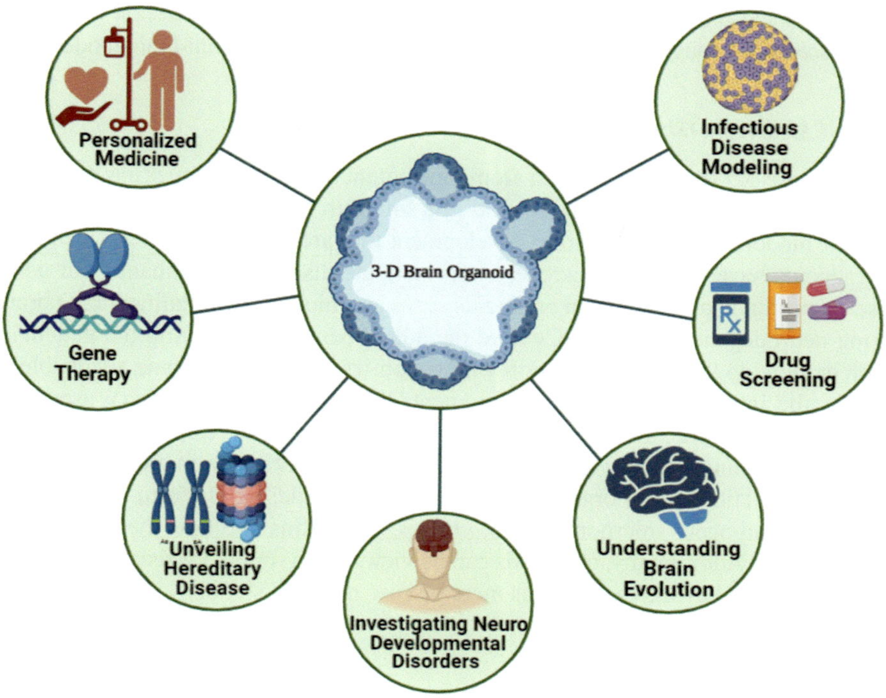

FIGURE 12.1 Significance of 3D mini-brains (Organoids).

12.7 FUTURISTIC APPROACH

The potential to study complex genetics and their progressive states in three-dimensional brain organoids provides many possibilities for the purposes of modeling aspects of human cerebral pathological development, but there is still a lack of understanding in the brain organoids, due to their variations [48]. Three-dimensional organoids hold great potential for investigating, for example, in the formation of Alzheimer's disease and the pathogenic mechanisms associated with AD, such as gliosis and metabolism. In the coming years, this model system may serve as a foundation for identifying targets related to AD and the creation of innovative drug candidates for treating AD [1]. Organoids are autonomous 3D clusters of cells forming an organ, allowing us to model the original functions of organs in a close manner. To study the development of organs, diseases, and drug testing, a new approach is possible, which is to use microfluidic applications on organoids to produce high-quality human models. Microfluidics has unlocked the potential to replicate physiological conditions through precise control of fluid dynamics, enabling the regulation of nutrient delivery. This technology facilitates the incorporation of external stimuli like electrical and mechanical forces and enables real-time monitoring of crucial factors such as pH, oxygen levels, and metabolite concentrations within the system. Utilising microfluidics in conjunction with organoids has opened exciting possibilities for creating robust human models that facilitate the study of organ development, diseases, and drug testing [49].

Newly developed organoid intelligence (OI) is a concept that seeks to leverage the inherent biological capabilities of these brain organoids, interfacing them with computer technology to enable biocomputing and synthetic intelligence. This involves inputting information, observing how the organoids process and execute tasks, and studying the resulting impact on their structural and functional networks. Moreover, delving into how learning processes contribute to the formation and alteration of connectivity patterns within these organoids gives valuable information into the early stages of cognitive development in the human brain [50]. With ongoing progress in the field, scientists and researchers are actively tackling challenges such as the oversimplified portrayal of brain intricacies. Promisingly, the integration of cutting-edge technologies like single-cell transcriptomics and the development of organoid-cerebral systems-on-a-chip offer potential solutions to augment the accuracy and realism of these 3D mini-brain models.

12.8 CONCLUSION

Brain organoids are 3D structures created from pluripotent stem cells mimicking organisation and containing various cell types like neurons and glia found in the human brain. They offer unique advantages for modeling aspects of brain development that are not yet well-replicated in animals. Notably, organoids have allowed researchers to investigate diseases and enhance our comprehension of disease mechanisms [51]. Brain organoids serve as a biologically significant 3D brain model for exploring neurodegeneration. hiPSCs have the added benefit of retaining the patient's genetic traits. When employing these in conjunction with genomics, organoids become a valuable platform for examining the intricacies of brain development, diseases, and evolutionary aspects [51–55]. Brain organoids play a crucial role in researching human brain development and investigating neuro-disorders including epilepsy, dementia, schizophrenia or other brain abnormalities resulting from Zika virus infection. Brain organoids offer a disease-modeling platform to explore effective factors, including pathogens microorganisms, viruses, and toxins and how they interact with and impact the brain [51]. Additionally, organoids can be viewed as a source for therapeutic applications in treatment, extending beyond their role as mere disease models [51, 56].

ACKNOWLEDGMENTS

The authors wish to thank the principal, S. S. Khanna Girls' Degree College (A Constituent College of the University of Allahabad, Prayagraj) and the DST-CURIE Grant of the Department of Science and Technology, New Delhi GOI (DST/CURIE-PG/2022/10 G), for providing all the necessary facilities in the college.

REFERENCES

1. Chang, Y., Kim, J., Park, H, Choi, H, and Kim, J. 2020. Modelling neurodegenerative diseases with 3D brain organoids. *Biological Reviews of the Cambridge Philosophical Society*, 95(5), 1497–1509. https://doi.org/10.1111/brv.12626
2. Jiang, X., Oyang, L., Peng, Q., Liu, Q., Xu, X., Wu, N., and Liao, Q. 2023. Organoids: Opportunities and challenges of cancer therapy. *Frontiers in Cell and Developmental Biology*,11, 1232528. https://doi.org/10.3389/fcell.2023.1232528
3. Wilson, H. V. 1907. On some phenomena of coalescence and regeneration in sponges. *Journal of Elisha Mitchell Science Society*, 23, 245–258. https://doi.org/10.1002/jez.1400050204
4. Lancaster, M. A., Renner, M., Martin, C. A., Wenzel, D., Bicknell, L. S., Hurles, M. E., Homfray, T., Penninger, J. M., Jackson, A. P., and Knoblich, J. A. 2013. Cerebral organoids model human brain development and microcephaly. *Nature*, 501(7467), 373–379. https://doi.org/10.1038/nature12517
5. Soldner, F., and Jaenisch, R. 2012. iPSC disease modeling. *Science*, 338(6110), 1155–1156. https://doi.org/10.1126/science.1227682
6. Muzio, L., and Consalez, G. G. 2013. Modeling human brain development with cerebral organoids. *Stem Cell Research & Therapy*, 4(6), 154. https://doi.org/10.1186/scrt384
7. McKenzie, A. T., Wang, M., Hauberg, M. E., Fullard, J. F., Alexey, K., Alexandra, K., et al. 2018. Brain cell type-specific gene expression and co-expression network architecture. *Scientific Reports*, 8(1), 8868. https://doi.org/10.1038/s41598-018-27293-5
8. Bose, R., Banerjee, S., and Dunbar, G. L. 2021. Modeling neurological disorders in 3D organoids using human-derived pluripotent stem cells. *Frontiers in Cell and Developmental Biology*, 9, 640212. https://doi.org/10.3389/fcell.2021.640212
9. Louit, A., Galbraith, T., and Berthod, F. 2023. In vitro 3D modeling of neurodegenerative diseases. *Bioengineering (Basel)*, 10(1), 93. https://doi.org/10.3390/bioengineering10010093
10. Zhuang, P., Sun, A. X., An, J., Chua, C. K., and Chew, S. Y. 2018. 3D neural tissue models: From spheroids to bioprinting. *Biomaterials*, 154, 113–133. https://doi.org/10.1016/j.biomaterials.2017.10.002

11. Kilian, K. A., Bugarija, B., Lahn, B. T., and Mrksich, M. 2010. Geometric cues for directing the differentiation of mesenchymal stem cells. *Proceedings of the National Academy of Sciences of the United States of America*, 107, 4872–4877. https://doi.org/10.1073/pnas.0903269107
12. Luca, A. C., Mersch, S., Deenen, R., Schmidt, S., Messner, I., Schafer, K. L., et al. 2013. Impact of the 3D microenvironment on phenotype, gene expression, and EGFR inhibition of colorectal cancer cell lines. *PLoS ONE*, 8, e59689. https://doi.org/10.1371/journal.pone.0059689
13. Kapalczynska, M., Kolenda, T., Przybyla, W., Zajaczkowska, M., Teresiak, A., Filas, V., and Lamperska, K. 2018. 2D and 3D cell cultures—A comparison of different types of cancer cell cultures. *Archives of Medical Science*, 14, 910–919. https://doi.org/10.5114/aoms.2016.63743
14. Lv, D., Hu, Z., Lu, L., Lu, H., and Xu, X. 2017. Three-dimensional cell culture: A powerful tool in tumor research and drug discovery. *Oncology Letters*, 14, 6999–7010. https://doi.org/10.3892/ol.2017.7134
15. Reynolds, B. A., Tetzlaff, W., and Weiss, S. 1992. A multipotent EGF-responsive striatal embryonic progenitor cell produces neurons and astrocytes. *Journal of Neuroscience*, 12, 4565–4574. https://doi.org/10.1523/JNEUROSCI.12-11-04565.1992
16. Phan, D. T. T., Bender, R. H. F., Andrejecsk, J. W., Sobrino, A., Hachey, S. J., George, S. C., and Hughes, C. C. W. 2017. Blood-brain barrier-on-a-chip: Microphysiological systems that capture the complexity of the blood-central nervous system interface. *Experimental Biology and Medicine (Maywood)*, 242, 1669–1678. https://doi.org/10.1177/1535370217694100
17. Qian, X., Song, H., and Ming, G. L. 2019. Brain organoids: Advances, applications, and challenges. *Development*, 146(8), dev166074. https://doi.org/10.1242/dev.166074
18. Lee, C. T., Bendriem, R. M., Wu, W. W., and Shen, R. F. 2017. 3D brain organoids derived from pluripotent stem cells: Promising experimental models for brain development and neurodegenerative disorders. *Journal of Biomedical Science*, 24(1), 59. https://doi.org/10.1186/s12929-017-0362-8
19. Hogberg, H. T., Bressler, J., Christian, K. M., Harris, G., Makri, G., O'Driscoll, C., et al. 2013. Toward a 3D model of human brain development for studying gene/environment interactions. *Stem Cell Research & Therapy*, 4(Suppl 1), S4. https://doi.org/10.1186/scrt365
20. Campion, D., Dumanchin, C., Hannequin, D., Dubois, B., Belliard, S., Puel, M., et al. 1999. Early-onset autosomal dominant Alzheimer disease: Prevalence, genetic heterogeneity, and mutation spectrum. *American Journal of Human Genetics*, 65(3), 664–670. https://doi.org/10.1086/302553
21. Blennow, K., de Leon, M. J., and Zetterberg, H. 2006. Alzheimer's disease. *Lancet*, 368(9533), 387–403. https://doi.org/10.1016/S0140-6736(06)69113-7
22. Clevers, H. 2016. Modeling development and disease with organoids. *Cell*, 165(7), 1586–1597. https://doi.org/10.1016/j.cell.2016.05.082
23. Qian, X., Nguyen, H. N., Song, M. M., Hadiono, C., Ogden, S. C., Hammack, C., and Ming, G. L. 2016. Brain-region-specific organoids using mini-bioreactors for modeling ZIKV exposure. *Cell*,165(5), 1238–1254. https://doi.org/10.1016/j.cell.2016.04.032
24. Huch, M., Gehart, H., van Boxtel, R., Hamer, K., Blokzijl, F., Verstegen, M. M., and Clevers, H. 2015. Long-term culture of genome-stable bipotent stem cells from adult human liver. *Cell*, 160(1–2), 299–312. https://doi.org/10.1016/j.cell.2014.11.050
25. Nantasanti, S., Spee, B., Kruitwagen, H. S., Chen, C., Geijsen, N., Oosterhoff, L. A., and Schotanus, B. A. 2015. Disease modeling and gene therapy of copper storage disease in canine hepatic organoids. *Stem Cell Reports*, 5(5), 895–907. https://doi.org/10.1016/j.stemcr.2015.09.002
26. Tang, X. Y., Wu, S., Wang, D., Chu, C., Hong, Y., Tao, M., and Liu, Y. 2022. Human organoids in basic research and clinical applications. *Signal Transduction and Targeted Therapy*, 7(1), 168. https://doi.org/10.1038/s41392-022-01024-9
27. Li, Y., Tang, P., Cai, S., Peng, J., Hua, G. 2020. Organoid-based personalized medicine: From bench to bedside. *Cell Regeneration*, 9, 21. https://doi.org/10.1186/s13619-020-00059-z
28. Dekkers, J. F., van der Ent, C. K., and Beekman, J. M. 2013. Novel opportunities for CFTR-targeting drug development using organoids. *Rare Diseases*, 1, e27112. https://doi.org/10.4161/rdis.27112
29. Dedhia, P. H., Bertaux-Skeirik, N., Zavros, Y., and Spence, J. R. 2016. Organoid models of human gastrointestinal development and disease. *Gastroenterology*, 150, 1098–1112. https://doi.org/10.1053/j.gastro.2015.12.042
30. Broutier, L., Mastrogiovanni, G., Verstegen, M. M., Francies, H. E., Gavarró, L. M., Bradshaw, C. R., Allen, G. E., et al. 2017. Human primary liver cancer-derived organoid cultures for disease modeling and drug screening. *Nature Medicine*, 23, 1424–1435. https://doi.org/10.1038/nm.4438
31. Vlachogiannis, G., Hedayat, S., Vatsiou, A., Jamin, Y., Fernández-Mateos, J., Khan, K., et al. 2018. Patient-derived organoids model treatment response of metastatic gastrointestinal cancers. *Science*, 359, 920–926. https://doi.org/10.1126/science.aao2774

32. Takasato, M., Er, PX., Chiu, HS., Maier, B., Baillie, GJ., Ferguson, C, et al. 2015. Kidney organoids from human iPS cells contain multiple lineages and model human nephrogenesis. *Nature*, 526, 564–568. https://doi.org/10.1038/nature15695
33. Skardal, A., Murphy, SV., Devarasetty, M., Mead, I., Kang, HW., Seol, YJ., et al. 2017. Multi-tissue interactions in an integrated three-tissue organ-on-a-chip platform. *Scientific Reports*, 7, 8837. https://doi.org/10.1038/s41598-017-08879-x
34. Huch, M., Boj, S. F., and Clevers, H. 2013. Lgr5(+) liver stem cells, hepatic organoids, and regenerative medicine. *Regenerative Medicine*, 8, 385–387. https://doi.org/10.2217/rme.13.39
35. Yui, S., Nakamura, T., Sato, T., Nemoto, Y., Mizutani, T., Zheng, X., et al. 2012. Functional engraftment of colon epithelium expanded in vitro from a single adult Lgr5+ stem cell. *Nature Medicine*, 18, 618–623. https://doi.org/10.1038/nm.2695
36. Geurts, M. H., de Poel, E., Amatngalim, GD., Oka, R., Meijers, FM., Kruisselbrink, E., et al. 2020. CRISPR-based adenine editors correct nonsense mutations in a cystic fibrosis organoid biobank. *Cell Stem Cell*, 26, 503–510.e507. https://doi.org/10.1016/j.stem.2020.01.019
37. Nie, J., and Hashino, E. 2017. Organoid technologies meet genome engineering. *EMBO Reports*, 18, 367–376. https://doi.org/10.15252/embr.201643732
38. Watson, C. L., Mahe, M. M., Múnera, J., Howell, J. C., Sundaram, N., Poling, H. M., et al. 2014. An in vivo model of human small intestine using pluripotent stem cells. *Nature Medicine*, 20, 1310–1314. https://doi.org/10.1038/nm.3737
39. Nie, Y. Z., Zheng, Y. W., Ogawa, M., Miyagi, E., Taniguchi, H. 2018. Human liver organoids generated with single donor-derived multiple cells rescue mice from acute liver failure. *Stem Cell Research & Therapy*, 9, 5. https://doi.org/10.1186/s13287-017-0749-1
40. Varzideh, F., Pahlavan, S., Ansari, H., Halvaei, M., Kostin, S., et al. 2019. Human cardiomyocytes undergo enhanced maturation in embryonic stem cell-derived organoid transplants. *Biomaterials*, 192, 537–550. https://doi.org/10.1016/j.biomaterials.2018.11.033
41. Chan, W. K., Fetit, R., Griffiths, R., Marshall, H., Mason, J. O., and Price, D. J. 2021. Using organoids to study human brain development and evolution. *Developmental Neurobiology*, 81(5), 608–622. https://doi.org/10.1002/dneu.22819
42. Eiraku, M., Watanabe, K., Matsuo-Takasaki, M., Kawada, M., Yonemura, S., Matsumura, M., Wataya, T., Nishiyama, A., Muguruma, K., and Sasai, Y. 2008. Self-organized formation of polarized cortical tissues from ESCs and its active manipulation by extrinsic signals. *Cell Stem Cell*, 3, 519–532.
43. Kadoshima, T., Sakaguchi, H., Nakano, T., Soen, M., Ando, S., Eiraku, M., and Sasai, Y. 2013. Self-organization of axial polarity, inside-out layer pattern, and species-specific progenitor dynamics in human ES cell-derived neocortex. *Proceedings of the National Academy of Sciences*, 110, 20284–20289.
44. Gonzalez, C., Armijo, E., Bravo-Alegria, J., Becerra-Calixto, A., Mays, C. E., and Soto, C. 2018. Modeling amyloid beta and tau pathology in human cerebral organoids. *Molecular Psychiatry*, 23(12), 2363–2374.
45 Yan, Y. W., Song, L. Q., Bejoy, J., Zhao, J., Kanekiyo, T., Bu, G. J., Zhou, Y., and Li, Y. 2018. Modeling neurodegenerative microenvironment using cortical organoids derived from human stem cells. *Tissue Engineering Part A*, 24, 1125–1137.
46. Kikuchi, S., Shinpo, K., Ogata, A., Tsuji, S., Takeuchi, M., Makita, Z., and Tashiro, K. 2002. Detection of N epsilon-(carboxymethyl)lysine (CML) and non-CML advanced glycation end-products in the anterior horn of amyotrophic lateral sclerosis spinal cord. *Amyotrophic Lateral Sclerosis and Other Motor Neuron Disorders*, 3, 63–68. https://doi.org/10.1080/146608202760196020
47. Singh, A., Kukreti, R., Saso, L., and Kukreti, S. 2019. Oxidative stress: A key modulator in neurodegenerative diseases. *Molecules*, 24(8), 1583. https://doi.org/10.3390/molecules24081583
48. Quadrato, G., and Arlotta, P. 2017. Present and future of modeling human brain development in 3D organoids. *Current Opinion in Cell Biology*, 49, 47–52. https://doi.org/10.1016/j.ceb.2017.11.010
49. Saorin, G., Caligiuri, I., and Rizzolio, F. 2023. Microfluidic organoids-on-a-chip: The future of human models. *Seminars in Cell & Developmental Biology*, 144, 41–54. https://doi.org/10.1016/j.semcdb.2022.10.001. PMID: 36241560.
50. Morales Pantoja, I. E., Smirnova, L., Muotri, A. R., Wahlin, K. J., Kahn, J., Boyd, J. L., and Hartung, T. 2023. First organoid intelligence (OI) workshop to form an OI community. *Frontiers in Artificial Intelligence*, 6, 1116870. https://doi.org/10.3389/frai.2023.1116870. PMID: 36925616; PMCID: PMC10013972.
51. Kim, S. H., and Chang, M. Y. 2023. Application of human brain organoids-opportunities and challenges in modeling human brain development and neurodevelopmental diseases. *International Journal of Molecular Sciences*, 24(15), 12528. https://doi.org/10.3390/ijms241512528

52. Sun, A. X., Yuan, Q., Fukuda, M., Yu, W., Yan, H., Lim, G. G. Y., et al. 2019. Potassium channel dysfunction in human neuronal models of Angelman syndrome. *Science (New York, N.Y.)*, 366(6472), 1486–1492. https://doi.org/10.1126/science.aav5386
53. Wulansari, N., Darsono, W. H. W., Woo, H. J., Chang, M. Y., Kim, J., Bae, E. J., Sun, W., Lee, J. H., Cho, I. J., Shin, H., Lee, S. J., and Lee, S. H. 2021. Neurodevelopmental defects and neurodegenerative phenotypes in human brain organoids carrying Parkinson's disease linked *DNAJC6* mutations. *Science Advances*, 7(8), eabb1540. https://doi.org/10.1126/sciadv.abb1540
54. Notaras, M., Lodhi, A., Dündar, F., Collier, P., Sayles, N. M., Tilgner, H., and Colak, D. 2022. Schizophrenia is defined by cell-specific neuropathology and multiple neurodevelopmental mechanisms in patient-derived cerebral organoids. *Molecular Psychiatry*, 27(3), 1416–1434.
55. Zagare, A., Barmpa, K., Smajic, S., Smits, L. M., Grzyb, K., Grünewald, A., Skupin, A., Nickels, S. L., and Schwamborn, J. C. 2022. Midbrain organoids mimic early embryonic neurodevelopment and recapitulate LRRK2-p. Gly2019Ser-associated gene expression. *American Journal of Human Genetics*, 109(2), 311–327. https://doi.org/10.1016/j.ajhg.2021.12.009
56. Kim, S. W., Woo, H. J., Kim, E. H., Kim, H. S., Suh, H. N., Kim, S. H., et al. 2021. Neural stem cells derived from human midbrain organoids as a stable source for treating Parkinson's disease: Midbrain organoid-NSCs (Og-NSC) as a stable source for PD treatment. *Progress in Neurobiology*, 204, 102086. https://doi.org/10.1016/j.pneurobio.2021.102086

13 Recent Advances in Therapeutics of Rare Neurological Disorders

Faiza Parvez, Atifa Haseeb Ansari, Durgesh Singh, and Sippy Singh
S.S. Khanna Girls' Degree College (A Constituent College of University of Allahabad), Prayagraj, India

13.1 INTRODUCTION

Rare neurological disorders encompass a diverse group of uncommon conditions that impact various components of the nervous system, including the brain, spinal cord, and peripheral nerves. A disorder is classified as rare if it affects less than five individuals out of every 10,000 people [1]. One of the defining features of these disorders is their limited prevalence within the general population, and they often manifest with intricate, distinctive symptoms [2]. These disorders are also referred to as orphan diseases, signifying their lesser-known status. Characterized by their heterogeneity and potential life-threatening nature, rare neurological disorders present a unique set of challenges owing to their scarce occurrence and the relative lack of research attention compared to more prevalent neurological conditions.

The intricacy of the central nervous system (CNS), encompassing the brain and spinal cord, introduces complexities in both diagnosis and treatment of rare neurological disorders [3]. Therapeutic interventions face hindrances due to several physiological barriers that closely regulate the exchange of molecules between the bloodstream and the brain tissue. These barriers are predominantly comprised of specialized glial and endothelial cells within the brain. The central nervous system is safeguarded by two primary barriers: the blood-brain barrier (BBB) and the blood-cerebrospinal fluid barrier (BCFSB). Among these, the blood-brain barrier plays a pivotal role, acting as the primary gateway for the passage of drugs and other therapeutic agents into the targeted regions of the brain [4]. This intricacy underscores the necessity for the ongoing development of innovative therapeutic strategies tailored specifically for addressing rare neurological disorders.

The inherent complexity of the brain's architecture further compounds the challenges in devising effective treatments for these disorders. The intricate interplay of neural networks, cellular diversity, and functional specialisation presents a formidable hurdle in the pursuit of therapeutic breakthroughs for rare neurological diseases [3]. Nevertheless, the ever-evolving landscape of neuroscientific research continuously unveils the complex structures and underlying molecular mechanisms of these disorders. This deeper understanding serves as a foundation for the development of novel therapeutics aimed at mitigating disease progression and alleviating symptoms. Noteworthy therapeutic avenues encompass gene therapy, stem cell therapy, immunotherapy, antisense therapy, and drug targeting via nanoparticles. In the realm of rare neurological disorders, ongoing research and innovative therapies are pivotal, offering promise to individuals grappling with these intricate conditions (**Table 13.1**).

13.2 THERAPEUTICS FOR RARE NEUROLOGICAL DISORDERS

The central nervous system is a sensitive and complex system that creates complications in the diagnosis and treatment of rare neurological disorders [3]. There is need of deep understanding of the causes and mechanisms, which is necessary to treat these disorders

DOI: 10.1201/9781032620442-13

TABLE 13.1
Some Rare Neurological Diseases and their Causes and Symptoms

S.No.	Diseases	Causes	Symptoms	References
1.	Amyotrophic lateral sclerosis (ALS)	Mutation in SOD1 and C9orf72 genes	Loss of motor function, weakness, paralysis, and respiratory failure	[5]
2.	Batten disease	Accumulation of lipopigments in lysosomes	Cognitive impairment, seizures, vision loss, and motor difficulties	[6]
3.	Huntington's disease (HD)	Mutations in Huntingtin (HTT) gene	Motor disturbances, cognitive decline, and psychiatric issues	[7]
4.	Guillain-Barré syndrome (GBS)	Aberrant immune response damages peripheral nerves	Viral or bacterial infection, weakness, numbness, and paralysis	[8]
5.	Rasmussen's encephalitis (RE)	Autoimmune T cells response to antigenic epitopes	Neurological deterioration, seizures, cognitive decline, and weakness	[9]
6.	Stiff person syndrome (SPS)	Autoimmune response against CNS	Muscle rigidity, sporadic muscle spasms, and chronic muscle pain	[10]
7.	Creutzfeldt-Jakob disease (CJD)	Accumulation of prion proteins in brain	Cognitive decline, memory problems, personality changes, muscle stiffness, and twitching	[11]
8.	Niemann-Pick disease	Genetic mutation in enzyme involved in lipid metabolism	Impaired motor skills, seizures, and cognitive decline, enlarged liver and spleen	[12]
9.	Rett syndrome	Genetic mutation in the MECP2 gene	Motor disturbance, repetitive hand movements, seizures, and breathing abnormalities	[13]
10.	Narcolepsy	Autoimmune response and dysfunction in brain regulation	Sleepiness, cataplexy, hallucinations, and sleep paralysis	[14]
11.	Friedreich's Ataxia	Mutation in frataxin gene leads to impaired mitochondrial function	Ataxia, muscle weakness, speech difficulties, heart problems, and impaired vision and hearing	[15]
12.	Frontotemporal dementias (FTD)	Single gene mutation in MAPT, GRN, and C9orf72 genes	Obsessive, and aggressive behaviors, loss of inhibitions and speech difficulties	[16]
13.	Gerstmann-Straussler disease (GSD)	Mutations in the prion protein gene (PRNP)	Memory loss, dementia, ataxia, and pathologic deposition of amyloid-like plaques in the brain	[17]
14.	Progressive supranuclear palsy (PSP)	Accumulation of tau protein in brain cells	Motor problems, balance difficulties, eye movement abnormalities, and cognitive changes	[18]
15.	Spinocerebellar ataxias	Gene mutation in ATXN, SCA, and CACNA1A genes	Tremors, muscle weakness, speech difficulties, and cerebellar and neurological degeneration	[19]

effectively [20]. Some of the therapeutics that may be used in treating neurological disorders are as follows:

13.2.1 Gene Therapy

Gene therapy can change genes by replacing the disease-causing gene mutation with a disease-free copy of the gene, eliminating the abnormal gene mutation, or inserting a new gene into the body to help fight diseases. Gene therapy can reverse genetic abnormalities in disease pathogenesis at the molecular level [21]. The CNS is bounded by the BBB and blood-cerebrospinal fluid. The cellular component is a layer of endothelial cells connected by tight junctions by pericytes and perivascular glial processes, all surrounded by a basal lamina to form a barrier [4].

To overcome these barriers, different gene delivery techniques have been employed for efficient gene therapy.

- *Amplification*: Enhancing the expression of the target gene is achieved through gene amplification. This technique proves valuable in cases where the incorporation of a missense mutation leads to a protein with impaired functionality, or when specific gene translations experience downregulation. Utilising gene amplification techniques enables the augmentation of protein levels. Research has indicated the application of this approach to elevate genes associated with neuroprotection and neuro-restoration, particularly in the context of conditions like ALS, Neiman-Pick disease, and Huntington's disease [22]. In specific neurological disorders, a reduction in neurotrophic levels, including brain-derived neurotrophic factor (BDNF), glial cell-derived neurotrophic factor (GDNF), and nerve growth factor (NGF), has been found. Addressing this issue, the possibility of restoring these proteins through gene amplification-based therapeutic approaches has been under investigation [23].
- *Silencing*: The technique of gene silencing holds the potential to diminish the synthesis of dysfunctional proteins or to decrease the presence of neurotoxic proteins responsible for provoking neuroinflammation. This method offers a prospective avenue for addressing neuropathological conditions arising from nonsense mutations, as it enables the post-transcriptional control of genes through the application of RNA interference (RNAi) technology [24]. Employing RNA interference (RNAi) to achieve gene silencing encompasses the utilisation of lentivirus-mediated reduction in glial fibrillary acidic protein (GFAP) and vimentin expression within astrocytes. Reactive astrocytes exhibit heightened levels of these two intermediate filament proteins, contributing to the formation of the glial scar that hampers axonal regeneration following injuries to the central nervous system. The application of RNAi-mediated knockdown targeting GFAP and vimentin resulted in decreased astrocyte reactivity and consequent reduction in glial scarring [25].
- *Editing*: Gene editing is a novel gene therapy approach involving engineered nucleases that possess sequence-specific DNA binding domains, coupled with a technique for non-specific DNA cleavage [26]. These nucleases precisely and efficiently alter DNA by initiating targeted double-strand breaks, thus prompting the cellular DNA repair mechanisms [27]. These endonucleases consist of zinc finger nucleases (ZFN), transcription activators like effector nucleases (TALEN), and clusters regulatory interspaced short palindromic repeats (CRISPR) [28]. ZFNs and TALENs result in the generation of double-strand breaks, prompting either inaccurate non-homologous end joining or precise homology-directed repair at the targeted genomic locus [29]. In Huntington's disease, the HTT gene mutation involves an elevated number of CAG repeats at the 5' end of gene. Zinc finger and TALEN nucleases offer a potential route for in vivo mutation correction or degradation of the mutated proteins [30].

13.2.2 Gene Delivery Techniques

Viruses are inherently structured to invade human cells using their genetic material, which facilitates their capacity to modify gene delivery systems. The main advantage of the viral vector is its unique and reliable properties for certain cell types. Different types of viral vectors that have been used in gene delivery systems are adenovirus, adeno-associated virus (AAV), lentivirus, herpes simplex virus (HSV), and various retroviruses.

- *Adenovirus (ADV)*: This virus is an unenveloped double-stranded DNA virus. However, the application of ADV is restricted due to its high virulence and the potential for toxicity within the central nervous system [31]. The modified helper-dependent ADV (HD-ADV) is used to remove all viral genes and therefore eliminate virulence [32]. Employing HD-ADV to manipulate DNA within human pluripotent stem cells offers an opportunity for gene therapy-oriented stem cell applications [33].

- *Adeno-associated virus (AVV)*: This virus is encapsulated by proteins and consists of single-stranded DNA. It is non-pathogenic and lacks the ability to replicate independently, thereby preventing significant immune responses or superinfections. Due to its favorable characteristics, it is the most frequently used viral vector [34]. It is feasible to modify this virus to contain minimal or zero viral genetic material, enabling the gene of interest to be transferred without the risk of causing any disease during transmission [34]. The AVV functions by silencing superoxide dismutase-1 (SOD1) to treat ALS or silencing the Huntingin (HTT) gene to treat Huntington's disease [35, 36].
- *Lentivirus*: This virus is an enveloped retrovirus containing protein-coated, single-stranded RNA. It could express a wide array of effector molecules [37]. This virus is harnessed for genetic material delivery and achieves genetic expression by integrating into the genetic makeup of the host. A novel variant called integrase-deficient lentivirus (IDLV) has been developed to avoid integration, thereby enhancing safety by lowering the occurrence of viral mutations and genome instability [38].
- *Herpes simplex virus*: This virus is an enveloped double-stranded DNA virus with a strong affinity for transfecting central nervous system neurons, particularly sensory neurons. Following infection, it maintains a dormant state as a stable nuclear episome within these neurons [39].

The use of viral vectors in delivering genes shows great promises in gene therapy but due to their toxicity and immunogenicity limits their widespread uses [40]. Non-viral vectors present an enhanced safety approach and have the potential to selectively target distinct populations of brain cells [41] which includes the use of including cationic polymers, cationic lipids, and engineered polypeptides. The cationic polymer effectively condenses nucleic acids into nanoparticles called 'polyplexes'. Coating negatively charged particles with cationic polymers like siRNA-thiolated gold nanoparticles has been shown to significantly improve the cytoplasmic delivery of siRNA. Cationic polymers may act as carriers of DNA or RNA transported via electrostatic interactions through ligands containing cell-targeted molecules attached to the functional groups of the polymer [42]. Cationic lipids are the most efficient non-viral vectors as they can carry large sizes of nucleic acids [43]. Several commercial combinations of cationic lipids are available, such as 1,2-dioleoyl-3-trimethylammonium propane (DOTAP), N-methyl-4- (dioleyl)methylpyridinium (SAINT-2), 3b- [N-(N0, N0 dimethylaminoethane)-carbamoyl] cholesterol (DC-Chol), or GS1, as well as the neutral lipid dioleoylphosphatidylethanolamine (DOPE) and cholesterol [44]. These positively charged lipids have demonstrated effective facilitation of nucleic acid delivery to neurons within the central nervous system (CNS) [45].

Engineered polypeptides possess specialized abilities for targeting distinct cell subpopulations, facilitating endosomal escape, and facilitating the nuclear entry of recombinant proteins. These engineered polypeptides have been designed to utilize recombinant fusion proteins based on the fragment C of the tetanus toxin for neuronal targeting, supplemented by a region that promotes endosomal escape, as well as DNA condensation enabled by the translocation site of the diphtheria toxin and the DNA-binding site of the GAL4 transcription factor [42].

Gene therapy shows promise for tackling conditions like ALS, Huntington's disease, Batten disease, and other rare neurological disorders caused by genetic defects. This avenue could potentially reverse disease progression. Advances in induced pluripotent stem cells, micro-RNA interventions, and CAR-T cell therapy are noteworthy. Despite progress, challenges remain including vector toxicity, immunogenicity, oncogenic risks, and translating findings from mice to humans.

13.2.3 Stem Cell Therapy

Stem cell therapy is widely used to treat neurological disorders. Stem cell therapy replaces and repairs damage and defective neurons of the brain. Hence, the objective of stem cell therapy is to facilitate the restoration of impaired or harmed nervous system cells, promote tissue repair, perform upkeep via paracrine influences, and conceivably foster the production of novel neurons. Different types of stem cells are used in stem cell therapy.

- *Neural stem cell (NSC)*: These are versatile precursor cells capable of generating neurotrophic elements and specialising into cells of both the central and peripheral nervous systems, encompassing neurons and supportive glial cells [46]. The embryonic brain harbors a wealth of neural stem cells, particularly during the initial phases of maturation. NSCs can release angiogenic factors to stimulate local tissue formation [47] and it has the potential to substitute impaired oligodendrocytes, facilitating the regeneration of myelin sheaths through the discharge of trophic factor cytokines [48]. These may also treat spinal cord injury by reducing inflammation by producing neurotrophic factors [49].
- *Mesenchymal stem cell (MSC)*: Sometimes referred to as mesenchymal stromal cells or therapeutic signaling cells, possess the ability to autonomously renew and transform into mesodermal cell types such as bone, adipose tissue, and cartilage cells [50]. Mesenchymal stem cells (MSCs) sourced from bone marrow are extensively employed for addressing uncommon neurological conditions like Huntington's disease and various other neurodegenerative ailments [51]. These MSCs deliver neuroprotection through their immunomodulatory and anti-inflammatory attributes [52].
- *Hematopoietic stem cells (HSCs)*: These are multipotent stem cells originating from bone marrow, peripheral blood, or umbilical cord blood (UCB). UCB has emerged as a valuable reservoir of HSCs for therapeutic interventions in neurological disorders [53]. Upon transplantation, these cells foster cell viability, stimulate the repair of impaired brain cells, mitigate inflammation, and facilitate the growth of new blood vessels (angiogenesis) [54]. HSCs provide immunity, reduce inflammation, and promote cognitive and motor function in neuron. HSCd also control the progression of diseases [55]. Embryonic stem cells (ESCs) play a pivotal role in addressing neurological disorders. These remarkable stem cells are harnessed for therapeutic applications. Upon implantation, ESCs possess the unique capability to integrate seamlessly into the striatum region, where they transform into tyrosine hydroxylase (TH) neurons, effectively reinstating motor functionality [56]. Beyond this, ESCs exhibit anti-inflammatory properties, fostering a reduction in inflammation, while simultaneously ensuring the preservation of ventral myelin segments within neurons [57].
- *Induced pluripotent stem cells (iPSCs)*: These cell possess a unique ability to generate distinct developmental cell strata under uniform circumstances, encompassing all three fundamental germ layers (endoderm, mesoderm, and ectoderm) [58]. These stem cells are produced by activating four transcription gene factors: OCT4 (octamer-4), SYR (sex determining region-Y, Box-2), KLF4 (Kruppel-like factor 4), and C-MYC (cellular myelocytomatosis oncogenes) [59]. iPSCs also function to reduce inflammation and restore or replace damaged neurons. Stem cell therapy is rapidly emerging as a beacon of hope in the realm of treating rare neurological disorders. This innovative approach holds the potential to revolutionize the management of conditions such as ALS, HD, Batten disease, and others. By harnessing the versatility of different types of stem cells, researchers are striving to unlock new avenues of healing for these complex disorders.

13.2.4 Immunotherapy

Immunotherapies stand out as potent tools for managing persistent diseases. The immune system has s natural capacity to repair damaged tissue and promote healing [60]. Microglia, specialized immune cells residing in the brain, play a vital role in this process. These cells are an important target of immunotherapy for neurological diseases [61]. Astrocytes, the most prevalent type of glial cells in the CNS, offer essential mechanical support to nerve cells in the brain. When activated, these astrocytes release substances likeinterleukin-6(IL-6) and tumor necrosis factor-alpha (TNFα), fostering a shielding environment [62]. Moreover, these cells actively scan the peripheral and central nervous system for immune cells, identifying regions of nerve damage and decline [63]. There are different types of immunotherapies:

- Checkpoint inhibitors play a crucial role in halting the activity of various checkpoint proteins that contribute to diseases. Two significant inhibitors are TNFα inhibitor and NOD-like receptor pyrin 3(NLRP3) inflammasome. TNFα inhibitors demonstrate their effectiveness by reducing inflammation in affected regions [64]. On the other hand, NLRP3 inflammasome, specifically MCC950 (a specific small molecule inhibitor of the NLRP3 inflammasome) obstructs the downstream generation of interleukin-1 beta (IL-1β), which regulates inflammatory reactions in the central nervous system (CNS). NLRP3 is also vital in detecting aggregated misfolded proteins in the brain, a factor that contributes to inflammation [65].
- Cytokines are a cluster of natural proteins within the body that aid in enhancing the immune system's effectiveness. Granulocyte macrophage colony-stimulating factor (GM-CSF), peroxisome proliferator-activated receptor gamma (PPAR-γ), and glucagon-like peptide 1 (GLP-1) are important cytokines. GM-CSF is an immunomodulatory growth factor that influences dendritic cells to manage the generation and growth of regulatory T cells. It also affects microglial proliferation and helps regulate inflammation [66]. GM-CSF hasanti-inflammatory, immunomodulatory, and neuroprotective properties. GLP-1 and PPAR-γ reduce neuroinflammation and improve motor functioning [67].
- Prasinezumab (PRX002) is a monoclonal antibody that targets the C-terminus, while BIIB054 targets the N-terminus of alpha-synuclein (α-syn) aggregates [68]. Gantenerumab and solaneuzumab are monoclonal antibodies designed to specifically target aggregated amyloid beta (Aβ) in the brain. These monoclonal antibodies reduce neuroinflammation and improve motor functioning [69]. A monoclonal antibody against phosphorylated tau effectively lessens the presence of insoluble or soluble tau variations in both the cortex and hindbrain [70]. Additionally, a monoclonal antibody engineered to target tau oligomers (known as TOMA) demonstrates notable efficacy in diminishing harmful tau oligomers, leading to improvements in memory deficits, cognitive decline, and a deceleration in disease progression [71].
- Vaccines provide active immunity against diseases. AADvacl and ACI-35 vaccines target tau oligomers and trigger anti-tau antibody production [72]. Mutant SOD1 and wt-SOD1 vaccinations suppress the expression of SOD1 gene, which cause disease [73]. AN-1792 vaccine, targeting amyloid β (Aβ40, Aβ42, and Aβ43), demonstrated favorable immunological outcomes and a decrease in disease-related pathology [74].

Immunotherapy is at the forefront of the fight against rare neurodegenerative disorders; its multipronged approach, spanning vaccines, gene-based interventions, and amyloid-targeted strategies, holds promise for shaping the future landscape of therapeutic interventions in these complex and challenging conditions.

13.2.5 Antisense Therapy

Antisense therapy relies on customized nuclear sequences, known as antisense oligonucleotides (ASO). These sequences are typically short, ranging from 12 to 25 nucleotides, and can be single-stranded DNA or RNA molecules, or double-stranded interfering RNA (iRNA) equipped with diverse chemical modifications. This innovative approach originated in 1978 when Zamecnik and

Stephenson first utilized it to therapeutically hinder gene expression [75]. Presently, antisense therapy is emerging as a promising contender for treating rare neurological disorders. These specialized molecules hold the potential to identify and access target neurons.

However, a challenge lies in their inability to efficiently cross the blood-brain barrier (BBB), necessitating modifications to enhance their efficacy. The spliceosome enzyme complex removes the intronic region from the pre-mRNA sequence. It can also remove exons at different rates; this is a process called alternative splicing, which results in more protein isoforms. This makes it possible to increase the protein encoded by the gene. Binding of ASO to the core of the pre-mRNA processing may disrupt normal splicing, induce intron protection in mature mRNA, or promote alternative splicing [76]. Some ASO scan attach to mRNA in a unique way. Unlike other ASOs, they do not break down the mRNA. When they stick to the mRNA, they prevent a specific enzyme complex in the cell from using the mRNA to build proteins. This process depends on the activity of enzymes, and it is designed to avoid overwhelming the normal enzyme functions of the cell. These special molecules can also be designed to target specific parts of mRNA that are involved in making proteins. This targeting helps the mRNA to be used more efficiently for creating proteins. Hence, these unique molecules have different ways of influencing how cells work to achieve better outcomes [77]. The coupling of ASO with mRNA generates a DNA-RNA pattern reminiscent of RNA H activity. This sequence resembles the functionality of the RNA H enzyme. As the target mRNA undergoes degradation, the ASO, characterized by its chemical stability, dissociates and becomes available to engage with another mRNA molecule. Presently, ASOs are harnessed to initiate the degradation of mRNA via a process that leverages the action of RNase H, aligning with the all-gap model. For instance, the drug Inotersen showcases a distinctive architectural configuration featuring a 5-10-5 cavity structure. This intricate arrangement entails a nucleus of ten base deoxynucleotides ensconced between five oligonucleotides spanning from the 5' to the 3' terminus. This unique structure effectively triggers the activation of RNase H, facilitating the targeted degradation of the intended mRNA substrate [78].

Small interfering RNA (siRNA) is designed to specifically target mature mRNA molecules. Within the RNA-induced silencing complex (RISC), the antisense strand of iRNA is loaded. This antisense segment aligns with the intended target mRNA, leading to its cleavage mediated by the argonaute 2 enzyme within the RISC assembly. The RISC complex exhibits remarkable versatility by degrading a diverse array of mRNA species, all the while preserving the iRNA antisense strands. This retained antisense strand retains its functionality and can engage in complementary base pairing with mRNA molecules for replication. The scope of iRNA encompasses distinct categories, including small hairpin RNA (shRNA) and microRNA (miRNA), each contributing to the orchestration of intricate gene regulatory processes [79]. Antisense therapy is a groundbreaking approach for treating rare neurological conditions like ALS, CJD, and HD. Advances in precise gene regulation have enabled this therapy's emergence, offering substantial benefits for brain diseases. By targeting genetic triggers, it goes beyond symptom relief and modifies disease progression. Its tailored nature suits the diverse genetics of rare conditions. Antisense therapy encompasses varied strategies, including gene expression modulation and exon skipping. This innovative approach has the potential to reshape care for rare neurological disorders.

13.2.6 Nanomaterials

Nanotechnology utilizes engineered nanomaterials and nanoparticles, ranging from 1 to 100 nanometers, to address neurological disorders. These tiny particles possess distinct properties due to their small size, enabling them to interact with the body's molecular systems and even pass through the blood-brain barrier [20]. Their size and structure allow these nanomolecules to cross the blood-brain barrier passively, making them effective for potential treatments. Over the past decade, diverse nanomaterials in various shapes have gained prominence in treating different neurological disorders. Scientists have achieved remarkable progress in the realm of nanomedicine, particularly in neuroscience. Metallic nanoparticles (NPs) have demonstrated promising outcomes in addressing neurological disorders. These nanoparticles can be customized in terms of their shape and size and can be combined with diverse compounds like drugs, antibodies, polypeptides, and polymers through various ligands [80]. Numerous metallic nanoparticle variations serve as effective carriers for targeted therapy aimed at the brain. These nanoparticles include FeONPs, AgNPs, gadolinium metallofullerene NPs, ultra small gadolinium oxide NPs, Ce2O3 NPs, ZnONPs, AuNPs, and PtNPs [81]. Metallic nanoparticles are also used for the imaging of CNS [82]. Metallic nanoparticles or nanomaterials can easily cross the blood brain barrier. These particles travel through the blood and reach regions in the brain and develop therapeutics potential for treating neurological disorders [83].

Dendrimers represent nanostructured materials featuring layered branching reminiscent of an onion skin arrangement. These nanoparticles grow in tiered formations, like the branching seen in globular proteins. Particularly intriguing is the outermost layer, which boasts numerous functional groups. These distinctive features render dendrimers well-suited for applications in medical science, where they serve as nanocarriers for a range of substances, including drugs, proteins, polypeptides, and amino acids [84]. Some examples of dendrimers areglycol, PEGylated, poly(propylene imine), poly(amidoamine), polyether-polyester, and pH dendrimers [85].

Recent investigations into dendrimer mechanisms underscore their role as drug carriers. These molecules form complexes with drugs, facilitating their traversal across the crucial blood-brain barrier for effective delivery. Dendrimers exhibit migration toward the parenchyma, with pronounced localisation in glial cells within brain injury sites [86]. Their distinctive structure and compact size position dendrimers as potent nanocarriers for central nervous system drugs, yielding widespread applications in this field [87]. Quantum dots (QDs) stand out as distinctive nanoparticles extensively employed in the treatment of various neurological conditions. These particles possess unique attributes such as sensitivity and selectivity when utilized at the nanoscale. Due to their compact size, QDs can penetrate the blood-brain barrier or access the brain via circulation. Notably, a concern arises regarding neurotoxicity, encompassing oxidative stress, elevated cytoplasmic Ca^{2+} levels, autophagy-induced neural cell damage, synaptic transmission disruption, and diminished plasticity [88].

QDs emit strong, long-lasting fluorescent light. They help visualize brain structures and mechanisms and can be used for drug delivery due to their special properties. Therefore, quantum dots are also used in diagnosis of neurological disease like Gullain-Barré syndrome, Rasmussen encephalitis, etc. [89]. Polymeric micelles are tiny structures composed of amphiphilic block copolymers. Their unique angel chain configuration makes them exceptionally adept as drug delivery systems. This is crucial because drug molecules face challenges in reaching the brain due to protective barriers like the blood-brain barrier and blood-cerebrospinal fluid barrier. Nanocarriers play a pivotal role in treating central nervous system disorders [90].

Polymeric micelles possess distinct traits, including their diminutive size and shell-like structure. These micelles are amphiphilic by nature, readily self-assembling in water. This property shields the drug from off-target cells. Nanocarriers utilize natural proteins derived from organic sources, like lactoferrin (LF) sourced from milk. LF has a unique affinity for brain surfaces and attaches to brain receptors, facilitating its passage through the blood-brain barrier [91]. Exosomes, derived from internal cell membranes, represent natural nanomaterials offering distinct advantages. They possess non-immunogenic properties and the capacity to transport various payloads. Created via human mesenchymal stem cells (MSCs) treated with interferon-gamma (IFN) [92], exosomes hold potential in delivering siRNA to specific cells like microglia, neurons, and oligodendrocytes, leading to targeted gene suppression. With remarkable blood-brain barrier penetration capabilities, exosomes find utility in treating rare neurological disorders such as ALS [93]. Nano-based approaches, involving nanoparticles and nanomaterials, offer an innovative strategy with promise for treating and diagnosing diverse neurological disorders. Various types of nanomaterials are extensively employed in both imaging and treating neurological conditions. Notably, nanoparticles have emerged as a potential means to surmount the blood-brain barrier and cerebrospinal fluid barrier, facilitating drug delivery to the brain and its intended effects.

13.3 CONCLUSION

The landscape of therapeutics for rare neurological disorders is constrained by their infrequent occurrence, unique symptomatology, and intricate molecular mechanisms. To advance therapeutic strategies for these conditions, a comprehensive grasp of the intricate structure of the brain is imperative. Ongoing investigations in neurosciences continually unveil the labyrinthine structure and molecular intricacies underlying these disorders, pivotal for devising effective interventions. The CNS presents a formidable challenge due to the formidable defense mechanisms posed by the blood-brain barrier and the blood-cerebrospinal fluid barrier. These barriers restrict the penetration of treatments, warranting innovative strategies for effective intervention.

Encouragingly, several therapeutic avenues have surfaced, exhibiting potential promise in the management of rare neurological disorders. These therapeutic modalities span a spectrum of approaches, encompassing gene therapy, stem cell therapy, immunotherapy, antisense therapy, and nanoparticle-based drug targeting. Recently neurodegenerative diseases are a risk factor, hence the primary concern should be their therapeutics. Synthetic drugs have harmful impact on body organs. Nutraceutical products, including herbs such as ginkgo, ashwagandha, turmeric, and honey, have better compatibility with minimal side effects. Secondary metabolites and flavonoids play a crucial role in neuroprotection [94]. They collectively exhibit considerable efficacy in not only arresting the progression of the diseases but also ameliorating the burden of symptoms. These methodologies demonstrate a remarkable potential to modulate the course of these disorders, offering a glimmer of hope to afflicted individuals and their families. Although the horizon of rare neurological disorders remains void of definitive cures, these treatments stand as beacons of progress, mitigating symptoms and impeding the inexorable advancement of the diseases. As research endeavors persist and technologies evolve, the pursuit of curative strategies for these enigmatic disorders continues, fueling optimism for transformative breakthroughs on the horizon.

ACKNOWLEDGMENTS

The Authors wish to thank the principal, S. S. Khanna Girls' Degree College (A Constituent College of University of Allahabad, Prayagraj) and the DST-CURIE Grant of Department of Science and Technology, New Delhi GOI (DST/CURIE-PG/2022/10 G), for providing all the necessary facilities in the college.

REFERENCES

1. Farnaes L et al. 2018. Rapid whole-genome sequencing decreases infant morbidity and cost of hospitalization. *NPJ Genomic Medicine*.4, 3–10. Doi: 10.1038/s41525-018-0049-4.
2. Giunti P, Morris S, Relja M, Pastores G, Quoidbach V. 2019. Toward earlier diagnosis and treatment of rare neurological disorders: the value of coordinated care and specialist centers. *Croatian Medical Journal*. 60(2), 156–157. Doi: 10.3325/cmj.2019.60.156.
3. Ramos-Cabrer P, Agulla J, Argibay B, Pérez-Mato M, Castillo J. 2011. Serial MRI study of the enhanced therapeutic effects of liposome-encapsulated citicoline in cerebral ischemia. *International Journal of Pharmaceutics*. 405(1-2), 228–233. Doi: 10.1016/j.ijpharm.2010.12.014.
4. Abbott NJ, Patabendige AA, Dolman DE, Yusof SR, Begley D. 2010. J. Structure and function of the blood–brain barrier. *Neurobiology of Disease*. 37(1), 13–25. Doi: 10.1016/j.nbd.2009.07.030.
5. Brown RH, Al-Chalabi A. 2017. Amyotrophic lateral sclerosis. *The New England Journal of Medicine*. 377, 162–172. Doi: 10.1056/NEJMra1603471.
6. Johnson TB, Cain JT, White KA, Ramirez-Montealegre D, Pearce DA, Weimer JM. 2019. Therapeutic landscape for batten disease: current treatments and prospects. *Nature Reviews Neurology*.15(3), 161–178. Doi: 10.1038/s41582-019-0138-8.
7. Wheelock VL, Tempkin T, Marder K, Nance M, Myers RH, Zhao H, Kayson E, Orme C, Shoulson I. 2003. Huntington Study group: predictors of nursing home placement in Huntington disease. *Neurology*. 60(6), 998–1001. Doi: 10.1212/01.wnl.0000052992.58107.67.

8. Willison HJ, Jacobs BC, van Doorn PA. 2016. Guillain-Barré syndrome. *Lancet Neurology*. 388(10045), 717–727. Doi: 10.1016/S0140-6736(16)00339-1.
9. Bien CG, Kruse CA, Jensen FE, Bauer J, Pardo CA, Vincent A, Mathern GW, Cross JH. 2014. Rasmussen's encephalitis: clinical features, pathobiology, and treatment advances. *Lancet Neurology*. 13(2), 195–205. Doi: 10.1016/S1474-4422(13)70260-6.
10. Stayer C, Meinck HM. 1998. Stiff-man syndrome: an overview. *Neurologia*. 83–8. Doi: 10.1212/wnl.49.6.1591
11. Sitammagari KK, Masood W. 2023. Creutzfeldt Jakob Disease. *StatPearls Publishing*. 234–240. Doi: 10.53347/rid-163058.
12. Bajwa H, Azhar W. 2023. Niemann-pick disease. *StatPearls. Treasure Island (FL): StatPearls Publishing*. PMID: 32310589 Doi: 10.7759/cureus.40822.
13. Downs J, Leonard H, Wong K, Newton N, Hill K. 2017. Quantification of walking-based physical activity, and sedentary time in individuals with rett syndrome. *Developmental Medicine & Child Neurology*. 59(6), 605–11. Doi: 10.1111/dmcn.13398.
14. Bornstein A, Hedström A, Wasling P. 2021. Actigraphy measurement of physical activity and energy expenditure in narcolepsy type 1, narcolepsy type 2 and idiopathic hypersomnia: a sensewear armband study. *The Journal of Sleep Research*. 30(2):e13038. Doi: 10.1111/jsr.13038
15. Delatycki MB, Williamson R, Forrest SM, Friedreich. 2000. Ataxia: an overview. *Journal of Medical Genetics*. 37(1), 1–8. Doi: 10.1136/jmg.37.1.1.
16. Bang J, Spina S, Miller BL. 2015. Frontotemporal dementia. *Lancet (London, England)*. 386(10004), 1672–1682. Doi: 10.1016/S0140-6736(15)00461-4.
17. Liberski PP. 2012. Gerstmann-Sträussler Scheinker Disease. In: Ahmad, S.I. (eds) neurodegenerative diseases. Advances in Experimental Medicine and Biology, 724. *Springer, New York, NY*. Doi:10.1007/978-1-4614-0653-2_10.
18. Taghdiri F, Sato C, Ghani M, Moreno D, Rogaeva E, Tartaglia MC. 2016. Novel GRN mutations in patients with corticobasal syndrome. *Scientific Reports*. 6, 22913. Doi: 10.1038/srep22913.
19. Paulson HL. 2009. The spinocerebellar ataxias. *Journal of Neuro-Ophthalmology*. 29(3), 227–237. Doi: 10.1097/WNO0b013e3181b416de.
20. Huang M, Gu X, Gao X. 2019. Nanotherapeutic strategies for the treatment of neurodegenerative diseases. In Brain Targeted Drug Delivery System; *Elsevier: Amsterdam, The Netherlands*. 321–356. Doi: 10.1016/B978-0-12-814001-7.00013-5
21. Goncalves GAR, Paiva RMA. 2017. Gene Therapy: advances, challenges, and perspectives. *Einstein (Sao Paulo, Brazil)*. 15(3), 369–375. Doi: 10.1590/S1679-45082017RB4024.
22. Gray SJ. 2013. Gene Therapy and neurodevelopmental disorders. *Neuropharmacology*. 68, 136–142. Doi: 10.1016/j.neuropharm.2012.06.024.
23. Wang B, Yuan J, Xu J, et al. 2017. Brain-derived and glial cell line-derived neurotrophic factor fusion protein immobilization to laminin. *Experimental and Therapeutic Medicine*. 13(1), 178–186. Doi: 10.3892/etm.2016.3925.
24. Valdmanis PN, Kay MA. 2017. Future of rav Gene therapy: platform for RNAi, Gene editing, and beyond. *Human Gene Therapy*. 28(4), 361–372. Doi: 10.1089/hum.2016.171.
25. Desclaux M, Teigell M, Amar L, et al. 2009. A novel and efficient gene transfer strategy reduces glial reactivity and improves neuronal survival and axonal growth in vitro. *PLoS One*. 4(7),6227. Doi: 10.1371/journal.pone.0006227.
26. Joung JK, Sander JD. 2013. TALENs: a widely applicable technology for targeted genome editing. *Nature Reviews Molecular Cell Biology*. 49–55. Doi: 10.1038/nrm3486.
27. Memi F, Ntokou A, Papangeli I. 2018. CRISPR/Cas9 gene-editing: research technologies, clinical applications, and ethical considerations. *Seminars in Perinatology*. 42(8), 487–500. Doi: 10.1053/j.semperi.2018.09.003.
28. Gaj T, Gersbach CA, Barbas CF. 2013. ZFN, TALEN, and CRISPR/Cas-based methods for genome engineering. *Trends in Biotechnology*. 31(7), 397–405. Doi: 10.1016/j.tibtech.2013.04.004.
29. Chandrasegaran S, Carroll D. 2016. Origins of programmable nucleases for genome engineering. *Journal of Molecular Biology*. 428(5 Pt B), 963–989. Doi: 10.1016/j.jmb.2015.10.014.
30. Aronin N, DiFiglia M. 2014. Huntingtin-lowering strategies in Huntington's disease: antisense oligonucleotides, small RNAs, and gene editing. *Movement Disorders*. 29(11), 1455–1461. Doi: 10.1002/mds.26020.
31. Germano IM, Fable J, Humayun Gultekin S, et al. 2003. Adenovirus/herpes simplex-thymidine kinase/ganciclovirud complex: preliminary results of a phase i trial in patients with recurrent malignant gliomas. *Journal of Neuro-Oncology*. 65(3), 279–289. Doi:10.1023/b:neon.0000003657.95085.56.

32. Thomas CE, Ehrhardt A, Kay MA. 2003. Progress and problems with the use of viral vectors for gene therapy. *Nature Reviews Genetics*. 4(5), 346–358. Doi: 10.1038/nrg1066.
33. Mitani K. 2014. Gene Targeting in human-induced pluripotent stem cells with adenoviral vectors. *Methods in Molecular Biology*. 1114, 163–167. Doi: 10.1007/978-1-62703-761-7_10.
34. Chen X, Goncalves MAFV. 2016. Engineered viruses as genome editing devices. *Molecular Therapy*. 24(3), 447–457. Doi: 10.1038/mt.2015.164.
35. Saraiva J, Nobre RJ, Pereira de Almeida L. 2016. Gene Therapy for the CNS using AAVs: the impact of systemic delivery by AAV9. *Journal of Controlled Release*. 241, 94–109. Doi: 10.1016/j.jconrel.2016.09.011.
36. Thomsen GM, Gowing G, Latter J, et al. 2014. Delayed disease onset and extended survival in the SOD1G93A rat model of amyotrophic lateral sclerosis after suppression of mutant SOD1 in the motor cortex. *Journal of Neuroscience*. 34(47), 15587–1560. Doi:10.1523/JNEUROSCI.2037-14.2014.
37. Matrai J, Chuah MK, Vanden Driessche T. 2010. Recent advances in lentiviral vector development and applications. *Molecular Therapy*. 18(3), 477–490. Doi: 10.1038/mt.2009.319.
38. Sengupta R, et al. 2016. An optimized protocol for packaging pseudotyped integrase defective lentivirus. *Biological Procedures Online*. 18, 4–14. Doi:10.1186/s12575-016-0044-z.
39. Lentz TB, Gray SJ, Samulski RJ. 2012. Viral vectors for gene delivery to the central nervous system. *Neurobiology of Disease*. 48(2), 179–188. Doi: 10.1016/j.nbd.2011.09.014.
40. Chira S, Jackson CS, Oprea I, et al. 2015. Progresses towards safe and efficient gene therapy vectors. *Oncotarget*. 6(31), 30675–30703. Doi: 10.18632/oncotarget.5169.
41. Guo X, Huang L. 2012. Recent advances in nonviral vectors for gene delivery. *Accounts of Chemical Research*. 45(7), 971–979. Doi: 10.1021/ar200151m.
42. Jayant RD, Sosa D, Kaushik A, et al. 2016. Status of non- viral gene therapy for CNS disorders. *Expert Opinion on Drug Delivery*. 13(10), 1433–1445. Doi: 10.1080/17425247.2016.1188802.
43. Zhao Y, Huang L. 2014. Lipid nanoparticles for gene delivery. *Advanced Genetics*. 88, 13–36. Doi:10.1016/B978-0-12-800148-6.00002-X.
44. Bergen JM, Park I-K, Horner PJ, et al. 2008. Nonviral approaches for neuronal delivery of nucleic acids. *Pharmaceutical Research*. 25(5), 983–998. Doi: 10.1007/s11095-007-9439-5.
45. Suk JS, Xu Q, Kim N, et al. 2016. PEGylation as a strategy for improving nanoparticle-based drug and gene delivery. *Advanced Drug Delivery Reviews*. 99(Pt A), 28–51. Doi: 10.1016/j.addr.2015.09.012.
46. Christian KM, Song H, Ming GL. 2014. Functions, and dysfunctions of adult hippocampal neurogenesis. *Annual Review of Neuroscience*. 37, 243–262. Doi: 10.1146/annurev-neuro-071013-014134.
47. Ryu S, Lee SH, Kim SU, Yoon BW. 2016. Human neural stem cells promote proliferation of endogenous neural stem cells and enhance angiogenesis in the ischemic rat brain. *Neural Regeneration Research*. 11(2), 298–304. Doi: 10.4103/1673-5374.177739.
48. Pluchino S, Quattrini A, Brambilla E, Gritti A, Salani G, Dina G, Galli R, Del Carro U, et al. 2003. Injection of adult neurospheres induces recovery in a chronic model of multiple sclerosis. *Nature*. 422(6933), 688–694. Doi:10.1038/nature01552.
49. Bottai D, Madaschi L, Di Giulio AM, Gorio A. 2008. Viability-dependent promoting action of adult neural precursors in spinal cord injury. *Molecular Medicine*. 14(9-10), 634–644. Doi: 10.2119/2008-00077. Bottai.
50. Caplan AI. 2017. Mesenchymal stem cells: time to change the name! Stem cells. *Translational Medicine*. 6(6),1445–1451. Doi: 10.1002/sctm.17-0051.
51. Karussis D, Karageorgiou C, Vaknin-Dembinsky A, et al. 2010. Safety and immunological effects of mesenchymal stem cell transplantation in patients with multiple sclerosis and amyotrophic lateral sclerosis. *Archives of Neurology*. 67(10), 1187–1194. Doi:10.1001/archneurol.2010.248.
52. Dahbour S, Jamali F, Alhattab D, et al. 2017. Mesenchymal stem cells and conditioned media in the treatment of multiple sclerosis patients: clinical, ophthalmological, and radiological assessments of safety and efficacy. *CNS Neuroscience & Therapeutics*. 23(11), 866–874. Doi:10.1111/cns.12759.
53. Dessels C, Alessandrini M, Pepper MS. 2018. Factors influencing the umbilical cord blood stem cell industry: an evolving treatment landscape. *STEM CELLS Translational Medicine*. 7(9), 643–650. Doi: 10.1002/sctm.17-0244.
54. McDonald CA, Fahey MC, Jenkin G, Miller SL. 2018. Umbilical cord blood cells for treatment of cerebral palsy; Timing and treatment options. *Pediatric Research*. 83(1-2), 333–344. Doi: 10.1038/pr.2017.236.
55. Chez M, Lepage C, Parise C, Dang-Chu A, Hankins A, Carroll M. 2018. Safety, and observations from a placebo-controlled, crossover study to assess use of autologous umbilical cord blood stem cells to improve symptoms in children with autism. *STEM CELLS Translational Medicine*. 7(4), 333–341. Doi: 10.1002/sctm.17-0042.

56. Liu S, Qu Y, Stewart TJ, Howard MJ, Chakraborty S, Holekamp TF, McDonald JW. 2000. Embryonic stem cells differentiate into oligodendrocytes and myelinate in culture and after spinal cord transplantation. *The Proceedings of the National Academy of Sciences*. 97(11), 6126–6131. Doi: 10.1073/pnas.97.11.6126.
57. Bottai D, Cigognini D, Madaschi L, Adami R, Nicora E, et al. 2010. Embryonic stem cells promote motor recovery and affect inflammatory cell in filtration in spinal cord injured mice. *Experimental Neurology*. 223(2),452–463. Doi: 10.1016/j.expneurol.2010.01.010.
58. Abad M, Mosteiro L, Pantoja C, Cañamero M, Rayon T, Ors I, Graña O, et al. 2013. Reprogramming in vivo produces teratomas and IPS cells with totipotency features. *Nature*. 502(7471), 340–345. doi:10.1038/nature12586.
59. Takahashi K, Tanabe K, Ohnuki M, Narita M, Ichisaka T, Tomoda K, Yamanaka S. 2007. Induction of pluripotent stem cells from adult human fibroblasts by defined factors. *Cell*. 131(5), 861–872. Doi: 10.1016/j.cell.2007.11.019.
60. Schwartz M, Kipnis J. 2005. Protective autoimmunity and neuroprotection in inflammatory and noninflammatory neurodegenerative diseases. *Journal of the Neurological Sciences*. 233(1–2), 163–166. Doi: 10.1016/j.jns.2005.03.014.
61. Klegeris PLMG. 2005. non-steroidal anti-inflammatory drugs (NSAIDs) and other anti-inflammatory agents in the treatment of neurodegenerative disease. *Current Alzheimer Research*. 2(3), 355–365. Doi: 10.2174/1567205054367883.
62. Mattotti M, Alvarez Z, Ortega JA, Planell JA, Engel E, Alcántara S. 2012. Inducing functional radial glia-like progenitors from cortical astrocyte cultures using micropatterned PMMA. *Biomaterials*. 33(6), 1759–1770. Doi: 10.1016/j.biomaterials.2011.10.086.
63. Maragakis NJ, Rothstein JD. 2006. Mechanisms of disease: astrocytes in neurodegenerative disease. *Nature Reviews Neurology*. 2(12), 679–689. Doi: Doi:10.1038/ncpneuro0355.
64. Barnum CJ, Chen X, Chung J, Chang J, Williams M, Grigoryan N, et al. 2014. Peripheral administration of the selective inhibitor of soluble tumor necrosis factor (TNF) XProR 1595 attenuates nigral cell loss and glial activation in 6-OHDA hemiparkinsonian rats. *The Journal of Parkinson's Disease*. 4(3), 349–360. Doi:10.3233/JPD-140410.
65. Coll RC, Robertson AA, Chae JJ, Higgins SC, Muñoz-Planillo R, Inserra MC, et al. 2015. A small-molecule inhibitor of the NLRP3 inflammasome for the treatment of inflammatory diseases. *Nature Medicine*. 21(3), 248–255. Doi:10.1038/nm.3806.
66. Kiyota T, Machhi J, Lu Y, Dyavarshetty B, Nemati M, Yokoyama I, et al. 2018. Granulocyte-macrophage colony-stimulating factor neuroprotective activities in Alzheimer's disease mice. *Journal of Neuroimmunology*. 319, 80–92. Doi: 10.1016/j.jneuroim.2018.03.009.
67. Bernardo A, Minghetti L. 2006. PPAR-gamma agonists as regulators of microglial activation and brain inflammation. *Current Pharmaceutical Design*. 12(1), 93–109. Doi: 10.2174/138161206780574579.
68. Jankovic J, Goodman I, Safirstein B, Marmon TK, Schenk DB, Koller M, et al. 2018. Safety, and tolerability of multiple ascending doses of PRX002/RG7935, an anti–α-synuclein monoclonal antibody, in patients with Parkinson disease: a randomized clinical trial. *JAMA Neurol*. 75(10), 1206–1214. Doi:10.1001/jamaneurol.2018.1487.
69. Klein G, Delmar P, Voyle N, Rehal S, Hofmann C, Abi-Saab D, et al. 2019. Gantenerumab reduces amyloid-β plaques in patients with prodromal to moderate Alzheimer's disease: a PET substudy interim analysis. *Alzheimer's Research & Therapy*. 11(1), 101. Doi:10.1186/s13195-019-0559-z
70. D'Abramo C, Acker CM, Jimenez H, Davies P. 2015. Passive immunization in JNPL3 transgenic mice using an array of phospho-tau specific antibodies. *PLoS ONE*. 10(8), e0135774. Doi: 10.1371/journal.pone.0135774.
71. Gerson JE, Farmer KM, Henson N, Castillo-Carranza DL, Murillo MC, Sengupta U, et al. 2018. Tau oligomers mediate α-synuclein toxicity and can be targeted by immunotherapy. *Molecular Neurodegeneration*. 13(1), 13. Doi:10.1186/s13024-018-0245-9.
72. Theunis C, Crespo-Biel N, Gafner V, Pihlgren M, López-Deber MP, Reis P, et al. 2013. Efficacy and safety of a liposome-based vaccine against protein tau, assessed in tau. P301L mice that model tauopathy. *PLoS ONE*. 8(8), e72301. Doi: 10.1371/journal.pone.0072301.
73. Takeuchi S, Fujiwara N, Ido A, Oono M, Takeuchi Y, Tateno M, et al. 2010. Induction of protective immunity by vaccination with wild type aposuperoxide dismutase 1 in mutant SOD1 transgenic mice. *Journal of Neuropathology & Experimental Neurology*. 69(10), 1044–1056. Doi:10.1097/NEN.0b013e3181f4a90a.
74. Jäkel L, Boche D, Nicoll JA, Verbeek MM. 2019. Aβ43 in human Alzheimer's disease: effects of active Aβ42 immunization. *Acta Neuropathologica Communications*. 7(1), 141. Doi 10.1186/s40478-019-0791-6.

75. Zamecnik PC, Stephenson ML. 1978. Inhibition of *Rous sarcoma* virus replication and cell transformation by a specific oligodeoxynucleotide. *The Proceedings of the National Academy of Sciences.* 75(1), 280–284. Doi: 10.1073/pnas.75.1.280.
76. Lee Y, Rio DC. 2015. Mechanisms and regulation of alternative pre mRNA splicing. *Annual Review of Biochemistry. 84*, 291–323. Doi: 10.1146/annurev-biochem-060614-034316.
77. Good L. 2003. Translation repression by antisense sequences. *Cellular and Molecular Life Sciences.* 60(5), 854–861. Doi: 10.1007/s00018-003-3045-4.
78. Liang X-H, Shen W, Sun H, Migawa MT, Vickers TA, Crooke ST. 2016. Translation efficiency of mRNAs is increased by antisense oligonucleotides targeting upstream open reading frames. *Nature Biotechnology.* 34(8), 875–880. Doi: 10.1038/nbt.3589.
79. Treiber T, Treiber N, Meister G. 2019. Regulation of microRNA biogenesis and its crosstalk with other cellular pathways. *Nature Reviews Molecular Cell Biology.* 20(1), 5–20. Doi: 10.1038/s41580-018-0059-1.
80. Dong J, Tao L, Abourehab MA, Hussain Z. 2018. Design, and development of novel hyaluronate-modified nanoparticles for combo-delivery of curcumin and alendronate: fabrication, characterization, and cellular and molecular evidence of enhanced bone regeneration. *International Journal of Biological Sciences.* 116, 1268–1281. Doi: 10.1016/j.ijbiomac.2018.05.116.
81. Waris A, Din M, Ali A, Afridi S, Baset A, Khan AU, Ali M. 2021. Green fabrication of co and co3O4 nanoparticles and their biomedical applications: a review. *Open Life Sciences.* 16(1), 14–30. Doi: 10.1515/biol-2021-0003.
82. Mody VV, Siwale R, Singh A, Mody HR. 2010. Introduction to metallic nanoparticles. *Journal of Pharmacy and Bioallied Sciences.* 2(4), 282–289. Doi: 10.4103/0975-7406.72127.
83. Ayaz M, Ovais M, Ahmad I, Sadiq A, Khalil AT, Ullah F. 2020. Biosynthesized metal nanoparticles as potential Alzheimer's Disease therapeutics. In Metal Nanoparticles for Drug Delivery and Diagnostic Applications; *Elsevier: Amsterdam, The Netherlands.* 31–42. Doi:10.1016/B978-0-12-816960-5.00003-3.
84. Boas U, Heegaard PM. 2004. Dendrimers in drug research. *Chemical Society Reviews.* 33(1), 43–63. Doi: 10.1039/B309043B.
85. Beg S, Samad A, Alam MI, Nazish I. 2011. Dendrimers as novel systems for delivery of neuropharmaceuticals to the brain. *CNS & Neurological Disorders – Drug Targets.* 10(5), 576–588. Doi: 10.2174/187152711796235023.
86. Nance E, Zhang F, Mishra MK, Zhang Z, Kambhampati SP, Kannan RM, Kannan S. 2016. Nanoscale effects in dendrimer mediated targeting of neuroinflammation. *Biomaterials.* 101, 96–107. Doi: 10.1016/j.biomaterials.2016.05.044.
87. Leiro V, Duque Santos S, Lopes CD, Paula Pêgo A. 2018. Dendrimers as powerful building blocks in central nervous system disease: headed for successful nanomedicine. *Adv. Funct. Mater.* 28(12), 1700313. Doi: 10.1002/adfm.201700313.
88. Wu T, Zhang T, Chen Y, Tang M. 2016. Research advances on potential neurotoxicity of quantum dots. *Journal of Applied Toxicology.* 36(3), 345–351. Doi: 10.1002/jat.3229.
89. Aswathy RG, Yoshida Y, Maekawa T, Kumar DS. 2010. Near-infrared quantum dots for deep tissue imaging. *Analytical and Bioanalytical Chemistry.* 397(4), 1417–1435. Doi: 10.1007/s00216-010-3643-6.
90. Monteiro AFM, Viana JDO, Nayarisseri A, Zondegoumba EN, Mendonça Junior FJB, Scotti MT, Scotti L. 2018. Computational studies applied to flavonoids against Alzheimer's and Parkinson's diseases. *Oxidative Medicine and Cellular Longevity.* 2018, 7912765. Doi: 10.1155/2018/7912765.
91. Patel MM, Patel BM. 2017. Crossing the blood–brain barrier: recent advances in drug delivery to the brain. *CNS Drugs.* 31(2), 109–133. Doi: 10.1007/s40263-016-0405-9.
92. Haney MJ, Klyachko NL, Zhao Y, Gupta R, Plotnikova EG, He Z, et al. 2015. Exosomes as drug delivery vehicles for Parkinson's disease therapy. *Journal of Controlled Release.* 207, 18–30. Doi: 10.1016/j.jconrel.2015.03.033.
93. Wang G, Rayner S, Chung R, Shi B, Liang X. 2020. Advances in nanotechnology-based strategies for the treatments of amyotrophic lateral sclerosis. *Materials Today Bio.* 6, 100055. Doi: 10.1016/j.mtbio.2020.100055.
94. Ansari AH, Singh D, Singh S. 2024. Effect of plant-based diets on the brain. *Nutraceutical Fruits and Foods for Neurodegenerative Disorders.* 249–260. Doi: 10.1016/B978-0-443-18951-7.00014-1.

14 Bioinformatics-Based Modeling in Different Neurodegenerative Diseases and Associated Therapeutic Strategies

Atifa Haseeb Ansari, Swayam Prabha Rai, Shivani Pathak, and Sippy Singh
Department of Zoology, S.S. Khanna Girls' Degree College
(A Constituent College of University of Allahabad), Prayagraj, India

14.1 INTRODUCTION

The nervous system, including brain and spinal cord, consists of neuronal structures that are involved in electrical impulse transmission. Neurodegenenerative diseases are becoming a major threat as they are increasingly prevalent and affecting billions of people all over the world. There are 18 major disorders. These diseases are associated with the functionally impaired nerve cells within the brain. Examples of neurodegenerative diseases are Alzheimer's disease (AD), Parkinson's disease (PD), Huntington's disease (HD), amyotrophic lateral sclerosis (ALS), dementia, and ataxia. Neurodegeneration can be of either hereditary or sporadic type. In neurodegenerative disorders, nerve cells die, and this leads to incurable disease. It has been reported that 400,000 people have sclerosis and 30,000 are affected by Huntington's disease. Impairment and loss of nerve cells with respect to malfunctioning in structural and functional is termed as neurodegeneration; these conditions are recognised by progressive loss of neurons, which may be caused by sudden injury or is sometimes inherited in nature [1]. Disease can lead to neuronal and mental disabilities as it affects both the central as well as peripheral nervous system. Oxidative stress is one of the major causes of neuronal degeneration [2].

- *Alzheimer's disease*: This is a dementia, which is linked with mental disabilities associated with neuronal dysfunction in the hippocampus region of the brain. The main symptom of this disorder involves memory loss. It is characterised by deposition of beta-amyloid granules and microglial activation.
- *Parkinson's disease*: It is associated with dopamine, acetylcholine, and nor-epinephrine [3]. This is the second most common type of dementia which involves mitochondrial dysfunction [4]. The main salient feature of this disorder is loss of dopaminergic neurons.
- *Huntington's disease*: This disease is also referred as Huntington's chorea, which is hereditary in nature and may result in severe conditions like brain cell death. Mutation is also one reason for Huntington's disease. It is characterised by jerky body movements.
- *Schizophrenia*: This neurodegenerative disease is a common type of mental disorder and is identified by symptoms such as intellectual disabilities and alteration in perception of reality.

DOI: 10.1201/9781032620442-14

TABLE 14.1
Different Types of Neurodegenerative Diseases, their Mechanism of Action, Symptoms, Effects, Diagnostic Markers, and Associated Herbs

Diseases	Mechanisms of Action	Symptoms	Effects	Diagnostic Markers	Herbs
Alzheimer's Disease	Accumulation of amyloid granules	Memory abnormalities, neuronal death, seizure, allergic dermatitis, headache, dizziness, and stress	Behavioral variations	Presinilin-1, presinilin-2, CSF	Beta-carotene, lavandula, flavonoids, carotenoids
Huntington's Disease	Release of dopamine at higher level, mutated gamma-aminobutyric acid (GABA)	Psychological disorders, symptoms, and stress	Jerky-body movement	HTT, cytokines, glutathione	Turmeric, ginseng
Parkinson's Disease	Abnormalities in dopaminergic neurons	Anxiety, confusion, neuro-inflammation, and ataxia	Deficiency in dopaminergic activity	PARKIN, NR4A2, loss of dopamine transporter	Soyabean, brahmi, vitamin balanced diet

- *Amyotrophic lateral sclerosis*: It is a type of disability that is associated with the cerebral cortex and motor neurons and is characterised by twitching muscles.
- *Stroke*: It is a condition that involves neuronal loss and is also referred to as a brain attack. The main cause is hypertension [5] and interrupted blood supply (**Table 14.1**). **Table 14.1** shows the different types of neurodegenerative diseases, their mechanisms of action, symptoms, effects, diagnostic markers, and associated herbs.

14.2 BIOINFORMATICS TOOLS ACT AS EMERGING THERAPEUTICS IN NEURODEGENERATIVE DISEASES

Bioinformatics plays vital role in many biological areas. The scope and importance of bioinformatics tools in drug development is emerging and the field is also useful in the identification of targets, biomarkers, and potential therapeutics for treatment of NDDs; it will evolve soon. Recently bioinformatics models have been emerging for genotyping data in blood cell samples and for healthy controls consistently in both coding and non-coding. Neuroprotection is the process used for preservation of the neuronal structure. Although there are different types of medication sources available for neurodegenerative disorders, including allopathy, therapy, and herbal products, nutraceutical products such as *ginkgo biloba, Melissa officinalis,* and *Passiflora incarnate* are widely used in the treatment. Beyond this, bioinformatics tools or biomarkers are also emerging as new medication sources in neuronal disorders. Such as different assays such as ELISA, MAPT, and PRNP are used for AD. Microarrays, next generation sequencing (NGS), and data based on genotype are evolving among the best methods for identification of neurodegenerative disease based on blood samples [6, 7]. Microarray datasets [8] play a crucial role in identification of stages in nerve impulse transmission, protein modification, and mRNA splicing (**Figure 14.1**).

Previous research suggested that Cytoscape [9] is a key factor in molecular and biological pathways. Gene set enrichment analysis (GSEA) is used for identification of genes that are linked to diseases [10]. The growth in bioinformatics opened up a wide area for clinical applications [11, 12]. Machine learning (ML) provides advanced forms of algorithms to solve complex problems and is used in tumor identification. An artificial neural network (ANN) is used to identify biomarkers for AD. Machine learning is used to identify diseases from the brain, especially the frontal,

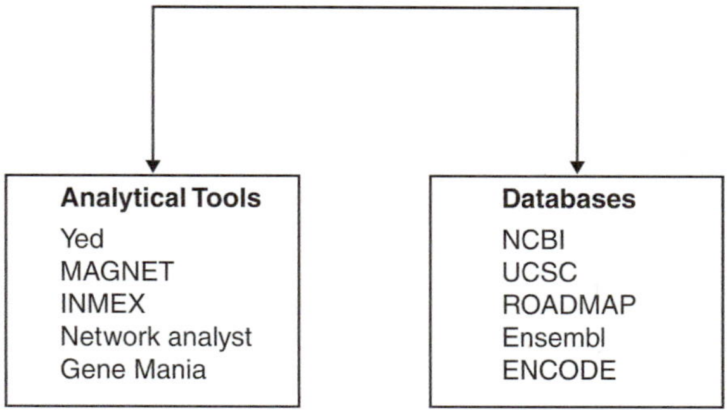

FIGURE 14.1 Analytical tools to visualise and databases for creation of biological networks in neurodegenerative diseases.

temporal, and hippocampal regions. Genetic algorithms identify biomarkers that help in treatment of Alzheimer's disease. The Gene MANIA Cytoscape regulates the function of markers ORA12 and TPII. The intrinsic modeling plays a crucial role in the genome information. Genetics-based algorithm analytical methods are used to identify the solutions for biological operations including mutations [13] from the frontal and temporal regions. It has been reported that studies based on genes for AD and PD are used in the molecular pathways [14].

Bioinformatics is the branch of applied science that plays a significant role in neurological research, with respect to new drug discovery and possible treatment [15]. Genomic-wide association (GWAS) is widely used in sequencing [16] and is linked with PD [17, 18]. Recently, new research studies have emerged on coding and non-coding proteins using microarray and genotypic data. MicroRNAs are the most abundant and noncoding ranges up to a few nucleotides binding to the 3′-untranslated region [19, 20], widely used in neurodegenerative diseases [21, 22]. Omic data is used in analysis of risk factors of diseases [23]. In recent decades, bioinformatics has had a profound impact on molecular and biological pathways. Tools such as transcriptomics, genomics, and proteomics are generated for understanding the interlinking networks of disabilities and pathways to reduce risk factors, and they are also advantageous in the development of drugs to treat neurodegenerative disorders.

14.3 PARKINSON'S DISEASE

Parkinson's disease a is long-term neurodisorder that affects the motor neurons [24]. Biomarkers associated with PD have been widely emerging in the form of proteomics and bioinformatics techniques using plasma samples. Quantitative analysis occurs on the basis of the chromatography. There are a few drugs that have potential for prevention and treatment of disease by using data mining and analysis based on genomics which provide new research and development in the field of drug delivery.

- *Mechanism*: Bioinformatics analysis acts as potent molecular markers for neurodegenerative disorders. It has been reported that 1997 marked the initial phase of PD genomics, which plays an integral role in the development of therapeutics. Expression data upstream analysis (EDUSA) provides bioinformatics analysis for signaling cascades of genes and determining hierarchy process. Next-generation sequencing (NGS) datasets are utilised to identify differentially expressed genes (DEGs) in PD. Gene ontology (GO) is used to understand the functional role of genes. A receiver operating curve (ROC) explores the diagnostic status of hub genes. There are numerous biomarkers, including neurotropic factor (a derivative of the brain) (BDNF) [25], histone deacetylase 4 (HDAC4) [26], leucine-rich repeat kinase2 (LRRK2) [27], and JAK/STAT signaling pathway [28]. It has been reported that GEO2R online analysis was useful in identification of DEGs between PD patients and other normal organisms. GO analysis can be categorised into biological process (BP), molecular function (MF), and cellular component (CC). In terms of MF, genes are abundant in ATP and GTP binding, whereas in the case of CC, these genes are significantly enriched in mitochondria, golgi bodies, neuronal cells, and microtubules.

14.4 ALZHEIMER'S DISEASE

The pathological mechanism of this disease is not still understood; hence, the drugs have led to poor therapeutics; thus, machine learning methods such as least absolute shrinkage and selection operator (LASSO)and Random-forest are emerging as new methodologies for treatment. Analysis based upon miRNA utilises databases including target scanning and visualisation and integrated discovery (DAVID), commonly used in AD. miRNAs were integrated in tissues such as brain and blood; it has been observed that miRNAs targets the nucleus and transcription [29]. Next-generation techniques based on sequencing proved to be a potent tool for identification of gene mutations and their linkage with the diseases.

- *Mechanism*: It has been reported that excessive alcohol consumption leads to increased risk factor. Ferroptosis is another form of cell death that is dependent on iron [30], and this pathway is linked with alcohol and AD. Gene Expression Omnibus (GEO) databases were used and the ferroptosis genes were obtained from Kyoto Encyclopedia of Genes and Genomes (KEGG). The protein interaction network (PPI) was made by means of gene-based MANIA databases for gene list analysis [31].

14.5 HUNTINGTON'S DISEASE

Huntington's is an autosomal neurodisorder which occurs by mutation in the huntingtin (HTT) gene due to repetitive CAG sequences. The gene plays a crucial role in the synaptic functions. It has been reported that gene profiling is rapidly increasing for identification of mRNA; bioinformatics mining is considered to be the central method to understand the pathological condition of neurodisorders. Detection of biomarkers may be used for drug development.

- *Mechanism*: It has been reported that hubgenes and messenger RNAs are linked with HD and pre-HD. KEGG analysis demonstrated that ubiquitin-mediated proteolysis plays an important role in HD. Gene expression analysis is commonly used for complex disease etiology. Gene set enrichment is an important tool for identification of chromosome location and interconnection with disorders.

14.6 SCHIZOPHRENIA

Schizophrenia is a neurodegenerative disease which involves group of symptoms, including perception, thinking, emotion, behavior, and mental disabilities. Generally, the analysis is based upon the GEO database. Hub genes were screened out by NFKBIA and CDKNIA. Several genetic algorithms are widely used for categorising gene variants, based on genomics and transcriptomics data. These transcriptomic data are used for demonstration of alteration in genes during pathogenesis. Genome-wide association is used to identify dysfunction in synapses of patients [32].

- *Mechanism*: Bioinformatics analysis of schizophrenia is dependent on GEO datasets. It has been reported that three different types of datasets and hub genes were analyzed by using bioinformatics tools. The three datasets include DEGs. Out of these, 30 upregulated and 31 downregulated which were enriched in nuclei and protein construction. Tools such as NFKBIA, CDKNIA, BTG2, and GADD45B are widely used to screen out four hub genes. Experiments concluded that (GO) analysis indicates that hub genes are mainly enriched in DNA damage response, regulation of cell cycle, protein sorting into nucleus, and ubiquitin binding.

14.7 MULTIPLE SCLEROSIS

Multiple sclerosis is an autoimmune neurodegeneration of the central nervous system which is recognised by oligodendrocyte loss [33]. Oligodendrocytes are necessary for myelin cells [34];a myelinated sheath is protective in nature, made up of fat, and is involved in transmission of nerve impulses [35]. The 21st century marked technological and innovative advancement in many fields. Data may be in the form of genomic sequences, medical data, and molecular and biological pathways; bioinformatics tools allow the exploitation of such data, which are widely used in genome annotation, gene expression analysis, and drug targeting.

- *Mechanism*: Genome-wide association studies (GWAS) are emerging as advancement in the field of genetics, which is capable of testing genetic variations in order to identify the genotypic and phenotypic associations of patients [36]. It also compares the allelic frequency between healthy and diseased person [37]. Microarray is another technology used to identify differences of gene expression in healthy and non-healthy persons [38]. Various algorithms can also provide sufficient information for datasets, which are useful for researchers. Biomarkers such as neurofilaments and antibodies can perform several applications which are advantageous in monitoring diseases and treatment response. Computational biology provides potent markers for multiple sclerosis, including DNA arrays, proteomics, and antibody arrays, which can be powerful tools in neurodegenerative diseases.

14.8 AMYOTROPHIC LATERAL SCLEROSIS

This is a fatal neurodisorder that is produced by neuronal degeneration, which especially displays deposition of cytoplasmic molecules. It is a progressive neurological disorder marked by muscle twitching and hallucination and it involves two group of neurons – upper and lower motor neurons – which are adversely effected in this disorder.

- *Mechanism*: Recently, Sanger sequencing is a common method useful in the screening of mutagens in the ALS gene [39]. Nanopore sequencing uses motor protein and single-stranded DNA molecules through a nanopore particle. RNA sequencing is being used to identify transcriptomic alterations in cells. Bioinformatics tools have now been explored including HipSTR and GangSTR, as potent markers to detect structural variation in data [40, 41].

14.9 STROKE

Stroke is described as a hemorrhagic disease, which is also referred to as an attack of the brain and cerebrovascular circulatory reduction. It involves cerebral and neuronal loss and is characterised by high blood pressure, hemorrhage and blood vessel blockage. Stroke can be defined as the most devastating neuronal disabilities and may cause physiological and psychological impairments [42].

- *Mechanism*: It has been reported that GO and the Kyoto Encyclopedia of Genes and Genomes pathways were analyzed with respect to DEGs and bioinformatics tools were used for the pathogenesis and key genes, which provide treatment strategies. Cytoscape software is widely used to identify differences in miRNA-genes. The main mechanism behind stroke involves inflammation, oxidative damage, programmed cell death pathways, and angiogenesis. CD19 molecules were the most significant genes and belong to immunoglobulin superfamily; they are expressed on B-cell regulate kinases activity [43] (**Figure 14.2**).

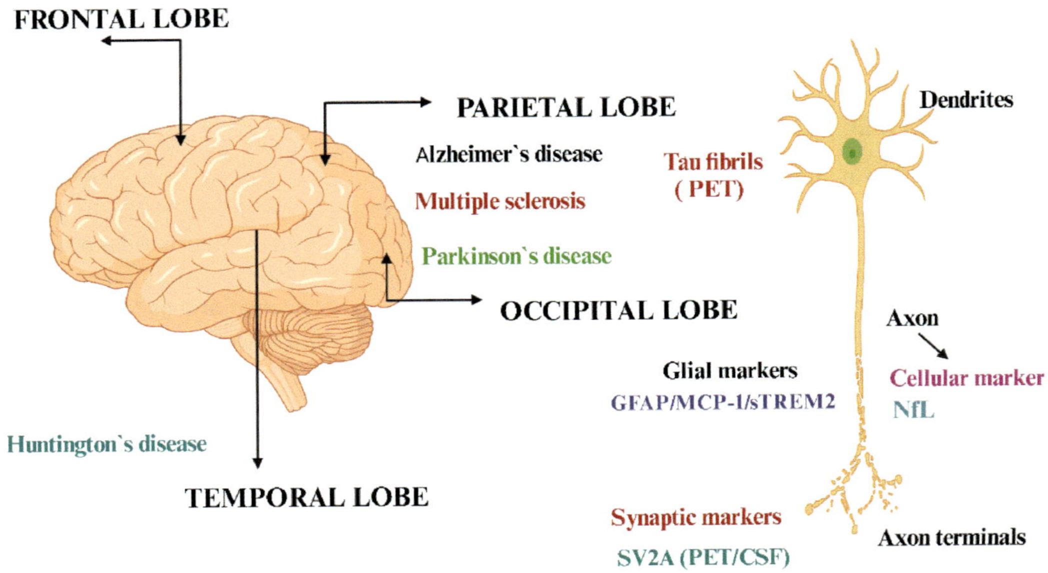

FIGURE 14.2 Biomarkers used in neurodegenerative diseases.

14.10 CONCLUSION

Bioinformatics research, including genomics and databases, is emerging as rapidly increasing technological evidence in the advancement of medication fields. Such global analysis to identify altered proteins and genes in patients suffering from neuronal injuries will further help in preventing as well as curing many diseases such as AD, PD, and HD. Molecular pathways lead to development of new methodologies. Utilising biomarkers such as databases and analytical tools is one of the challenging protocols for treatment of neurodegeneration. Molecular signatures and therapeutic approaches are used in modification of drug discoveries; bio-analytical tools have established a significant role in the biological aspects, which can be act as potent inhibitors for neurodegenerative disorders. It has been reported that there is a center for genomics to study neurological diseases, which deals with the mutation cases. These approaches can provide new discoveries in the field of neuroscience. Bioinformatics analysis aids in genomic studies and obtains information about molecular biology at evolutionary stages. This chapter aims at highlighting bioinformatics model for treatment of neurodegenerative disease.

ACKNOWLEDGEMENTS

The authors wish to thank the principal, S. S. Khanna Girls' Degree College (A Constituent College of University of Allahabad, Prayagraj) and the DST-CURIE Grant of Department of Science and Technology, New Delhi GOI (DST/CURIE-PG/2022/10 G), for providing all the necessary facilities in college.

REFERENCES

1. Bredesen DE, Rao RV, Mehlen P. (2006). Cell death in the nervous system. Nature. 443(7113):796–802. doi: 10.1038/nature05293
2. Pandit MK. (2011). Neuroprotective properties of some Indian medicinal plants. Int J Pharm Biol Arch. 2(5):1374–9.
3. Rios JL, Onteniente M, Picazo D, Montesinos MC. (2016). Medicinal plants and natural products as potential sources for anti-Parkinson drugs. Planta Med. 82:942–51.

4. Mythri R, Kumar A, Mms B. (2015). Nutraceuticals and Other Natural Products in Parkinson's Disease Therapy: Focus on Clinical Applications. In Bioactive Nutraceuticals and Dietary Supplements in Neurological and Brain Disease. Academic Press: Cambridge, MA, USA, pp. 421–431. doi: 10.1016/B978-0-12-411462-3.00044-8
5. Mota AH (2016). A review of medicinal plants used in therapy of cardiovascular disease. IJPPR. 8:572–91.
6. Karni S, Soreq H, Sharan R. (2009). A network-based method for predicting disease-causing genes. J Comput Biol. 16:181–9. doi: 10.1089/cmb.2008.05TT
7. Zhang B, Gaiteri C, Bodea LG, Wang Z, McElwee J, Podtelezhnikov AA, et al. (2013). Integrated systems approach identifies genetic nodes and networks in late-onset Alzheimer's disease. Cell. 135:707–20. doi: 10.1016/j.cell.2013.03.030
8. Barbash S, Soreq H. (2012). Threshold-independent meta-analysis of Alzheimer's disease transcriptomes shows progressive changes on hippocampal functions, epigenetics and microRNA regulation. Curr Alzheimer Res. 9:425–35. doi: 10.2174/156720512800492512
9. Saito R, Smoot ME, Ono K, Ruscheinski J, Wang PL, Lotia S, et al. (2012). A travel guide to Cytoscape plugins. Nat Methods. 9:1069–76. doi: 10.1038/nmeth.2212
10. Subramanian A, Tamayo P, Mootha VK, Mukherjee S, Ebert BL, Gillette MA, et al. (2005). Gene set enrichment analysis: A knowledge-based approach for interpreting genome-wide expression profiles. Proc Natl Acad Sci. 102(43):15545–50. doi: 10.1073/pnas.0506580102
11. Doss CGP, Zayed H. (2017). Comparative computational assessment of the pathogenicity of mutations in the Aspartoacylase enzyme. Metab Brain Dis. 32(6):2105–18. doi: 10.1007/s11011-017-0090-5
12. Zaki OK, Krishnamoorthy N, El Abd HS, Harche SA, Mattar RA, Nofal MY, Zayed H. (2017). Two patients with Canavan disease and structural modeling of a novel mutation. Metab Brain Dis. 32(1):171–7. https://link.springer.com/article/10.1007/s11011-016-9896-9
13. Holland JH. (1992). Genetic algorithms. Sci Am. 267(1):44–50. doi: 10.1038/scientificamerican0792-66
14. Tanzi E. (2012). "The genetics of Alzheimer disease. Cold Spring Harb Perspect Med. 2:a006296. doi: 10.1101/cshperspect.a006296
15. Uhlen M, Hallstro MBM, Lindskog C, Mardinoglu A, Ponten F, Nielsen J. (2016). "Transcriptomics resources of human tissues and organs. Mol Syst Biol. 12:862. doi: 10.15252/msb.20155865
16. Scott LJ, Mohlke KL, Bonnycastle LL, Li CJ, Willer Y, Duren WL. (2007). A genome-wide association study of type 2 diabetes in Finns detects multiple susceptibility variants. Science. 316(80):1341–5. doi: 10.1126/science.1142382
17. Nalls MA, Blauwendraat C, Vallerga CL, Heilbron K, Bandres-Ciga S, Chang D, Tan M, Kia DA, Noyce AJ, Xue A. (2019). Identification of novel risk loci, causal insights, and heritable risk for Parkinson's disease a meta-analysis of genome-wide association studies. Lancet Neurol. 18:1091–102. doi: 10.1016/S1474-4422(19)30320-5
18. Foo JN, Chew EGY, Chung SJ, Peng R, Blauwendraat C, Nalls MA, Mok KY, Satake W, Toda T, Chao Y. (2020). Identification of risk loci for Parkinson disease in Asians and comparison of risk between Asians and Europeans: A genome-wide association study. JAMA Neurol. 77:746–54. doi: 10.1016%2FS1474-4422(19)30320-5
19. Wei CY, Zhu MX, Lu NH, Peng R, Yang X, Zhang PF, Wang L, Gu JY. (2019). Bio-informatics-based analysis reveals elevated MFSD12 as a key promoter of cell proliferation and a potential therapeutic target in melanoma. Oncogene. 38(11):1876–91. doi: 10.1038/s41388-018-0531-6
20. Lee RC, Feinbaum RL, Ambros V. (1993). The C. elegans heterochronic gene lin-4 encodes small RNAs with antisense complementarity to lin-14. Cell. 75(5):843–54. doi: 10.1016/0092-8674(93)90529-y
21. Reed ER, Latourelle JC, Bockholt JH, Bregu J, Smock J, Paulsen JS, Myers RH. (2018). MicroRNAs in CSF as prodromal biomarkers for Huntington disease in the PREDICT-HD study. Neurology. 90(4):e264–72. doi: 10.1212/wnl.0000000000004844
22. Bagnoli M, Canevari S, Califano D, Losito S, Maio MD, Raspagliesi F, Carcangiu ML, Tofoli G, Cecchin E, Sorio R, et al. (2016). Development and validation of a microRNA-based signature (MiROvaR) to predict early relapse or progression of epithelial ovarian cancer: A cohort study. Lancet Oncol. 17(8):1137–46. doi: 10.1016/s1470-2045(16)30108-5
23. Mathew JP, Ullman-Cullere MH. (2011). Emerging landscape of genomics in the electronic health record for personalized medicine. Human Mutation. 32(5):512–6.doi: 10.1002/humu.21456
24. Ma XW, Guo RY. (2017). Dose-dependent effect of Curcuma longa for the treatment of Parkinson's disease. Exp Ther Med. 13:1799–805.
25. Palasz E, Wysocka A, Gasiorowska A, Chalimoniuk M, Niewiadomski W, Niewiadomska G. (2020). BDNF as a promising therapeutic agent in Parkinson's disease. Int J Mol Sci. 21(3):1170. doi: 10.3390/ijms21031170

26. Lang C, Campbell KR, Ryan BJ, Carling P, Attar M, Vowles J, Perestenko OV, Bowden R, Baig F, Kasten M, et al. (2019). Single-cell sequencing of iPSC-dopamine neurons reconstructs disease progression and identifies HDAC4 as a regulator of Parkinson cell phenotypes. Cell Stem Cell. 24(1):93–106.e6. doi: 10.1016/j.stem.2018.10.023
27. Hur EM, Lee BD. (2021). LRRK2 at the crossroad of aging and Parkinson's disease. Genes (Basel). 12(4):505. doi: 10.3390/genes12040505
28. Lashgari NA, Roudsari NM, Momtaz S, Sathyapalan T, Abdolghaffari AH, Sahebkar A. (2021). The involvement of JAK/STAT signaling pathway in the treatment of Parkinson's disease. J Neuroimmunol. 361:577758. doi: 10.1016/j.jneuroim.577758
29. Sun C, Liu J, Duan F, Cong L, Qi X. (2022). The role of the microRNA regulatory network in Alzheimer`s disease: A bioinformatics analysis. Arch Med Sci. 18(1):206–22. doi: 10.5114/aoms/80619
30. Dixon SJ, Lemberg KM, Lamprecht MR, Skouta R, Zaitsev EM, Gleason CE, et al. (2012). Ferrroptosis: An iron-dependent form of nonapoptotic cell death. Cell. 149:1060–72. doi:10.1016?j.cell.2012.03.042
31. Franz M, Rodriguez H, Lopes C, Zuberi K, Montojo J, Bader GD, et al. (2018). GeneMANIA update. Nucleic Acids Res. 46, W60–W64. doi:10.1093?nar?gky311
32. Gandal MJ, et al. (2018). Transcriptome-wide isoform-level dysregulation in ASD, schizophrenia, and bipolar disorder. Science (New York, N.Y.). 14, 362(6420):eaat8127. doi: 10.1126/science.aat8127.
33. Titus HE, Chen Y, Podojil JR, Robinson AP, Balabanov R, Popko B, Miller SD. (2020). Pre-clinical and clinical implications of inside-out vs outside-in paradigms in multiple sclerosis etiopathogenesis. Front Cell Neurosci.14:599717. doi: 10.3389/fncel.2020.599717
34. Dulamea AO. (2017). Role of oligodendrocycte dysfunction in demyelination, remyelination and neurodegeneration in multiple scleross. Adv Exp Med Biol.958:91–127. doi: 10.1007/978-3-319-47861-6_7
35. Stys P, Tsutsui S. (2019). Recent advances in understanding multiple sclerosis. F1000Res. 8:F1000 Faculty Rev-2100. doi: 10.12688/f1000research.20906.1
36. Tam V, Patel N, Turcotte M, Bosse Y, Pare G, Meyre D. (2019). Benefits and limitations of genome-wide association studies. Nat Rev Genet. 20:467–48. doi: 10.1038/s41576-019-0127-1
37. Cotsapas C, Mitrovic M. (2018). Genome-wide association studies of multiple sclerosis. Clin Transl Immunol. 7:e1018. doi: 10.1002/cti2.1018
38. Rodriguez-Esteban R, Jiang X. (2017). Differential gene expression in disease: A comparison between high-throughput studies and the literature. BMC Med Genomics. 10:59. doi: 10.1186/s12920-017-0293-y
39. Luigetti M, Lattante S, Zollino M, Conte A, Marangi G, Del Grande A, et al. (2011). SOD1 G93D sporadic amyotrophic lateral sclerosis (SALS) patient with rapid progression and concomitant novel ANG variant. Neurobiol Aging. 32(10):1924.e15–8. doi: 10.1016/j.neurobiolaging.2011.04.004
40. Mousavi N, Shleizer-Burko S, Yanicky R, Gymrek M. (2019). Profling the genome-wide landscape of tandem repeat expansions. Nucleic Acids Res. 47(15):e90. doi: 10.1093/nar/gkz501
41. Willems T, Zielinski D, Yuan J, Gordon A, Gymrek M, Erlich Y. (2017). Genome wide profling of heritable and de novo STR variations. Nat Methods. 14(6):590–2. doi: 10.1038/nmeth.4267
42. Zhang S, Cheng S, Zhang Z, et al. (2021). Related risk factors associated with post-stroke fatigue: A systematic review and meta-analysis. Neurol Sci. 42(4):1463–147. doi: 10.1007/s10072-020-04633-w
43. Li X, Ding Y, Zi M, Sun Li, Zhang W, Chen S, Xu Y. (2017).CD19, from bench to bedside. Immunol Lett. 183:86–95. doi: 10.1016/j.imlet.2017.01.010

15 Holistic Management of Diverse Neurodegenerative Diseases in Different Translational Models

Prabhash Kumar Pandey, Ashutosh Shrivastava*, Anshuman Srivastava, and Shiva Shukla*
Centre for Advance Research, Faculty of Medicine, King George's Medical University
Lucknow, Uttar Pradesh, India

Mohit
Department of Prosthodontics, Faculty of Dental Sciences, King George's Medical University
Lucknow, Uttar Pradesh, India

Amita Jain and Vimala Venkatesh
Department of Microbiology, Faculty of Medicine, King George's Medical University
Lucknow, Uttar Pradesh, India

Soni Tiwari
Department of Microbiology, Dr. Rammanohar Lohia Avadh University
Ayodhya, Uttar Pradesh, India

Farrukh Jamal
Department of Biochemistry, Dr. Rammanohar Lohia Avadh University
Ayodhya, Uttar Pradesh, India

15.1 INTRODUCTION

Neurodegenerative illnesses are a serious risk to human well-being. The senior population has been growing, and this has led to an increase in the prevalence of many age-related illnesses. Worldwide, neurodegenerative illnesses impact millions of individuals [1, 2]. The nerve cells in the brain or peripheral nervous system gradually lose their ability to function and eventually die, which causes

* These authors contributed equally to the corresponding authorship of this chapter.

DOI: 10.1201/9781032620442-15

these illnesses. While several therapies may assist in decreasing some of the physical or psychological symptoms of neurodegenerative illnesses, these conditions are chronic, progressive, and ultimately deadly. When the patient reaches the final stages of the illness and is physically and intellectually incapable, his condition worsens irreversibly.

Although neurodegenerative disease research has increased over the past few decades, patients still do not have access to efficient treatments [3, 4]. Several medications with the potential to treat these diseases failed their phase III clinical trials. A critical revaluation of previous efforts to produce medicines led to recommendations for significant challenges to overcome and strategies to improve future research [5].

Neurodegenerative disorders have a wide range of etiologies; some result in memory and cognitive issues, while others impair a person's mobility, speech, and breathing. Effective treatments are desperately needed, but they won't be accessible until the fundamental mechanisms of each disease are thoroughly comprehended. We know very little about the basic biology behind aging and neurodegeneration [6]. Even though researchers have identified many important components of human neurodegeneration, we still don't fully understand the process and must make a lot of unfounded assumptions. Popular therapeutic theories have, out of necessity, relied more on assumptions than on factual information [7]. Consequently, to create sensible therapies, significant scientific gaps in our understanding of the fundamental causes of neurodegenerative illness must be filled.

Building a model system that replicates the traits that set a disease apart is one way to learn how the illness functions [8]. Researchers have long used a variety of powerful experimental model species, such as the fruit fly, nematode worm, mouse, and even baker's yeast, to study neurodegenerative illnesses [9]. These studies have produced significant insights into the disease processes. The holistic management of different neurodegenerative diseases in different translational models is a difficult task that includes figuring out how these diseases work, creating and testing possible treatments, and taking into account how each disease model is different [10]. We have underlined in this chapter the significance of model-based research as well as diverse approaches to managing and treating the various neurodegenerative disorders using different translational models.

15.2 TRANSLATIONAL MODELS USED FOR THE STUDY OF NEURODEGENERATIVE DISEASES

Preclinical models called 'translational models' are utilised in biomedical research to examine and comprehend the mechanisms underlying various neurodegenerative disorders. These models seek to close the gap between *in vitro* or cell culture-based basic research and human clinical trials. They support the identification of biomarkers for diagnosis and monitoring, the testing of prospective treatments, and deeper knowledge of the diseases by researchers [11, 12]. A comprehensive strategy for handling various neurodegenerative diseases in various translational models necessitates a thorough comprehension of the illnesses, a multidisciplinary strategy, and a dedication to ongoing research and patient care. Collaboration among researchers, doctors, patients, and caregivers is essential to solving the numerous problems that these deadly diseases present [13]. The advancement of our knowledge of neurodegenerative disorders and the creation of possible therapeutics depend heavily on translational models. They provide a setting for testing therapeutic approaches and evaluating their efficacy and safety before implementing them in clinical trials involving human subjects [14]. Listed below are a few typical translational models for the investigation of neurodegenerative diseases.

15.2.1 Animal Models

To investigate neurodegenerative illnesses, mice, rats, and non-human primates are widely utilised as animal models. To mimic the genetic and clinical characteristics of these conditions, transgenic rats and mice with mutations associated with Huntington's, Parkinson's, and Alzheimer's diseases have been created. Researchers use these models to test potential treatment strategies and to comprehend how diseases develop [15]. Animal models are frequently created to imitate the early stages of neurodegenerative disorders. However, late-stage diseases cannot be accurately modeled in these circumstances. It is difficult to mimic disease progression in the late stages when symptoms are at their worst. Researchers frequently combine *in vitro* experiments, human-derived cell models, and animal models to get around these constraints [15]. Additionally, technological advancements and a deeper comprehension of disease causes are enhancing the creation of more precise and pertinent animal models. Even so, it's crucial to take these limitations into account when analysing the outcomes of animal studies and thinking about how they apply to neurodegenerative disorders in people [16].

15.2.2 Cell Culture Models

To learn more about the cellular and molecular causes of neurodegenerative diseases, it is important to use *in vitro* models with neuronal cell cultures, especially induced pluripotent stem cells made from patient cells. They enable scientists to control, examine, and test possible drug candidates in biological processes [17].

15.2.3 *Drosophila* Models

Because of their straightforward neural system, brief lifespan, and genetic tractability, fruit flies (*Drosophila melanogaster*) are employed as model organisms to investigate neurodegenerative illnesses. They have been crucial in revealing the basic molecular mechanisms behind various illnesses [18].

15.2.4 *Caenorhabditis elegans* Models

This nematode worm is another straightforward model organism used to research neurological illnesses. The capacity to research age-related neurodegeneration and quick genetic tinkering are advantages they provide [19].

15.2.5 Zebrafish Models

Zebrafish share a great deal of genetic resemblance with humans and have a neurological system that is well understood. They can be used as a platform for investigating the genetics of neurodegenerative diseases and for medication screening [20].

15.2.6 *In Vivo* Imaging

Modern neuroimaging techniques such as functional magnetic resonance imaging (FMRI), positron emission tomography (PET), and magnetic resonance imaging (MRI) allow researchers to monitor the course of a disease and evaluate the effects of treatment on actual patients. These techniques provide valuable translational data for clinical studies [21].

15.2.7 Human Brain Organoids

Brain organoids are three-dimensional cell culture models that duplicate some aspects of brain development and function. They are created from human induced pluripotent stem cells (iPSCs). They are employed to investigate the pathophysiology of neurodegenerative illnesses and evaluate prospective medications [22].

15.2.8 Genetic Models

Genetic engineering advancements have made it possible to produce transgenic animals or cells that have particular disease-related mutations. The genetic underpinnings of neurodegenerative illnesses can be studied using these models [23].

15.2.9 Neurobehavioral Models

To evaluate cognitive and motor performance in animal models and to understand the functional impact of neurodegeneration and test prospective treatments, researchers use a variety of behavioral tests [24].

15.3 NEURODEGENERATIVE DISEASES AND TRANSLATIONAL MODELS

15.3.1 Alzheimer's Disease

Translational models in Alzheimer's disease (AD) research serve as the essential connection between fundamental scientific discoveries and practical applications in patient care [25]. These models, which include genetically modified mice and advanced cellular systems, are very important for figuring out how AD works and what treatments might work. Genetically modified mice expressing human genes associated with AD have been instrumental in elucidating the intricate processes underlying the disease [26]. With their usefulness in developing amyloid plaques and neurofibrillary tangles, these mice provide invaluable information about disease progression. This replication of key AD characteristics allows us to test experimental therapies and gain crucial insights into their efficacy, thereby paving the way for future clinical trials [27].

Non-human primates, particularly macaques, and humans have a lot in common genetically and physiologically. Studying AD in these primates offers a unique perspective, considering their complex brain structures and cognitive capabilities. By observing a cognitive decline in primates, we can gain a deeper understanding of AD-related changes in brain function, shedding light on the disease's impact on memory, learning, and behavior [28]. These insights are invaluable for developing interventions that target specific cognitive impairments in AD patients.

Fruit flies, specifically *Drosophila melanogaster*, have proven surprisingly useful in AD research. Despite their simple nervous systems, fruit flies have orthologs of human AD-related genes. This similarity enables large-scale genetic screening, allowing us to identify genes and pathways influencing neurodegeneration. These tiny insects have provided essential information about cellular processes involved in AD, often serving as a starting point for more complex studies in mammalian models.

Caenorhabditis elegans, a transparent roundworm, offers unique advantages for studying neuronal function and pathology [29]. Its short lifespan and well-defined neural circuitry enable it for rapid genetic studies. Further, we can observe the effects of specific genes or compounds on neuronal health and longevity. Even though it is easier to study than mammalian systems, *C. elegans* has helped us to learn a lot about basic biological processes that may play a role in AD, like how proteins misfold and how mitochondria don't work properly [30].

iPSCs represent a ground-breaking technology in AD research. These cells are derived from adult tissues and reprogrammed into a pluripotent state, allowing scientists to generate patient-specific neuronal cells [31]. iPSCs mirror the genetic background of individual patients, providing a personalised platform for studying disease mechanisms and screening potential drug compounds. By observing how neurons derived from AD patients respond to various treatments, we can identify therapies tailored to specific genetic profiles, ushering in the era of precision medicine for AD [32].

Brain organoids, three-dimensional cultures derived from iPSCs, replicate the complex cellular architecture of the human brain. These miniature brain-like structures offer a unique opportunity to study disease pathology in a controlled environment [33]. Researchers can observe the development of amyloid plaques and neurofibrillary tangles, providing insights into the early stages of AD. Brain organoids also help researchers try out new ways to treat brain problems because they are more like real brain tissue than two-dimensional cell cultures [34].

In the realm of computational models, advanced algorithms and simulations are revolutionising our approach to understanding AD. By integrating diverse biological data, these models predict disease pathways, identify potential drug targets, and analyze large-scale datasets. Computational models offer a bird's-eye view of AD, allowing us to make predictions and test hypotheses efficiently [35]. These models complement experimental research, providing a virtual platform for exploring the vast landscape of AD-related data.

From genetically modified mice providing insights into disease progression to iPSCs enabling personalised medicine approaches, these models are revolutionising the way we study and treat AD. Fruit flies and roundworms, despite their simplicity, offer fundamental insights into AD-related processes, acting as essential building blocks for more complex studies [36]. Advanced computational models provide a holistic perspective, integrating vast datasets and guiding experimental research. Patient-derived xenografts, on the other hand, bridge the translational gap, ensuring that laboratory discoveries have direct relevance to individual patients [37]. As these models continue to evolve, they hold the promise of unlocking new therapeutic avenues, bringing us closer to effective treatments and, ultimately, a cure for AD.

15.3.2 Parkinson's Disease

Parkinson's disease (PD) continues to challenge the medical community due to its complex nature and the limited efficacy of current treatments. Translational models have emerged as crucial tools in understanding the underlying mechanisms of PD and testing potential therapies before they reach clinical trials [38]. These models, ranging from animal systems to cell-based cultures, provide valuable insights into the disease's progression and aid in the development of holistic management strategies.

One significant area of research in translational models focuses on the role of alpha-synuclein aggregation, a hallmark characteristic of PD. Scientists using transgenic mice that have too much alpha-synuclein have shown that clumped proteins hurt dopaminergic neurons, which leads to motor problems like those that people with Parkinson's disease have [39]. Researchers have used these models to test different anti-aggregation compounds, such as antibodies and small molecules, and the results look good for getting rid of harmful alpha-synuclein deposits.

iPSCs from PD patients have transformed the field of cell-based models and PD research [40]. These cells can be differentiated into dopaminergic neurons, allowing scientists to observe disease-specific cellular changes in a controlled environment. Scientists have recently used CRISPR-Cas9 to change genes in iPSCs to create isogenic cell lines that are different in only some genetic mutations linked to Parkinson's disease [41]. This approach enables researchers to pinpoint the exact genetic alterations responsible for PD pathogenesis, leading to the identification of novel drug targets.

Animal models, particularly primates like macaques, have also played a pivotal role in PD research. These models closely mimic human brain structures and enable scientists to study the complex motor and non-motor symptoms of PD [42]. Researchers have also used these monkeys to look into how well deep brain stimulation (DBS) therapy works. DBS is a surgical method that involves putting electrodes into certain parts of the brain to change how neurons fire. Studies on macaques have given us important information that has helped us to improve the parameters of DBS and make it work better for people with Parkinson's disease [43]. Additionally, zebrafish models have gained popularity in PD research due to their genetic tractability and transparent embryos, allowing for real-time visualisation of neuronal development. Drug screening has benefited greatly from the use of zebrafish models, which allow compounds to be evaluated for their potential to restore dopaminergic neurons and enhance motor behavior [44]. Potential neuroprotective drugs have been found as a result of these studies, opening the door for additional research in mammalian models and, eventually, clinical trials.

Another area of translational research involves the gut-brain axis, exploring the connection between the gut microbiota and PD pathology. Emerging evidences suggest that alterations in gut microbiota composition may influence PD development. Researchers have utilised germ-free mice, devoid of any gut microbiota, to study the impact of microbial colonisation on PD-related neuroinflammation and motor dysfunction [45]. These findings have created new opportunities for treatment approaches, such as probiotics and dietary modifications, aimed at restoring healthy gut microbiota to mitigate PD symptoms. In that way, translational models have significantly advanced our understanding of Parkinson's disease and paved the way for innovative therapeutic strategies [45].

These models have taught us a lot about how PD starts, from studying alpha-synuclein aggregation in transgenic mice to using iPSCs for personalised medicine and looking into the gut-brain axis in germ-free mice. Continued research in translational models holds the promise of uncovering novel therapies, bringing us closer to effective treatments for this debilitating neurological disorder [46].

15.3.3 Huntington's Disease

Translational models in Huntington's disease (HD) research have revolutionised our understanding of this devastating neurodegenerative disorder. These models, ranging from genetically modified animals to cell-based cultures, bridge the gap between basic research findings and clinical applications. Through these models, scientists have made significant strides in developing holistic management strategies for HD [47].

Scientists have used genetically modified animals, especially transgenic mice that express mutant huntingtin (HTT) protein, to help them study HD. Investigators have utilised these mice to investigate the molecular underpinnings of the disease. For instance, studies in transgenic mice have revealed the toxic effects of mutant HTT, leading to impaired neuronal function and motor deficits [48]. This understanding has guided the development of therapeutic interventions, including antisense oligonucleotides (ASOs). ASOs are drugs that are meant to target and lower mutant HTT expression. They have shown promise in mouse models, lowering HTT levels and improving motor symptoms [49]. Such breakthroughs highlight the potential of ASOs as disease-modifying treatments for HD.

iPSCs from HD patients have been a major source of insights into the molecular mechanisms underlying the disease in cell-based models. iPSCs can be differentiated into neurons, allowing scientists to study disease-specific changes in neuronal function [50]. Using iPSC-derived neurons, researchers have uncovered aberrant signaling pathways associated with HD pathology. Also, CRISPR-Cas9 technology has made it possible to fix certain HD-related mutations in iPSCs, which has led to the creation of genetically corrected cells [51]. These cells serve as valuable tools for understanding the direct impact of mutations and for testing potential therapeutic compounds. Recent studies using iPSC-derived neurons have focused on enhancing protein clearance pathways [52]. Small molecules made to help to get rid of harmful HTT aggregates have been tested in these cellular models and have shown promise in lowering the formation of aggregates and the harmful effects on cells [53].

Beyond pharmacological approaches, non-pharmacological interventions have gained attention in translational studies. Enriched environments and physical exercise have been explored in animal models, such as rats, to assess their impact on HD progression [54]. Rats exposed to enriched environments, which feature more social interaction and physical activity, have demonstrated delayed disease onset and improved motor coordination. These findings emphasise the importance of environmental enrichment in promoting neuronal plasticity and mitigating motor deficits in HD [55].

Dietary interventions have also been investigated in animal models to identify potential neuroprotective effects. Mouse models fed with specific diets, such as the ketogenic diet, have demonstrated metabolic improvements and enhanced motor performance [56]. The ketogenic diet, low in carbohydrates and high in fats, has shown promise in ameliorating HD-related symptoms. These studies shed light on the therapeutic potential of dietary modifications for managing HD [57].

Translational models have significantly advanced our understanding of Huntington's disease and paved the way for innovative therapeutic strategies. From looking at mutant HTT toxicity in transgenic mice to looking at protein clearance pathways in neurons made from iPSCs, these models have taught us a lot about how HD starts. The integration of pharmacological and non-pharmacological interventions, guided by translational research, holds great promise for improving the lives of individuals affected by Huntington's disease [58]. As these translational studies continue to evolve, the prospects for effective treatments and, ultimately, a cure for HD become increasingly hopeful.

15.3.4 Amyotrophic Lateral Sclerosis

Lou Gehrig's disease, also known as amyotrophic lateral sclerosis (ALS), is a debilitating neurological illness marked by a gradual degeneration of motor neurons in the brain and spinal cord [59]. This degeneration leads to muscle weakness, paralysis, and ultimately respiratory failure, making ALS a fatal condition with no known cure [60]. Researchers worldwide are tirelessly working to understand the underlying mechanisms of ALS and develop effective treatments. One promising approach in this pursuit is the holistic management of ALS in various translational models [61]. Translational models play a crucial role in ALS research by allowing scientists to study the disease's progression and test potential therapies in a controlled environment. These models, ranging from cell cultures to animal models, serve as invaluable tools in the quest for a deeper understanding of ALS and the development of novel treatments [62]. Holistic management of ALS in translational models involves a multifaceted approach that addresses different aspects of the disease. One key aspect is understanding the genetic and molecular basis of ALS [63]. Genetic mutations, such as those in the C9, f72, and SOD1 genes, are associated with familial ALS cases. By studying these mutations in translational models, researchers gain insights into the disease's molecular pathways, enabling the development of targeted therapies [64]. In addition to genetic factors, oxidative stress and inflammation are prominent features of ALS pathology. Studies utilise translational models to explore the impact of oxidative stress on motor neurons and surrounding cells [65]. By identifying specific antioxidants and anti-inflammatory agents that mitigate these processes, scientists can potentially develop interventions that slow disease progression. Furthermore, excitotoxicity, a phenomenon where excessive stimulation of nerve cells leads to damage or death, is implicated in ALS [66]. Translational models help to investigate the role of excitotoxicity and develop drugs that modulate neurotransmitter levels, preserving motor neuron function. Holistic management also encompasses neuroprotective strategies [67]. Studies have explored the potential of neurotrophic factors, which promote the survival and function of neurons, in translational models. Scientists want to keep muscles working by making these factors more available or creating gene therapies that make more of them. This will keep motor neurons from dying and keep muscles working [68].

Addressing the damaged blood-brain barrier (BBB), a barrier that prevents dangerous substances from entering the brain, is another crucial component of comprehensive ALS care. In ALS, the BBB becomes permeable, allowing toxic molecules to enter the central nervous system [69]. Researchers used translational models to look into the integrity of the BBB and come up with ways to strengthen it. This would stop harmful substances from getting in and slow the progression of the disease. Furthermore, scientists concentrate on creating efficient symptomatic treatments to enhance the quality of life for those with ALS [70]. Translational models enable the testing of potential symptomatic treatments, such as muscle-strengthening exercises, respiratory therapies, and assistive devices. These interventions aim to enhance patients' mobility, respiratory function, and overall well-being. The holistic management of ALS in various translational models represents a comprehensive approach to understanding and treating this devastating disease. Researchers can learn more about ALS pathology and find different ways to help by looking into genetic factors, oxidative stress, inflammation, excitotoxicity, neuroprotection, BBB integrity, and symptomatic therapies

[71]. Through collaborative efforts and continued research in translational models, scientists strive to discover innovative therapies that may one day offer hope to ALS patients and their families.

15.4 INNOVATIVE MANAGEMENT STRATEGIES FOR NEURODEGENERATIVE DISEASES

Innovative methods for treating neurodegenerative diseases are essential, particularly given the rising prevalence of these illnesses, including amyotrophic lateral sclerosis, Parkinson's disease, Alzheimer's disease, and Huntington's disease. These illnesses result in a variety of cognitive and physical disabilities due to the progressive death of neurons [72]. An interdisciplinary strategy that incorporates medical, social, and technological advancements is necessary for the management of neurodegenerative illnesses. Here are a few cutting-edge approaches to treating neurodegenerative illnesses.

15.4.1 Precision and Personalised Therapies

Life expectancy has steadily increased during the past century. The aging population and the increasing prevalence of chronic diseases have had a substantial effect on healthcare costs. This highlights the fascinating conundrum of managing patients with long-term illnesses while boosting healthcare costs. Primary prevention that works could be a promising strategy. Considering this, precision and tailored medicine are useful instruments for investigating pathological processes before clinical symptoms appear and helping physicians to choose a suitable, focused treatment for patient care [73]. This will, of course, significantly reduce time commitments and healthcare costs while simultaneously improving patient quality of life and care. These circumstances are comparable in that they develop slowly over several decades. They therefore need a comprehensive, long-term response that is managed by different medical specialists and has access to the necessary equipment and materials. On the other hand, this can also offer a lot of chances to spot those who are vulnerable, offer protective measures, and start therapies early in the illness process [74]. In the world of medicine, precision and tailored medicines are showing promise in the treatment of neurodegenerative illnesses. Numerous debilitating symptoms are caused by neurodegenerative disorders, which are defined by the progressive death of neurons. These illnesses are incredibly diverse, with each patient experiencing a different disease course, set of symptoms, and underlying molecular causes [1]. The goal of precision and personalised therapeutics is to combat this heterogeneity by adjusting the course of treatment following a patient's unique illness features and genetic profile.

Some essential components of customised and targeted therapy for neurodegenerative disease treatment include the following.

15.4.1.1 Genetic Profile

The genetic profile of patients is a basic component of customised therapy. Knowing a patient's genetic background, including differences in certain genes linked to neurodegenerative illnesses, can be useful in predicting the course of the disease and identifying possible targets for treatment [75].

15.4.1.2 Biomarker Identification

Biomarkers that can show the course of a disease, how well a patient responds to treatment, or the many subtypes of a disease are key components of precision medicine. Genetic, proteomic, and imaging markers are examples of biomarkers that can be used to monitor a patient's response to medication and inform treatment decisions [76].

15.4.1.3 Targeted Medicines

Once a patient has particular genetic or molecular defects, targeted medicines can be created. The goal of these treatments is to either treat the anomalies or lessen their impact. For instance, monoclonal antibodies have been created to target the disease's hallmark, amyloid beta protein, in certain types of Alzheimer's disease [77].

15.4.1.4 Patient Stratification

Treatment outcomes for distinct subgroups of neurodegenerative disorders may vary. Treatments can be customised to specific subgroups by classifying patients according to their molecular and genetic profiles, which increases the chance of therapeutic success [78].

15.4.1.5 Repurposing Drugs

Using already-existing medications in novel ways is another aspect of precision medicine. It may be discovered that current medications offer potential benefits for neurodegenerative disorders by comprehending the molecular mechanisms behind these conditions as well as the hereditary elements impacting disease progression [79].

15.4.1.6 Tailored Treatment Plans

Tailored therapy goes beyond medication choice. It entails creating treatment programs that take into account each patient's particular situation, including comorbidities, way of life, and preferences. Along with medicine, this may entail dietary changes, physical therapy, and cognitive training.

15.4.1.7 Continuous Monitoring

To evaluate treatment response and make any required modifications, patients receiving personalised therapies frequently need to be followed up on. Wearable technology, telemedicine, and routine biomarker evaluations can all simplify the monitoring process [80].

15.4.1.8 Clinical Trials and Research

Clinical trials with targeted therapy are a common component of precision medicine in neurodegenerative diseases. Patients with certain genetic profiles may be selected for these trials, and their responses to therapy may help refine therapeutic approaches [81].

15.4.1.9 Obstacles

Despite their great potential, precision and personalised therapies face numerous challenges, including the cost and availability of genetic testing, governmental restrictions, and the need for collaboration among researchers, physicians, and pharmaceutical companies [80]. The gathering and application of genetic and health information give rise to ethical and privacy issues. A crucial component of putting tailored therapy into practice is making sure patient data is secure and used ethically [82].

In summary, precision and customised medications that take into account each patient's unique genetic and molecular traits are revolutionising the treatment of neurodegenerative illnesses. These treatments have the potential to significantly improve treatment outcomes and quality of life for individuals with these disabling conditions, even though they are still in their early stages of development. Innovative approaches to treating neurodegenerative diseases integrate medical, technical, and psychosocial modalities in a comprehensive and multidisciplinary manner. These methods aim to improve patients' quality of life, give them hope for the future, and ultimately aid in the hunt for a cure for these terminal diseases [83]. It's important to keep in mind that research on the long-term safety and efficacy of many of these techniques is still being conducted, as are experimental or exploratory clinical trials. **Figure 15.1** shows the different role of precision and personalised medicine in the management of neurodegenerative disorders. This figure illustrates the distinctive roles of precision and personalized medicine in the comprehensive management of neurodegenerative disorders. Precision medicine, represented on the left side of the figure, focuses on tailoring treatments based on specific molecular characteristics of the disease and the individual patient. This approach involves identifying genetic variations, biomarkers, and other molecular signatures that can inform targeted therapies aimed at addressing the underlying mechanisms driving the neurodegenerative process. Precision medicine enables more accurate diagnosis, prognosis, and treatment selection, potentially leading to better outcomes and improved patient care.

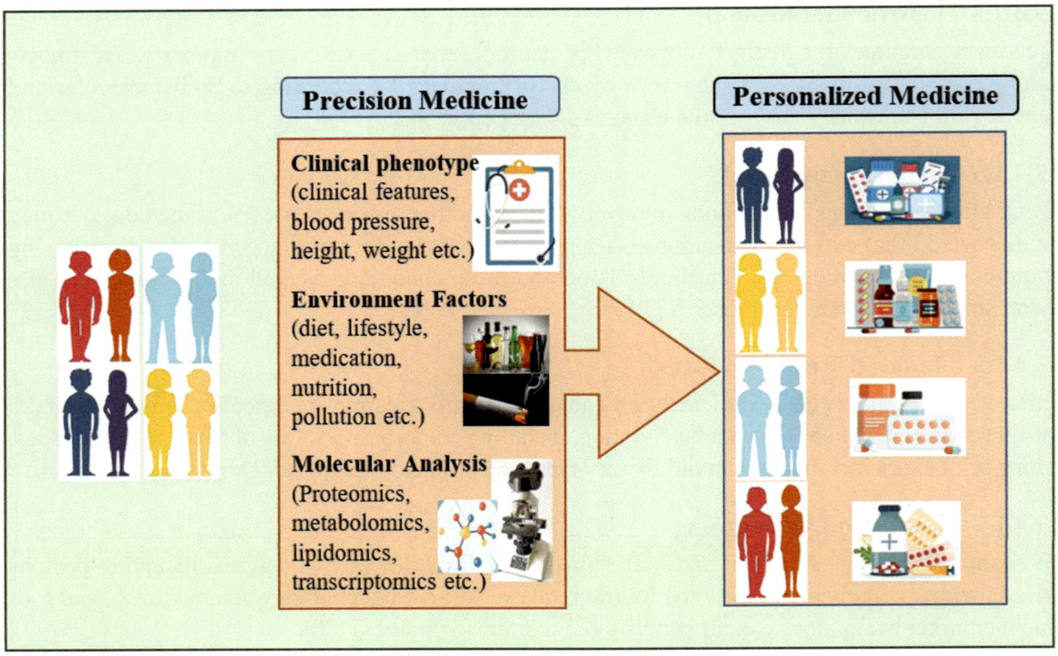

FIGURE 15.1 Role of precision and personalised medicine in the management of neurodegenerative disorders.

15.4.2 Immunotherapies

Immunotherapies, which harness the body's immune system to target and treat diseases, have gained attention in the field of neurodegenerative diseases. While the primary focus of immunotherapy has historically been in cancer treatment, researchers are exploring its potential for managing neurodegenerative conditions [84]. Immune system dysfunction may have an impact on the abnormal protein aggregation, chronic inflammation, and neuronal death that characterise these diseases. Immunotherapies are being explored in the following ways to treat neurodegenerative diseases.

15.4.2.1 Antibody-Based Therapies

Monoclonal antibodies that target specific proteins involved in neurodegenerative diseases have been developed. For example, Aducanumab is an FDA-approved monoclonal antibody for Alzheimer's disease that targets beta-amyloid plaques. These antibodies can help in the removal of pathological protein aggregates and reduce inflammation in the brain [85].

15.4.2.2 Anti-Inflammatory Therapies

Neurodegenerative diseases progress due in part to chronic neuroinflammation. Immunotherapies that target and regulate inflammatory responses have the potential to impede the development of illness. These can include anti-inflammatory drugs such as corticosteroids or more targeted anti-inflammatory agents [86].

15.4.2.3 Immune Cell Modulation

Certain strategies try to change how immune cells, such as microglia, function in the brain. The immune cells that live in the central nervous system are called microglia, and neurodegenerative illnesses may be exacerbated by their malfunction [87]. Researchers are exploring ways to regulate the activity of these cells to reduce inflammation and promote neuroprotection.

15.4.2.4 Vaccination Strategies

Vaccination against specific disease-related proteins, like beta-amyloid in Alzheimer's disease, is being investigated as a preventive strategy. This approach aims to stimulate an immune response against the pathological protein aggregates, leading to their clearance [88].

15.4.3 Stem Cell Therapies

Stem cell-based immunotherapies are being explored for neurodegenerative diseases. These approaches involve transplanting immune-modulating or regenerative stem cells to repair damaged neural tissue and modulate the immunological reaction in the central nervous system [89].

It's critical to remember that immunotherapies for neurodegenerative illnesses are still primarily being developed and researched. Numerous clinical trials are being conducted to assess their efficacy and safety. Additionally, there are challenges to overcome, such as the blood-brain barrier, which can limit the delivery of immunotherapeutic agents to the brain. Immunotherapies show promise in the management of neurodegenerative illnesses, but further study is required to completely comprehend their modes of action, long-term effects, and safety profiles [90]. Patients interested in these treatments should consult with healthcare professionals and consider participation in clinical trials when appropriate.

15.4.4 Gene Therapies

Gene therapy holds promise in the management of neurodegenerative diseases by targeting the genetic and molecular causes of these conditions. Huntington's disease, Amyotrophic lateral sclerosis, and certain types of Parkinson's disease are examples of neurodegenerative disorders that are frequently linked to particular genetic abnormalities or abnormal gene expression. Gene therapy approaches aim to correct, replace, or modulate the genes involved in these diseases [91]. Gene editing techniques, like CRISPR-Cas9, are being used to modify genes associated with neurodegenerative diseases. These techniques can potentially correct genetic mutations or modulate the expression of specific genes to reduce disease progression [92].

Here are some key ways gene therapies are being explored in the management of neurodegenerative diseases.

15.4.4.1 Gene Silencing

In diseases where the overexpression of a specific gene is implicated, gene therapy can be used to silence or downregulate the target gene. For example, small interfering RNA (siRNA) or antisense oligonucleotides (ASOs) can be designed to target and inhibit the expression of the mutant genes responsible for diseases like Huntington's disease [93].

15.4.4.2 Gene Replacement

In some cases, neurodegenerative diseases result from the absence or dysfunction of a specific gene. Gene therapy can be used to introduce a functional copy of the gene to restore normal cellular function [94]. This approach has been explored in clinical trials for diseases like spinal muscular atrophy.

15.4.4.3 Gene Editing

Advanced gene editing techniques, such as CRISPR-Cas9, are being investigated to correct specific genetic mutations associated with neurodegenerative diseases. This involves precise modification of the patient's DNA to repair or replace the faulty genes [95]. This approach is still in the experimental phase and faces challenges related to safety and accuracy.

15.4.4.4 Neuroprotection and Neuroregeneration

Gene therapy can be used to promote the survival and regeneration of damaged neurons. For example, neurotrophic factors or growth-promoting genes can be delivered to the affected brain regions to stimulate nerve growth and prevent further degeneration.

15.4.4.5 Targeting Protein Aggregates

In diseases like Alzheimer's, Parkinson's, and ALS, abnormal protein aggregation is a key pathological feature. Gene therapy can target the genes involved in the production or clearance of these misfolded proteins to reduce their accumulation [96].

15.4.4.6 Viral Vectors

Viral vectors, such as adeno-associated viruses (AAVs), are frequently utilised to transfer therapeutic genes to the central nervous system. These vectors are modified to carry the therapeutic gene and are designed to minimise immune responses and other potential side effects [97].

It's critical to remember that gene therapy for neurodegenerative illnesses is still a developing subject that is undergoing clinical trials and research. Challenges include ensuring the safety and long-term efficacy of gene therapy, the potential for immune responses to viral vectors, and ethical considerations related to genetic modification [98]. Regulatory approval and broader clinical implementation of these therapies will depend on further research and development. Patients and their healthcare providers should stay informed about the progress in gene therapy research for specific neurodegenerative diseases and consider participating in clinical trials when appropriate. The field of gene therapy is continually advancing, and it may offer new hope for individuals affected by these debilitating conditions in the future [91].

15.4.5 Neuroinflammation Management Strategies

The issue of neuroinflammation management in neurodegenerative disorders is a multifaceted and dynamic area of medical study and practice. Multiple sclerosis, Parkinson's disease, Alzheimer's disease, and other neurodegenerative illnesses are thought to progress more slowly as a result of neuroinflammation. Although the majority of neurodegenerative diseases have no known treatment, controlling neuroinflammation may be able to slow down the illness's course and enhance the quality of life for those who are impacted [99]. The following strategies can be used to control neuroinflammation in neurodegenerative illnesses.

15.4.5.1 Medications

Nonsteroidal anti-inflammatory drugs (NSAIDs), are frequently used to treat neuroinflammation in diseases like Alzheimer's disease because they can help to reduce inflammation in the brain. Neuroinflammation in multiple sclerosis is treated with immunomodulating medications such as glatiramer acetate and interferon beta [100]. Etanercept is one example of a tumor necrosis factor (TNF) inhibitor that has demonstrated promise in treating neuroinflammation in conditions like Parkinson's and Alzheimer's.

15.4.5.2 Antioxidants

Vitamins C and E are examples of antioxidants that may help lessen inflammation and oxidative stress in the brain. These antioxidants can be obtained by eating a diet high in fruits and vegetables.

15.4.5.3 Dietary Adjustments

Studies have indicated that the occurrence of neurodegenerative disorders may be reduced by following a Mediterranean diet rich in fruits, vegetables, whole grains, and healthy fats like olive oil. It is believed that this diet reduces inflammation in the brain [101].

15.4.5.4 Anti-Inflammatory Lifestyle and Exercise

Adopting an anti-inflammatory lifestyle, which includes giving up smoking and excessive alcohol use, keeping a healthy weight, and getting enough sleep, could halt the neuroinflammation. Consistent physical activity reduces inflammation and improves mental health in general. It has been demonstrated that exercise helps with neuroprotection and lowers inflammation in many neurodegenerative illnesses [102].

15.4.5.5 Stress Management

Persistent stress might make brain inflammation worse. Neuroinflammation may be managed with the aid of stress-reduction methods, including yoga, mindfulness meditation, and deep breathing exercises [103].

15.4.5.6 Omega-3 Fatty Acids

Omega-3 fatty acids are found in fish oil and may be helpful in the handling of neuroinflammation because of their anti-inflammatory properties.

15.4.5.7 Immunomodulation Therapies

Various novel approaches to controlling the immunological response and reducing neuroinflammation in neurodegenerative illnesses are being investigated, including stem cell transplantation and immunomodulatory medicines [104].

Seeking advice from a healthcare professional is essential for a customised strategy for neuroinflammation management in neurodegenerative illnesses. Depending on the particular disease, its stage, and unique patient characteristics, there may be differences in treatment options and recommendations [105]. Furthermore, further investigation is probably going to turn up fresh methods and treatments for treating neuroinflammation in these ailments.

15.4.6 Neuroprotective Approaches

Experiments have demonstrated the biological properties of natural products to include anti-inflammatory, anti-apoptotic, and antioxidant properties. More studies have demonstrated the efficacy of natural products in a range of *in vitro* and *in vivo* preclinical frameworks of neurodegenerative diseases. Fruits, vegetables, nuts, herbs, and fresh and marine flora are only a few examples of natural items that include phytoconstituents, such as polyphenolic antioxidants [106]. These phytoconstituents can improve memory and other brain functions while delaying neurodegeneration. They are also known to be crucial for the management and avoidance of many neurodegenerative diseases. Owing to the limitations of currently available pharmacological treatments for neurodegenerative disorders, extensive research has been conducted on novel substances and nonpharmacological approaches to modify the course of these diseases while reducing drug side effects. Preserving the number of neurons, averting cell death, and restoring the functionality of damaged neurons are all parts of the broad concept of neuroprotection [107]. It seems logical that the development of drugs that halt or slow the advancement of neurodegenerative diseases will result from a deeper comprehension of the etiology and pathophysiology of these conditions. Undoubtedly, both sectors have seen tremendous progress in the last several years, and there is now much more hope for the launch of 'neuroprotective' medications [108]. **Table 15.1** represents the various functions of natural products' constituents in the management of different neurodegenerative disorders.

15.4.7 Cognitive Approaches

Neurodegenerative disease management mostly relies on cognitive interventions. They can assist people in overcoming the emotional and cognitive difficulties brought on by these illnesses, even while they are unable to change the underlying pathology [131]. The following cognitive therapies are frequently employed in the treatment of neurodegenerative diseases.

15.4.7.1 Cognitive Stimulation Therapy (CST)

CST includes planned group exercises intended to improve social interaction, memory, and cognitive function. When used in conjunction with other treatments, such as Alzheimer's disease prevention, it can help to preserve cognitive function in affected patients [132].

TABLE 15.1
Various Functions of Natural Products' Constituents in the Management of Different Neurodegenerative Disorders

S. No.	Plants and Their Common Names	Role in the Management of Neurodegenerative Disease	Constituents	Functions	Refs.
1.	*Ananas comosus* (pineapple)	Alzheimer's disease	Bromelain	Breaks down amyloid-β42 oligomers and monomers in AD patients' CSF and breaks down artificial forms as well	[109]
2.	*Vaccinium angustifolium* (lowbush blueberry)	Alzheimer's disease	Anthocyanins	Reduction of oxidative damage and decline in age-related protein expression, including NF-kB	[110]
3.	*Juglans regia* (Persian walnut)	Alzheimer's disease	Vitamins, polyphenols, notably ellagic acid, alpha-tocopherol, and fatty acids	Defibrillation of preformed Aβ, inhibition of Aβ fibril formation, and amyloidogenic activity	[111]
4.	*Catharanthus roseus* (Madagascar periwinkle)	Alzheimer's disease	Citric acids, flavonoids, and caffeoylquinic acid	Inhibition of acetylcholine esterase enzyme	[112]
5.	*Arbutus unedo* (strawberry tree)	Alzheimer's disease	Carotenoids, gallic acid, tannins, vitamins C and E, anthocyanins	Antioxidant capacity to protect PC12 cells from hydrogen peroxide-induced neurotoxicity, enhance spatial learning, and increase memory	[113]
6.	*Vitis vinifera* (grape vine)	Alzheimer's disease	Resveratrol	Reduction of inflammation and oxidative stress by sirtuin 1 activation, which also lowers NF-kB activity and FOXO protein apoptotic activity.	[114]
7.	*Magnolia officinalis* (Houpo)	Alzheimer's disease	Magnolol, 4-O-methylhonokiol, honokiol, and obovatol	The capacity of AChE to inhibit and enhance memory	[115]
8.	*Cinnamomumverum* (cinnamon)	Alzheimer's disease	Cinnamyl alcohol, cinnamaldehyde, eugenol, and cinnamyl acetate	Inhibition of the production of oligomers and amyloid fibrils	[116]
9.	*Securinega suffruticosa* (yi ye qiu)	Parkinson's disease	Securinine	Anti-inflammatory effects contributing to neuroprotection	[117]
10.	Sex hormones	Parkinson's disease	17β-estradiol	Neuroprotection through the activation of receptors	[118]
11.	Estrogenic compounds	Parkinson's disease	Progesterone with 17β-estradiol	Neuroprotective effects	[119]
12.	Natural polyphenolic compounds	Parkinson's disease	Phytoestrogens	Neuroprotection through the activation of receptors	[118]
13.	*Genistein* (genistein)	Amyotrophic lateral sclerosis	4,5,7-trihydroxy isoflavone	Anti-angiogenic and anti-viral	[120]

(Continued)

TABLE 15.1 *(Continued)*
Various Functions of Natural Products' Constituents in the Management of Different Neurodegenerative Disorders

S. No.	Plants and Their Common Names	Role in the Management of Neurodegenerative Disease	Constituents	Functions	Refs.
14.	*Ginkgo biloba* (ginkgo)	Amyotrophic lateral sclerosis	Ginkgolides A, B, C, and J are terpenestrilactones Bilobalides, a sesquiterpene tri lactone, and myricetinkaem-pherol and quercetin, a flavonol glycoside	Cognitive/memory enhancement	[121]
15.	*Epigallocatechin Gallate (EGCG)* (catechin)	Amyotrophic lateral sclerosis	Flavan-3-ol	Antioxidant, anti-inflammatory	[122]
16	*Resveratrol* (resveratrol)	Amyotrophic lateral sclerosis	3,5,4-trihydroxy trans stilbene	Anti-aging, anti-ischemic	[123]
17.	*Panax ginseng and other species I* (ginseng)	Amyotrophic lateral sclerosis	Protopanaxadiol and protopanaxatriol (ginsenosides)	Antioxidant, anti-inflammatory	[124]
18.	*Tomatoes and tomato-based products*	Huntington's disease	Lycopene	Neuroprotection via an additional nitric acid route and antioxidant activity	[125]
19.	*Several fruits and vegetables*	Huntington's disease	Flavonoids	Nitric oxide synthase suppression	[126]
20.	*Bacopa monnieri* (Brahmi)	Huntington's disease	Leaf powder	Protective function against oxidative damage to the brain	[127]
21.	*Sesamum indicum (oil)* (sesame)	Huntington's disease	Sesamol	Effects that guard against 3-NP-induced HD	[128]
22.	*Curcuma longa* (turmeric)	Huntington's disease	Curcumin	Inflation of SDH activity and eradication of motor defects	[129]
23.	*Several plants, particularly grapevines*	Huntington's disease	Resveratrol	Enhancement of maze performance and locomotor activity	[123]
24.	*Panax ginseng* (five fingers)	Huntington's disease	Root	Preservation of the structural integrity of neurons and enhancement of cognitive function	[130]

15.4.7.2 Cognitive Rehabilitation

Programs for cognitive rehabilitation are individualised interventions to enhance particular cognitive abilities, including memory, attention, and problem-solving. Regular exercises and measures to adjust for cognitive deficiencies are frequently included in these programs [133].

15.4.7.3 Reality Orientation Therapy

This treatment aids people with cognitive impairments in keeping track of identity, location, and time. It can lessen disorientation and confusion, which are prevalent in illnesses like Alzheimer's [134].

15.4.7.4 Reminiscence Therapy

Reminiscence therapy helps people remember and talk about their prior experiences, which can improve their mood, strengthen their sense of self, and increase interaction and communication [135].

15.4.7.5 Errorless Learning

This method helps people with cognitive limitations feel less frustrated by reducing the possibility of making mistakes. It's frequently applied to assignments like imparting new talents or retrieving forgotten ones [136].

15.4.7.6 Montessori-based Interventions

These interventions modify the concepts of Montessori education to assist people suffering from neurodegenerative disorders in participating in meaningful activities that correspond to their cognitive capacities and interests [137].

15.4.7.7 Cognitive Behavioral Therapy (CBT)

CBT can assist people in coping with the behavioral and emotional side effects of neurodegenerative illnesses. It emphasises recognising and changing harmful mental patterns and actions [138].

15.4.7.8 Speech and Language Therapy

Speech and language therapists assist people with communication issues, such as issues with speech and comprehension as well as expression of language.

15.4.7.9 Occupational Therapy (OT)

Occupational therapists assist patients in preserving their capacity to carry out daily tasks, adjusting to cognitive modifications, and developing plans to increase their level of independence.

15.4.7.10 Art Therapy

People who struggle with verbal communication may find that art therapy offers them a creative alternative for self-expression and emotional processing.

15.4.7.11 Music Therapy

Music therapy can elevate mood, encourage social engagement and interaction, and stimulate cognitive and emotional reactions.

15.4.7.12 Mindfulness and Relaxation Techniques

These methods can assist people in coping with the usual emotional difficulties linked to neurodegenerative disorders, such as stress, anxiety, and depression.

15.4.7.13 Support Groups and Education

Individuals and their caregivers can benefit from emotional support, useful guidance, and a sense of community through participating in support groups and educational programs.

The needs, preferences, stages, and kinds of neurodegenerative disease all influence the choice of cognitive therapy. To address the condition's physical, emotional, and cognitive elements, a multidisciplinary team may be included in the integration of these therapies into an all-encompassing care plan [139].

15.4.8 Physical Rehabilitation

Strategies for physical rehabilitation are crucial parts of neurodegenerative disease management. These tactics seek to increase or preserve physical function, promote mobility, and lower the chance of problems related to these disorders [140]. The following are a few physical rehabilitation techniques for neurodegenerative disease management.

15.4.8.1 Physical Therapy

To help with strength, flexibility, balance, and mobility, physical therapists create customised exercise and movement regimens. Their main focus is on treating particular neurodegenerative disease symptoms such as muscle weakness, aberrant gait, and poor coordination.

15.4.8.2 Occupational Therapy

Occupational therapists assist people with neurodegenerative disorders in carrying out activities of daily living in a more autonomous manner. To preserve independence and enhance function, they suggest assistive technology and offer adaptive technique instruction.

15.4.8.3 Speech and Swallowing Therapy

Speech therapists assist patients who may have trouble speaking or swallowing as a result of illnesses like ALS or Parkinson's disease. The goals of these therapies are to enhance swallowing safety, vocal clarity, and articulation.

15.4.8.4 Assistance Devices

Those with neurodegenerative disorders can become more mobile and independent by using assistance devices such as wheelchairs, walkers, canes, and communication aids.

15.4.8.5 Range of Motion (ROM) Exercises

ROM exercises help to prevent contractures, which can restrict mobility, and maintain joint flexibility. They are particularly crucial for people who have restricted movement.

15.4.8.6 Strengthening Exercises

People with neurodegenerative disorders must maintain and increase their muscle strength. Specific exercises can enhance stability and reduce the risk of falling.

15.4.8.7 Balance and Coordination Training

Personalised balance activities, such as yoga or tai chi, can improve stability and lower the chance of falls.

15.4.8.8 Gait Training

Designed to help people with irregular gait keep their mobility and prevent falls, gait training aims to improve the pattern and quality of walking.

15.4.8.9 Hydrotherapy (Aquatic Therapy)

People with mobility issues may find that water-based workouts in a safe setting, like a pool, are very helpful. Water's buoyancy lessens the force on muscles and joints.

15.4.8.10 Aerobic Exercise

Regular aerobic exercise can enhance general health and cardiovascular fitness. Examples of this type of exercise include walking, swimming, and stationary cycling.

15.4.8.11 Breathing Therapy

People who have certain neurodegenerative disorders, such as ALS, may have a weakening in their breathing muscles. The therapies that respiratory therapists offer can help with breathing and lung health.

15.4.8.12 Integration of Cognitive and Motor Functions

Various neurodegenerative illnesses, such as Parkinson's disease, might impair the synchronisation of cognitive and motor abilities. Cognitive-motor therapy might be helpful when combined with physical activity [141].

15.4.8.13 Community-based Exercise Programs

Patients should be encouraged to take part in fitness classes or other community-based exercise programs that are specially created for those with neurodegenerative illnesses.

15.4.8.14 Education and Self-Management

Care teams can provide knowledge and skills to help people and others who care for them control their symptoms and keep their bodies functioning.

Physical rehabilitation techniques must be customised to meet the unique demands and capabilities of every individual with a neurodegenerative disease. A caregiver, therapist, and medical professional working together in a multidisciplinary approach can offer complete care that takes into account the physical difficulties brought on by these illnesses. As the illness worsens, regular evaluations and modifications to the rehabilitation program are frequently required [142].

15.4.9 Gut-brain Axis and Microbiota Modulation

The gut-brain axis and microbiome modulation have garnered significant attention in recent times owing to their possible application in the management of neurodegenerative diseases. Though more research is needed to fully understand the mechanisms involved, the gut microbiota may have an impact on brain health and be involved in the start and progression of neurodegenerative illnesses like ALS, Parkinson's disease, and Alzheimer's. The term 'gut microbiota' refers to the trillions of microorganisms that live within the gut. These microorganisms consist of many germs, viruses, bacteria, and fungi. The composition and variety of the gut microbiota can have an impact on many health issues, including those that affect the immune system, metabolism, and digestion [143].

Important information on the gut-brain axis and microbiome modification concerning neurodegenerative illnesses is as follows.

15.4.9.1 Gut-Brain Axis

The central nervous system (CNS) and the gastrointestinal tract (gut) are connected by this two-way gut-brain axis communication mechanism. This exchange takes place via immunological, endocrine, and neurological processes.

15.4.9.2 Gut Microbiota and Neurodegenerative Diseases

A new study points to a possible connection between changes in the microbiota of the gut and the onset or aggravation of neurodegenerative illnesses. An imbalance or disturbance in the gut microbiota, known as dysbiosis, may be a factor in oxidative stress, neuroinflammation, and the generation of toxic compounds that can have an impact on the brain.

15.4.9.3 Mechanisms of Influence

There are various ways in which the gut microbiota can affect the gut-brain axis: (A) the synthesis of neurotransmitters and neuroactive substances that influence mood and cognitive performance, such as serotonin, dopamine, and short-chain fatty acids; (B) immune system regulation, which may affect neuroinflammation; and (C) the production of metabolic waste products that can cross the blood-brain barrier and impact brain function [144].

15.4.9.4 Possible Techniques for Modulating the Microbiota

- *Probiotics*: Restoring a healthy gut microbiota balance can be facilitated by introducing helpful bacteria into the gut through probiotic supplements or fermented foods.
- *Prebiotics*: Prebiotics are non-digestible fibers that encourage the development of good gut bacteria. Foods, including fruits, vegetables, and whole grains, can provide them.
- *Diet*: A varied, well-balanced diet high in fruits, vegetables, and fiber can support healthy gut microbiota.

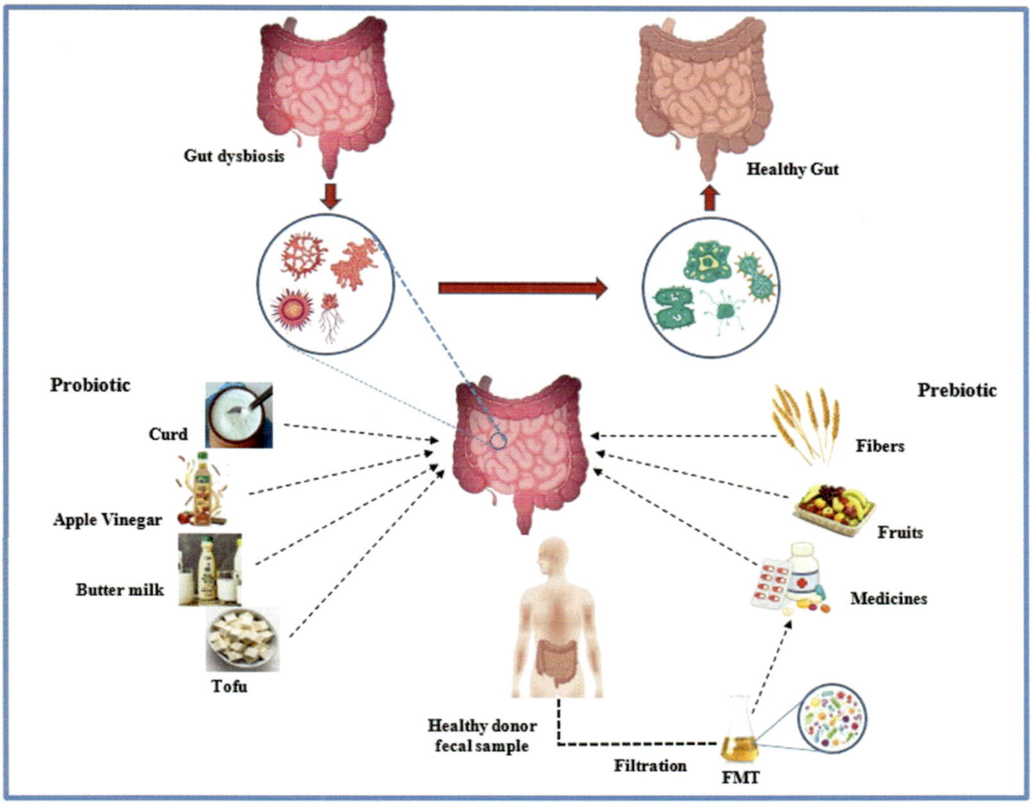

FIGURE 15.2 Potential treatments aimed at targeting the gut microbiota to modify the neurodegeneration caused by gut dysbiosis including fecal microbiota transplantation (FMT); prebiotics such as barley, fiber, and fruits; and probiotics like kimchi, yogurt/curd, pickles, buttermilk, and apple cider vinegar.

- *Fecal microbiota transplantation* (*FMT*): FMT is the process of giving a recipient unbalanced gut microbiota (fecal matter) from a healthy donor. In some gastrointestinal problems, it has demonstrated potential factor.

Although the gut-brain axis and microbiota regulation appear promising, it's crucial to understand that research in this field is still in its early stages and that the precise mechanisms and potential treatments are still being uncovered. Individualised strategies for microbiome modification might be required since different people may have different reactions in their gut microbiota concerning neurodegenerative illnesses [45]. Research on the gut-brain axis and microbiome regulation appears a promising approach for the treatment of neurodegenerative illnesses. Maintaining a healthy gut microbiota through a balanced diet, probiotics, and other therapies may have potential benefits for brain health and the management of various diseases; however, the precise linkages and therapeutic options are still being investigated. To fully comprehend and utilise the gut-brain axis therapeutic potential in the setting of neurodegenerative disorders, more research is required [145]. **Figure 15.2** represents the various treatment processes to modify the neurodegeneration by targeting the gut microbiota.

15.4.10 Ethical and Social Implications

15.4.10.1 Ethical

It might be difficult to handle moral conundrums in end-stage neurodegenerative disorders for both caregivers and medical professionals. These dilemmas often revolve around issues related to patient autonomy, quality of life, end-of-life care, and family involvement. When considering the patient as a whole, these illnesses harm their mental and emotional health, as well as their families and general quality of life. From a bioethical standpoint, these illnesses result in people losing their ability to make decisions for themselves [3, 146].

15.4.10.2 Respect for Autonomy

For as long as is feasible, make sure people suffering from neurodegenerative disorders are given the chance to make decisions about their treatment. Advanced care planning, which includes designating a healthcare proxy and using advance directives, may be necessary for this. When a patient is unable to make decisions for themselves, family members or surrogate decision-makers should be included in making decisions that correspond with the patient's prior desires [147].

15.4.10.3 Harmonising Palliative Care with Treatment

Deciding whether to switch to palliative care or stick with harsh treatments that might provide only temporary relief can present ethical challenges. Have frank conversations regarding the advantages, disadvantages, and risks of different therapy options with patients and their families. Promote the use of palliative care with an emphasis on comfort, quality of life, and symptom management, and make sure the patient's family is aware of this change in the care's objectives [148].

15.4.10.4 Withholding or Stopping Medical Intervention

In rare situations, it could be morally right to refuse or stop using certain medical treatments that are no longer helpful or in keeping with the patient's objectives. Adherence to the relevant legal and ethical requirements is necessary. When making these tough choices, involve a multidisciplinary team that includes the patient's relatives, ethicists, and palliative care specialists [149].

15.4.10.5 Family Conflicts and Involvement

Family members may dispute over how best to care for and treat the illness. It is the goal of healthcare providers to keep lines of communication open with all parties. Try to resolve disputes among the family while honoring the patient's previous desires and best interests.

15.4.10.6 Standard of Living and Pain

It is important to assess the patient's level of suffering and standard of living regularly. Depending on whether the patient's illness is causing them undue suffering, ethical considerations may need to be taken to determine the right course of medical care.

15.4.10.7 Cultural and Religious Considerations

Be aware that various religions and cultures may have different viewpoints on end-of-life care. As long as the patient's welfare is maintained, try your best to respect and take into account their beliefs.

15.4.10.8 Psychological Care and Emotional Assistance

It offers patients and their families' psychological and emotional assistance. Everyone who has a neurodegenerative disease may experience severe mental health and well-being declines.

15.4.10.9 Frequent Ethical Consultations

To offer direction and assistance in handling difficult ethical conundrums, think about consulting clinical ethicists or ethics committees. The patient's best interests and well-being ought to come first in every situation. To successfully navigate the moral conundrums that frequently occur in end-stage neurodegenerative diseases, open and empathetic communication with patients and their families is essential.

15.4.11 Social

People with neurodegenerative diseases have a variety of social challenges, including those relating to their families, communities, and caregivers. Improving the general quality of life for individuals impacted by these diseases requires managing their social ramifications. The following are some social ramifications of neurodegenerative illnesses and techniques for managing them.

15.4.11.1 Social Isolation and Stigma

To lessen the stigma attached to neurodegenerative illnesses, raise awareness, and encourage education. A sense of belonging and lessening social isolation can be provided via support groups and community organisations.

15.4.11.2 Stress and Burnout among Caregivers

Handle it by offering options for education, support groups, and respite care. To avoid burnout, encourage them to ask for assistance and take pauses.

15.4.11.3 Financial Burden

Help people and families obtain financial resources such as insurance, money for respite care, and disability benefits. Give guidance on financial planning to assist in controlling long-term care expenses.

15.4.11.4 Employment Challenges

Press companies to provide adequate accommodations and flexible work schedules for staff members suffering from neurodegenerative diseases. Encourage the use of vocational rehabilitation services to assist people in keeping their jobs or locating acceptable alternatives.

15.4.11.5 Disruption of Family Dynamics

Offer family education and counseling to assist family members in overcoming the practical and emotional difficulties associated with providing care. To keep supportive ties within the family, promote open communication.

15.4.11.6 Shifts in Social Roles

Support people with neurodegenerative diseases to continue participating in social activities and feeling purposeful. Encourage adjustments to their tasks and roles within their social networks.

15.4.11.7 Effects on Relationships

Encourage emotional support and communication in relationships. By offering services and counseling, you can address changes in emotional closeness and relationships.

15.4.11.8 Loss of Independence

To maximise independence and preserve a sense of autonomy for people with neurodegenerative illnesses, implement assistive equipment, home adaptations, and training.

15.4.11.9 Obstacles to Healthcare Access

Promote easily accessible healthcare services and modes of mobility. Assist people in navigating the healthcare system by providing information on nearby services.

15.4.11.10 Cognitive and Emotional Changes

Provide therapy, such as cognitive rehabilitation, counseling, and psychiatric treatment, to address cognitive and emotional needs. Encourage the use of stress-reduction and mindfulness practices.

15.4.11.11 Community Engagement

Motivate people to get involved in their communities by volunteering, attending social events, or joining support groups.

15.4.11.12 Legal and Ethical Concerns

Assist people and families with advanced care planning, including the use of power of attorney, advance directives, and, if required, legal guardianship.

15.4.11.13 Educational Support

Support students with neurodegenerative conditions and promote inclusive learning environments. Provide customised lesson plans and accommodations to guarantee students get a top-notch education.

15.4.11.14 Investigation and Protest

Encourage research into neurodegenerative illnesses to identify viable therapies and treatments. Urge people to join advocacy groups and clinical trials, as well as families.

A comprehensive strategy including social workers, healthcare providers, support groups, and communities is needed to address the social effects of neurodegenerative illnesses. It is possible to raise the quality of life for those impacted and their families while encouraging social inclusion and well-being by offering a network of support and services [132].

15.5 CHALLENGES AND FUTURE DIRECTIONS

The field of managing neurodegenerative illnesses is intricate, dynamic, and fraught with difficulties. Still, there are encouraging avenues for the future that might result in better patient outcomes and more robust medicines [150]. The following are some major issues and potential paths for the treatment of neurodegenerative diseases.

15.5.1 Challenges

15.5.1.1 Initial Diagnosis

Many neurodegenerative disorders are discovered at an advanced stage of the illness, when major damage has already occurred. To intervene in the illness process before permanent damage occurs, it is imperative to develop precise and easily accessible technologies for early diagnosis.

15.5.1.2 Disease Heterogeneity

It is difficult to create a single treatment that works for everyone since neurodegenerative illnesses frequently exhibit a broad variety of clinical and pathological characteristics. It is necessary to use personalised medical techniques that take into account each person's unique genetic, molecular, and clinical profiles.

15.5.1.3 Blood-Brain Barrier

This barrier makes it more difficult to administer treatment to the damaged parts of the brain by preventing medications and therapeutic substances from passing through. Novel approaches to medication administration, like targeted ultrasound and nanoparticles, are being investigated to get beyond this obstacle [151].

15.5.1.4 Absence of Disease-Modifying Therapies

Rather than targeting the underlying causes of neurodegenerative disorders, the majority of existing treatments concentrate on managing symptoms. The development of disease-modifying medications that target the underlying causes of these illnesses is a significant challenge.

15.5.1.5 Ethical Concerns

Utilising cutting-edge technologies to treat neurodegenerative diseases, such as brain-computer interfaces and gene editing, presents ethical dilemmas; therefore, it's critical to establish an equilibrium between the possible advantages of these technologies and ethical issues.

79. O. Strianese, F. Rizzo, M. Ciccarelli, G. Galasso, Y. D'Agostino, A. Salvati, C. Del Giudice, P. Tesorio, M.R. Rusciano, Precision and personalized medicine: How genomic approach improves the management of cardiovascular and neurodegenerative disease, Genes (Basel) 11 (2020) 747–770.
80. K.B. Johnson, W.Q. Wei, D. Weeraratne, M.E. Frisse, K. Misulis, K. Rhee, J. Zhao, J.L. Snowdon, Precision medicine, AI, and the future of personalized health care, Clin Transl Sci 14 (2021) 86–93.
81. H. Hampel, S.E. O'Bryant, J.I. Castrillo, C. Ritchie, K. Rojkova, K. Broich, N. Benda, R. Nisticò, R.A. Frank, B. Dubois, V. Escott-Price, S. Lista, Precision Medicine - The golden gate for detection, treatment and prevention of Alzheimer's disease, J Prev Alzheimers Dis 3 (2016) 243–259.
82. V. Chiruvella, A.K. Guddati, Ethical issues in patient data ownership, Interact J Med Res 10 (2021) e22269.
83. J. Kumar, T. Patel, F. Sugandh, J. Dev, U. Kumar, M. Adeeb, M.P. Kachhadia, P. Puri, F. Prachi, M.U. Zaman, S. Kumar, G. Varrassi, R.S. Syed, Innovative approaches and therapies to enhance neuroplasticity and promote recovery in patients with neurological disorders: A narrative review, Cureus 15 (2023) e41914.
84. I. Mortada, R. Farah, S. Nabha, D.M. Ojcius, Y. Fares, W.Y. Almawi, N.S. Sadier, Immunotherapies for neurodegenerative diseases, Front Neurol 12 (2021) 654739.
85. C. Song, J. Shi, P. Zhang, Y. Zhang, J. Xu, L. Zhao, R. Zhang, H. Wang, H. Chen, Immunotherapy for Alzheimer's disease: Targeting β-amyloid and beyond, Translational Neurodegeneration 11 (2022) 18.
86. D. Kempuraj, R. Thangavel, P.A. Natteru, G.P. Selvakumar, D. Saeed, H. Zahoor, S. Zaheer, S.S. Iyer, A. Zaheer, Neuroinflammation induces neurodegeneration, J Neurol Neurosurg Spine 1 (2016) 1003–1017.
87. L. Muzio, A. Viotti, G. Martino, Microglia in neuroinflammation and neurodegeneration: From understanding to therapy, Front Neurosci 15 (2021) 742065.
88. F. Mantile, A. Prisco, Vaccination against β-amyloid as a strategy for the prevention of Alzheimer's disease, Biology (Basel) 9 (2020) 425–445.
89. D.M. Hoang, P.T. Pham, T.Q. Bach, A.T.L. Ngo, Q.T. Nguyen, T.T.K. Phan, G.H. Nguyen, P.T.T. Le, V.T. Hoang, N.R. Forsyth, M. Heke, L.T. Nguyen, Stem cell-based therapy for human diseases, Signal Transduct Target Ther 7 (2022) 272.
90. D.H. Upton, C. Ung, S.M. George, M. Tsoli, M. Kavallaris, D.S. Ziegler, Challenges and opportunities to penetrate the blood-brain barrier for brain cancer therapy, Theranostics 12 (2022) 4734–4752.
91. W. Chen, Y. Hu, D. Ju, Gene therapy for neurodegenerative disorders: Advances, insights and prospects, Acta Pharm Sin B 10 (2020) 1347–1359.
92. J. Nouri Nojadeh, N.S. Bildiren Eryilmaz, B.I. Ergüder, CRISPR/Cas9 genome editing for neurodegenerative diseases, EXCLI J 22 (2023) 567–582.
93. J.K. Watts, D.R. Corey, Silencing disease genes in the laboratory and the clinic, J Pathol 226 (2012) 365–379.
94. D.G.T. Parambi, K.S. Alharbi, R. Kumar, S. Harilal, G.E. Batiha, N. Cruz-Martins, O. Magdy, A. Musa, D.S. Panda, B. Mathew, Gene Therapy approach with an emphasis on growth factors: Theoretical and clinical outcomes in neurodegenerative diseases, Mol Neurobiol 59 (2022) 191–233.
95. W. Liu, L. Li, J. Jiang, M. Wu, P. Lin, Applications and challenges of CRISPR-Cas Gene-editing to disease treatment in clinics, Precis Clin Med 4 (2021) 179–191.
96. P. Sweeney, H. Park, M. Baumann, J. Dunlop, J. Frydman, R. Kopito, A. McCampbell, G. Leblanc, A. Venkateswaran, A. Nurmi, R. Hodgson, Protein misfolding in neurodegenerative diseases: Implications and strategies, Transl Neurodegener 6 (2017) 6.
97. D. Wang, P.W.L. Tai, G. Gao, Adeno-associated virus vector as a platform for gene therapy delivery, Nat Rev Drug Discov 18 (2019) 358–378.
98. R. Martier, P. Konstantinova, Gene therapy for neurodegenerative diseases: Slowing down the ticking clock, Front Neurol 14 (2020) 580179–580204.
99. W.W. Chen, X. Zhang, W.J. Huang, Role of neuroinflammation in neurodegenerative diseases (review), Mol Med Rep 13 (2016) 3391–3396.
100. D.L. Krause, N. Müller, Neuroinflammation, microglia and implications for anti-inflammatory treatment in Alzheimer's disease, Int J Alzheimers Dis 2010 (2010) 1–9.
101. D.F. Romagnolo, O.I. Selmin, Mediterranean diet and prevention of chronic diseases, Nutr Today 52 (2017) 208–222.
102. E. Kip, L.C. Parr-Brownlie, Healthy lifestyles and wellbeing reduce neuroinflammation and prevent neurodegenerative and psychiatric disorders, Front Neurosci 17 (2023) 1092537.
103. D. Krishnakumar, M.R. Hamblin, S. Lakshmanan, Meditation and yoga can modulate brain mechanisms that affect behavior and anxiety-a modern scientific perspective, Anc Sci 2 (2015) 13–19.
104. C.P. Palanisamy, J. Pei, P. Alugoju, N.V.A. Anthikapalli, S. Jayaraman, V.P. Veeraraghavan, S. Gopathy, J.R. Roy, C.S. Janaki, D. Thalamati, M. Mironescu, Q. Luo, Y. Miao, Y. Chai, Q. Long, New strategies

of neurodegenerative disease treatment with extracellular vesicles (EVs) derived from mesenchymal stem cells (MSCs), Theranostics 13 (2023) 4138–4165.
105. L.K. Wareham, S.A. Liddelow, S. Temple, L.I. Benowitz, A. Di Polo, C. Wellington, J.L. Goldberg, Z. He, X. Duan, G. Bu, A.A. Davis, K. Shekhar, A. Torre, D.C. Chan, M.V. Canto-Soler, J.G. Flanagan, P. Subramanian, S. Rossi, T. Brunner, D.E. Bovenkamp, D.J. Calkins, Solving neurodegeneration: Common mechanisms and strategies for new treatments, Mol Neurodegener 17 (2022) 23.
106. X. Chen, J. Drew, W. Berney, W. Lei, Neuroprotective natural products for Alzheimer's disease, Cells 10 (2021) 1309–1346.
107. R. Venkatesan, E. Ji, S.Y. Kim, Phytochemicals that regulate neurodegenerative disease by targeting neurotrophins: A comprehensive review, Biomed Res Int 2015 (2015) 814068.
108. A. Nowacek, L.M. Kosloski, H.E. Gendelman, Neurodegenerative disorders and nanoformulated drug development, Nanomedicine (Lond) 4 (2009) 541–555.
109. G.M. Sancesario, M. Nuccetelli, A. Cerri, J. Zegeer, C. Severini, M.T. Ciotti, M. Pieri, A. Martorana, C. Caltagirone, R. Nistico, S. Bernardini, Bromelain degrades Aβ1-42 monomers and soluble aggregates: An in vitro study in cerebrospinal fluid of Alzheimer's disease patients, Curr Alzheimer Res 15 (2018) 628–636.
110. M.A. Papandreou, A. Dimakopoulou, Z.I. Linardaki, P. Cordopatis, D. Klimis-Zacas, M. Margarity, F.N. Lamari, Effect of a polyphenol-rich wild blueberry extract on cognitive performance of mice, brain antioxidant markers and acetylcholinesterase activity, Behav Brain Res 198 (2009) 352–358.
111. V. Chauhan, A. Chauhan, Oxidative stress in Alzheimer's disease, Pathophysiology 13 (2006) 195–208.
112. D.M. Pereira, F. Ferreres, J. Oliveira, P. Valentão, P.B. Andrade, M. Sottomayor, Targeted metabolite analysis of Catharanthus roseus and its biological potential, Food Chem Toxicol 47 (2009) 1349–1354.
113. H.J. Heo, C.Y. Lee, Strawberry and its anthocyanins reduce oxidative stress-induced apoptosis in PC12 cells, J Agric Food Chem 53 (2005) 1984–1989.
114. V. Vingtdeux, U. Dreses-Werringloer, H. Zhao, P. Davies, P. Marambaud, Therapeutic potential of resveratrol in Alzheimer's disease, BMC Neurosci 9 Suppl 2 (2008) S6.
115. Y.K. Lee, D.Y. Yuk, T.I. Kim, Y.H. Kim, K.T. Kim, K.H. Kim, B.J. Lee, S.Y. Nam, J.T. Hong, Protective effect of the ethanol extract of Magnolia officinalis and 4-O-methylhonokiol on scopolamine-induced memory impairment and the inhibition of acetylcholinesterase activity, J Nat Med 63 (2009) 274–282.
116. A. Frydman-Marom, A. Levin, D. Farfara, T. Benromano, R. Scherzer-Attali, S. Peled, R. Vassar, D. Segal, E. Gazit, D. Frenkel, M. Ovadia, Orally administrated cinnamon extract reduces β-amyloid oligomerization and corrects cognitive impairment in Alzheimer's disease animal models, PLoS One 6 (2011) e16564.
117. B. Kandinov, N. Giladi, A.D. Korczyn, Smoking and tea consumption delay onset of Parkinson's disease, Parkinsonism Relat Disord 15 (2009) 41–46.
118. M.A. Arevalo, I. Azcoitia, L.M. Garcia-Segura, The neuroprotective actions of oestradiol and oestrogen receptors, Nat Rev Neurosci 16 (2015) 17–29.
119. S. Callier, M. Morissette, M. Grandbois, D. Pélaprat, T. Di Paolo, Neuroprotective properties of 17beta-estradiol, progesterone, and raloxifene in MPTP C57Bl/6 mice, Synapse 41 (2001) 131–138.
120. K. Polkowski, A.P. Mazurek, Biological properties of genistein. A review of in vitro and in vivo data, Acta Pol Pharm 57 (2000) 135–155.
121. Z. Zhang, D. Peng, H. Zhu, X. Wang, Experimental evidence of Ginkgo biloba extract EGB as a neuroprotective agent in ischemia stroke rats, Brain Res Bull 87 (2012) 193–198.
122. N. Khan, F. Afaq, M. Saleem, N. Ahmad, H. Mukhtar, Targeting multiple signaling pathways by green tea polyphenol (-)-epigallocatechin-3-gallate, Cancer Res 66 (2006) 2500–2505.
123. M. Yáñez, L. Galán, J. Matías-Guiu, A. Vela, A. Guerrero, A.G. García, CSF from amyotrophic lateral sclerosis patients produces glutamate independent death of rat motor brain cortical neurons: Protection by resveratrol but not riluzole, Brain Res 1423 (2011) 77–86.
124. T.A. Nakaya, M. Kita, H. Kuriyama, Y. Iwakura, J. Imanishi, Panax ginseng induces production of proinflammatory cytokines via toll-like receptor, J Interferon Cytokine Res 24 (2004) 93–100.
125. G. Hsiao, T.H. Fong, N.H. Tzu, K.H. Lin, D.S. Chou, J.R. Sheu, A potent antioxidant, lycopene, affords neuroprotection against microglia activation and focal cerebral ischemia in rats, In Vivo 18 (2004) 351–356.
126. G.M. Raso, R. Meli, G. Di Carlo, M. Pacilio, R. Di Carlo, Inhibition of inducible nitric oxide synthase and cyclooxygenase-2 expression by flavonoids in macrophage J774A.1, Life Sci 68 (2001) 921–931.
127. G.K. Shinomol, Muralidhara, Bacopa monnieri modulates endogenous cytoplasmic and mitochondrial oxidative markers in prepubertal mice brain, Phytomedicine 18 (2011) 317–326.
128. P.F. Hsieh, C.W. Hou, P.W. Yao, S.P. Wu, Y.F. Peng, M.L. Shen, C.H. Lin, Y.Y. Chao, M.H. Chang, K.C. Jeng, Sesamin ameliorates oxidative stress and mortality in kainic acid-induced status epilepticus by inhibition of MAPK and COX-2 activation, J Neuroinflammation 8 (2011) 57.

129. S. Choudhary, P. Kumar, J. Malik, Plants and phytochemicals for Huntington's disease, Pharmacogn Rev 7 (2013) 81–91.
130. K. Radad, G. Gille, L. Liu, W.D. Rausch, Use of ginseng in medicine with emphasis on neurodegenerative disorders, J Pharmacol Sci 100 (2006) 175–186.
131. B.N. Dugger, D.W. Dickson, Pathology of neurodegenerative diseases, Cold Spring Harb Perspect Biol 9 (2017) 28035–28056.
132. M. Ruthirakuhan, A.C. Luedke, A. Tam, A. Goel, A. Kurji, A. Garcia, Use of physical and intellectual activities and socialization in the management of cognitive decline of aging and in dementia: A review, J Aging Res 2012 (2012) 384875.
133. A. Barman, A. Chatterjee, R. Bhide, Cognitive impairment and rehabilitation strategies after traumatic brain injury, Indian J Psychol Med 38 (2016) 172–181.
134. O. Zanetti, G.B. Frisoni, D. De Leo, M. Dello Buono, A. Bianchetti, M. Trabucchi, Reality orientation therapy in Alzheimer disease: Useful or not? A controlled study, Alzheimer Dis Assoc Disord 9 (1995) 132–138.
135. B. Woods, L. O'Philbin, E.M. Farrell, A.E. Spector, M. Orrell, Reminiscence therapy for dementia, Cochrane Database Syst Rev 3 (2018) CD001120.
136. M.M. de Werd, D. Boelen, M.G. Rikkert, R.P. Kessels, Errorless learning of everyday tasks in people with dementia, Clin Interv Aging 8 (2013) 1177–1190.
137. N. Chaudhry, S. Tofique, N. Husain, D. Couture, P. Glasgow, M. Husain, T. Kiran, R. Memon, S. Minhas, A. Qureshi, F. Shuber, I. Leroi, Montessori intervention for individuals with dementia: Feasibility study of a culturally adapted psychosocial intervention in Pakistan (MIRACLE), BJPsych Open 6 (2020) e69.
138. P.J. Tully, D.A. Turnbull, J.D. Horowitz, J.F. Beltrame, B.T. Baune, S. Sauer-Zavala, H. Baumeister, C.G. Bean, R.B. Pinto, S. Cosh, G.A. Wittert, Transdiagnostic cognitive-behavioral therapy for depression and anxiety disorders in cardiovascular disease patients: Results from the CHAMPS pilot-feasibility trial, Front Psychiatry 13 (2022) 741039.
139. R.W. Levenson, V.E. Sturm, C.M. Haase, Emotional and behavioral symptoms in neurodegenerative disease: A model for studying the neural bases of psychopathology, Annu Rev Clin Psychol 10 (2014) 581–606.
140. R. Bonanni, I. Cariati, U. Tarantino, G. D'Arcangelo, V. Tancredi, Physical exercise and health: A focus on its protective role in neurodegenerative diseases, J Funct Morphol Kinesiol 7 (2022) 38–55.
141. P. Borrione, E. Tranchita, P. Sansone, A. Parisi, Effects of physical activity in Parkinson's disease: A new tool for rehabilitation, World J Methodol 4 (2014) 133–143.
142. J.H. Grand, S. Caspar, S.W. Macdonald, Clinical features and multidisciplinary approaches to dementia care, J Multidiscip Healthc 4 (2011) 125–147.
143. P. Sun, L. Su, H. Zhu, X. Li, Y. Guo, X. Du, L. Zhang, C. Qin, Gut Microbiota regulation and their implication in the development of neurodegenerative disease, Microorganisms 9 (2021) 2281–2298.
144. J. Appleton, The gut-brain axis: Influence of Microbiota on mood and mental health, Integr Med (Encinitas) 17 (2018) 28–32.
145. B. Bicknell, A. Liebert, T. Borody, G. Herkes, C. McLachlan, H. Kiat, Neurodegenerative and neurodevelopmental diseases and the gut-brain axis: The potential of therapeutic targeting of the microbiome, Int J Mol Sci 24 (2023) 9577–9626.
146. M. Klietz, Caregiver burden in movement disorders and neurodegenerative diseases: Editorial, Brain Sci 12 (2022) 1184–1187.
147. A. Giordano, L. De Panfilis, M. Perin, L. Servidio, M. Cascioli, M.G. Grasso, A. Lugaresi, E. Pucci, S. Veronese, A. Solari, Advance care planning in neurodegenerative disorders: A scoping review, Int J Environ Res Public Health 19 (2022) 803–822.
148. B.K. Mohanti, Ethics in palliative care, Indian J Palliat Care 15 (2009) 89–92.
149. B. Varkey, Principles of clinical ethics and their application to practice, Med Princ Pract 30 (2021) 17–28.
150. J. Cummings, Disease modification and neuroprotection in neurodegenerative disorders, Transl Neurodegener 6 (2017) 25.
151. S. Zhang, L. Gan, F. Cao, H. Wang, P. Gong, C. Ma, L. Ren, Y. Lin, X. Lin, The barrier and interface mechanisms of the brain barrier, and brain drug delivery, Brain Research Bulletin 190 (2022) 69–83.

16 Gene Therapy, OMICS Approaches, and Translational Neurodegenerative Models

Recent Advancements and Future Perspectives

Aakriti Sharma
Endocrinology division, CDRI, Lucknow, India

Shashank Shekhar Mishra
Centre of Experimental Medicine and Surgery, Institute of Medical Sciences, Banaras Hindu University, Varanasi, India

Sandeep Singh
Centre of Biotechnology, University of Allahabad, Prayagraj, India

16.1 INTRODUCTION

Gene therapy, OMICS approaches, and translational neurodegenerative models collectively constitute a cutting-edge triumvirate that stands at the forefront of unravelling and treating the complexities of neurodegenerative diseases. Gene therapy, rooted in the manipulation of genetic material, has evolved from its conceptual inception to groundbreaking applications. Its diverse strategies, encompassing gene replacement, silencing, and editing, hold considerable promise in addressing the intricate genetic origins of debilitating ailments such as Alzheimer's and Parkinson's.

Simultaneously, OMICS technologies – encompassing genomics, transcriptomics, proteomics, and metabolomics – offer a panoramic and comprehensive view of the molecular intricacies that underlie the complex landscape of neurodegeneration. From identifying genetic variants to delineating changes in gene expression and protein profiles, these advanced approaches play a pivotal role in deepening our understanding of the multifaceted mechanisms at play in neurodegenerative diseases.

In tandem with gene therapy and OMICS, translational neurodegenerative models emerge as indispensable bridges connecting bench research with the potential for clinical breakthroughs. These models, ranging from in vitro cell cultures to sophisticated in vivo systems such as induced pluripotent stem cells and organoids derived from patients, authentically replicate the complexities observed in human diseases. While challenges persist, continual advances in genome editing and the development of 3D cultures significantly enhance the reliability and relevance of patient-specific models.

This chapter embarks on a comprehensive exploration of these groundbreaking domains, delving into the fundamental principles of gene therapy, the expansive landscape of OMICS technologies, and the critical significance of translational models. Collectively, these techniques not only hold the key to targeted therapeutic interventions and biomarker discovery but also contribute to a

DOI: 10.1201/9781032620442-16

holistic understanding of neurodegenerative diseases. By shaping the future landscape of neuroscience research and clinical interventions, this triumvirate opens doors to innovative approaches that have the potential to restructure the diagnosis and treatment of neurodegenerative conditions.

16.2 SIGNIFICANCE OF THESE APPROACHES IN THE CONTEXT OF NEURODEGENERATIVE DISEASES

The integration of gene therapy, OMICS approaches, and translational neurodegenerative models constitutes a significant role in the persistent combat against neurodegenerative diseases. These formidable neurological disorders, exemplified by the likes of Alzheimer's and Parkinson's, present intricate challenges deeply rooted in complex genetic and molecular factors. Gene therapy emerges as a targeted and revolutionary approach, seeking to address the genetic origins of these diseases at their core, thereby aiming to correct aberrations rather than merely alleviating symptomatic manifestations.

In tandem with gene therapy, OMICS approaches wield immense significance by providing a comprehensive and systematic lens to decode the intricate molecular landscape of neurodegeneration. These cutting-edge technologies, encompassing genomics, transcriptomics, proteomics, and metabolomics, play a pivotal role in unravelling the mysteries of neurodegenerative diseases. By identifying genetic variants associated with susceptibility, delineating alterations in gene expression patterns, and elucidating changes in protein profiles, OMICS technologies contribute to a holistic understanding of the intricate molecular mechanisms. This multifaceted comprehension is not only foundational for advancing precise diagnostics but also essential for the development of tailored interventions that target specific molecular aberrations.

Translational neurodegenerative models stand as indispensable contributors to this integrative approach, providing critical platforms to test therapeutic strategies in a controlled setting before progressing to complex and dynamic clinical trials. Patient-derived models, in particular, bring a personalised dimension to the research landscape, allowing for the study of individualised disease signatures. This nuanced understanding fosters the development of personalised approaches to treatment, recognising the diversity inherent in the manifestation and progression of neurodegenerative diseases among individuals.

Collectively, the convergence of these interdisciplinary approaches represents a paradigm shift in neurodegenerative disease research. It not only deepens our comprehension of disease mechanisms but also holds the promise of defined therapies, diagnostics, and a more targeted, patient-centric approach to addressing these challenging conditions. In the ongoing quest for effective treatments that can make a tangible impact on the lives of individuals grappling with the profound consequences of neurodegenerative diseases, the harmonious interplay of gene therapy, OMICS approaches, and translational models stands as a beacon of hope, illuminating the path toward innovative solutions and improved outcomes.

16.3 IMPORTANCE OF UNDERSTANDING THE MOLECULAR BASIS OF NEURODEGENERATION FOR THERAPEUTIC INTERVENTIONS

Understanding the molecular basis of neurodegeneration stands as a critical imperative for advancing therapeutic interventions in the field of neurological disorders. Neurodegenerative diseases, exemplified by conditions like Alzheimer's and Parkinson's, initiate their destructive processes at the molecular level long before clinical symptoms surface. A profound exploration of these intricate molecular mechanisms is essential, providing invaluable insights into the initiation, progression, and ultimately, the manifestation of these debilitating disorders.

At the prime of this understanding is the identification of specific genetic mutations, the unravelling of abnormal protein aggregations, and the dissection of disrupted signaling pathways. Such insights form the bedrock upon which targeted therapeutic strategies can be meticulously constructed. The precision afforded by these strategies, addressing molecular aberrations with exactitude, holds the promise of rendering interventions more effective and tailored to the specific intricacies of each disorder. This, in turn, has the potential to not only ameliorate symptoms but also to impede or halt the relentless progression of neurodegenerative diseases.

Beyond immediate treatment benefits, a comprehensive grasp of the molecular underpinnings of neurodegeneration has profound implications for the broader landscape of neurological healthcare. The discovery of biomarkers, intricately tied to the molecular intricacies of these disorders, becomes conceivable. These biomarkers, serving as indicators of disease presence, progression, and treatment response, pave the way for early diagnosis and continuous monitoring of therapeutic efficacy. In essence, they become invaluable tools in the clinician's arsenal, facilitating proactive and personalised management strategies.

Therefore, the unravelling of molecular intricacies in neurodegeneration exceeds academic curiosity; it is the foundation upon which innovative and precisely targeted therapies can be built. This holistic understanding not only offers hope for improved clinical outcomes but also holds the potential to enhance the overall quality of life for individuals grappling with the profound challenges imposed by neurodegenerative conditions. The journey toward effective therapeutic interventions in neurodegeneration is intricately intertwined with the depth of our understanding at the molecular level, guiding us toward a future where these disorders are not only managed but genuinely conquered.

16.4 GENE THERAPY FOR NEURODEGENERATIVE DISEASES

16.4.1 Definition and Principles of Gene Therapy

Gene therapy represents a cutting-edge technique in biomedical science, heralding a paradigm shift in the approach to managing or minimising the diseases by manipulating the genetic material within an individual's cells. At its core, gene therapy embodies a profound principle – addressing the root causes of diseases at the genetic level rather than merely alleviating symptoms. This gradual process involves the strategic introduction, alteration, or removal of genetic material to correct or replace faulty genes, with the overarching goal of restoring normal cellular function.

The underlying premise of gene therapy is grounded in the fundamental understanding that numerous diseases, spanning from genetic disorders to certain types of cancers, find their origins in abnormalities within the DNA sequence. Through the application of diverse techniques, such as gene replacement, gene silencing, or gene editing, gene therapy endeavors to provide functional copies of missing or defective genes, inhibit the expression of harmful genes, or directly modify the DNA to rectify genetic abnormalities. This precision-oriented and targeted approach holds immense promise for treating a vast array of medical conditions, surpassing the boundaries of traditional therapeutic modalities.

The potential of gene therapy to transform the landscape of modern medicine cannot be overstated. Its impact extends across various domains, offering unprecedented avenues for tackling diseases at their very roots. As research and technological advancements in gene therapy continue to unfold, the prospects of this field influencing the trajectory of healthcare are becoming increasingly tangible. From inherited genetic disorders to complex conditions like cancer, gene therapy emerges as a beacon of hope, promising not just symptomatic relief but fundamental corrections at the genetic level, paving the way for a new era in the quest for improved and personalised medical interventions.

16.4.2 Historical Perspective and Key Milestones in Neurodegenerative Disease Gene Therapy

The historical evolution of neurodegenerative disease gene therapy traces a compelling journey, unfolding from initial conceptualisation to groundbreaking applications. Emerging in the late 20th

century, pioneering efforts embarked on the exploration of viral vectors as vehicles for delivering therapeutic genes into affected cells. A pivotal moment in 1990 marked the initiation of the first gene therapy clinical trial for Parkinson's disease, performed by Dr. Kaplitt and signifying a substantial leap forward in the field's advancement.

As the years progressed, the development of safer and more efficient delivery systems, such as adeno-associated viruses (AAV) and lentiviruses, propelled neurodegenerative disease gene therapy into the limelight. Notably, increased understanding of the genetic underpinnings of diseases like Huntington's and amyotrophic lateral sclerosis (ALS) spurred targeted gene therapy research, pushing the boundaries of what was previously deemed possible.

With the FDA's approval of the first gene therapy for a neurodegenerative disorder, therapy specifically designed to treat spinal muscular atrophy followed in the 2010s. This approval underscored the therapeutic potential of gene interventions in addressing the root causes of neurological conditions. These collective milestones illuminate the progressive trajectory of neurodegenerative disease gene therapy, progressing from early conceptualisation to tangible clinical applications, promising novel avenues for treating conditions that were once considered incurable.

In subsequent years, the evolution of neurodegenerative disease gene therapy experienced a surge in precision and customisation. The advent of CRISPR-Cas9 technology revolutionised the landscape, empowering targeted and specific modifications of disease-associated genes. This breakthrough enabled scientists to fine-tune interventions with unprecedented accuracy, addressing genetic anomalies at a molecular level. Moreover, collaborative international efforts cultivated a wealth of genetic data, facilitating the identification of novel therapeutic targets and enhancing the design of tailored gene therapies. As the field expanded, ethical considerations and safety protocols gained prominence, ensuring the responsible and secure deployment of these cutting-edge treatments. The ongoing synergy among genetic research, technological innovation, and regulatory frameworks continues to propel neurodegenerative disease gene therapy toward increasingly sophisticated and impactful applications, promising a future marked by transformative breakthroughs in the sphere of neurological disorders.

16.4.3 Types of Gene Therapies: Gene Replacement, Gene Silencing, Gene Editing

Gene therapy encompasses a spectrum of innovative approaches, each tailored to address specific genetic abnormalities associated with various diseases. Three prominent types of gene therapies – gene replacement, gene silencing, and gene editing – hold significant promise for treating a diverse range of conditions.

1. *Gene replacement*: Gene replacement therapy aims to address disorders caused by the absence or malfunction of a particular gene. In this approach, a functional copy of the defective gene is introduced into the patient's cells. This can be achieved using viral vectors or other delivery systems, allowing the expression of the correct gene and restoring normal cellular function. Notable applications include treating genetic disorders like cystic fibrosis or certain types of inherited blindness.
2. *Gene silencing*: Gene silencing, also known as RNA interference (RNAi), involves inhibiting the expression of a specific gene to counteract its detrimental effects. This approach utilises small RNA molecules to target and degrade the messenger RNA (mRNA) of the problematic gene, preventing the synthesis of the corresponding protein. Gene silencing is particularly relevant in diseases where overactive or mutated genes contribute to pathology, such as in certain types of cancer and neurodegenerative disorders.
3. *Gene editing*: Gene editing techniques, notably CRISPR-Cas9, enable precise modification of the DNA sequence within a cell. This revolutionary approach allows scientists to edit, add, or delete specific genetic material, offering unparalleled control over the genome. In the context of gene therapy, gene editing holds promise for correcting disease-causing

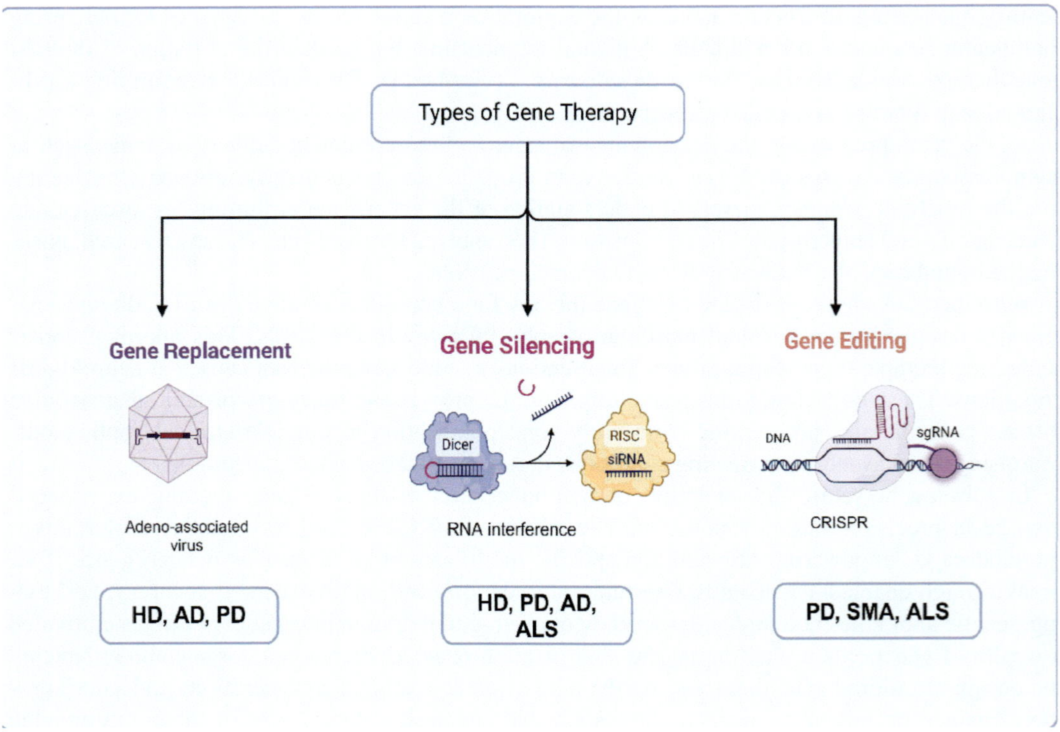

FIGURE 16.1 Different types of gene therapy – gene replacement, gene silencing, and gene editing – and their uses in respective neurodegenerative disease. *Abbreviations*: AD: Alzheimer's disease, ALS: Amyotrophic lateral sclerosis, HD: Huntington's disease, PD: Parkinson's disease. (Created with BioRender.com.)

mutations directly. It is actively investigated for disorders like sickle cell anemia and muscular dystrophy, where targeted modifications could potentially alleviate or eliminate genetic abnormalities.

These diverse gene therapy strategies collectively represent a transformative frontier in medicine, offering tailored solutions for a wide array of genetic and acquired diseases. The ongoing refinement of these techniques holds the potential to revolutionise the treatment landscape, providing more effective and personalised therapeutic interventions (**Figure 16.1**).

16.5 CASE STUDIES OF SUCCESSFUL GENE THERAPY APPLICATIONS IN NEURODEGENERATIVE DISORDERS

16.5.1 Gene Therapy for Leber Congenital Amaurosis

The narrative of Luxturna, as a groundbreaking gene therapy for Leber congenital amaurosis (LCA), serves as a compelling testament to the transformative impact that gene therapies can have on the intricate landscape of rare genetic disorders. LCA, an exceptionally rare genetic disorder, manifests with severe vision loss or blindness in infancy, primarily resulting from mutations in the RPE65 gene, crucial for producing a protein essential for normal vision.

Luxturna, a pioneering adeno-associated virus (AAV) vector-based gene therapy, represents a meticulously tailored approach to address LCA specifically caused by RPE65 mutations. Serving as a sophisticated delivery system, Luxturna facilitates the transportation of a functional RPE65 gene into retinal cells, thereby enabling the production of the essential protein required for vision.

The journey of Luxturna from conceptualisation to clinical reality is underscored by a series of rigorous clinical trials aimed at assessing both safety and efficacy. A pivotal phase III trial stands out, demonstrating significant improvements in vision among patients who received Luxturna. Notably, individuals exhibited enhanced abilities in navigating obstacle courses under low-light conditions, providing tangible evidence of the therapy's effectiveness.

The historic milestone achieved by Luxturna in December 2017 stands as a groundbreaking achievement that reverberates through the landscape of gene therapy, particularly in the expanse of inherited genetic diseases. The significance of Luxturna's accomplishment is multifaceted, extending far beyond its immediate impact on the treatment of Leber congenital amaurosis. As the first FDA-approved gene therapy for an inherited genetic disease, Luxturna carved an indelible mark in the history of medical advancements, signifying a paradigm shift in therapeutic approaches.

The approval of Luxturna proclaimed new era in the area of gene therapy, exemplifying the potential of this innovative approach to tackle the underlying genetic causes of a spectrum of neurodegenerative disorders. The ripple effects of Luxturna's approval resonated across the scientific and medical communities, sparking inspiration and catalysing subsequent developments and innovations in the burgeoning field of gene therapy. Luxturna, therefore, stands not only as a singular achievement for LCA treatment but as a catalyst for transformative progress in the broader landscape of genetic-based neurodegenerative conditions.

The narrative of Luxturna extends well beyond its initial approvals and the confines of clinical trials. Subsequent studies and long-term follow-ups have emerged as valuable chapters in Luxturna's story, providing crucial insights into the durability of its therapeutic effects. Patients undergoing Luxturna treatments have consistently demonstrated sustained improvements in vision over extended periods. This enduring impact not only reinforces the efficacy of Luxturna in addressing the specific challenges of LCA but also instills hope for the long-term viability and effectiveness of gene therapies as a whole.

Luxturna's success serves as a beacon of progress, casting a guiding light with broader implications for the field of gene therapy. It stands as a paradigmatic exemplar, showcasing the transformative potential of delivering functional genes to specific target tissues. Luxturna's success illuminates new avenues for developing treatments not only for LCA but also for a spectrum of genetic-based neurodegenerative conditions. The confidence instilled by Luxturna's achievements resonates across the scientific community, reinforcing the feasibility and promise of gene-based therapeutic approaches in reshaping the landscape of medical interventions for genetic disorders.

In conclusion, the success story of Luxturna in treating LCA encapsulates the transformative power of gene therapy. Its journey, from clinical trials to FDA approval and subsequent positive long-term outcomes, underscores the feasibility and promise of gene-based treatments. Luxturna not only stands as a groundbreaking therapy for LCA but also paves the way for continued advancements and breakthroughs in the evolving landscape of gene therapy for neurodegenerative disorders. The ripple effects of Luxturna's success resonate across the scientific community, inspiring optimism for further strides in the field of gene therapy.

16.5.2 Zolgensma for Spinal Muscular Atrophy (SMA)

In the intricate landscape of genetic disorders, spinal muscular atrophy (SMA) stands as a poignant challenge, casting its shadows through the gradual weakening of muscles due to the loss of motor neurons. The roots of SMA delve into genetic anomalies, particularly mutations in the survival motor neuron 1 (SMN1) gene, where insufficiency in the production of the SMN protein becomes the harbinger of motor neuron dysfunction.

Here, the narrative unfolds to reveal Zolgensma, a beacon of hope in the field of gene therapy crafted to confront the complexities of SMA. Through the elegant deployment of an AAV vector, Zolgensma intricately delivers a functional rendition of the SMN1 gene to the beleaguered motor neurons, initiating a cascade of events that promises sustained production of the vital SMN protein.

Picture the clinical stage, a crucible where Zolgensma undergoes meticulous trials, notably the phase III START trial. The canvas of this trial captures a symphony of success as infants treated with Zolgensma exhibit milestones in motor function surpassing the natural trajectory of SMA progression. A few precious ones gain the ability to sit, stand, and embark on the journey of walking independently.

The chronicle takes a significant turn in May 2019 when Zolgensma receives the imprimatur of the FDA, a testament to its transformative efficacy. This moment etches itself in history as the first gene therapy accorded approval for addressing SMA. The approval echoes the clarion call of a new era in neurodegenerative disorder treatment, one that directly engages with the genetic underpinnings of the condition.

As the story progresses, the narrative extends beyond the trial halls to the branches of long-term follow-up studies. These chapters unfold to reveal a tale of endurance, where the therapeutic effects of Zolgensma exhibit a remarkable longevity. Treated patients, like characters in an unfolding drama, continue to showcase improvements in motor function, underlining the enduring impact of this gene therapy on the relentless march of SMA.

Yet, the significance of Zolgensma's success surpasses the boundaries of SMA, resonating as a profound melody in the symphony of gene therapy for neurodegenerative disorders. It becomes a metaphorical lighthouse, illuminating a path toward a treatment paradigm that exceeds symptomatic relief, directly confronting the genetic anomalies dictating the course of the disease. Zolgensma's triumph not only inspires further exploration in the domain of SMA but sparks a cascade of research and development in gene therapies targeting the genetic origins of diverse neurodegenerative conditions.

In the tapestry of medical advancements, the case study of Zolgensma emerges as a compelling narrative, where the threads of science and innovation weave together to paint a portrait of hope and transformation. The tale of Zolgensma for SMA is not just a story of a therapy; it's a testament to the potential of gene therapy to redefine the narrative of neurodegenerative disorders, ushering in an era where precision and efficacy intertwine to script a brighter future for patients and the scientific community alike.

16.6 OMICS APPROACHES IN NEURODEGENERATIVE DISEASE RESEARCH

16.6.1 Overview of OMICS Technologies

OMICS technologies collectively constitute a transformative suite of high-throughput approaches that have become indispensable tools in the arena of biological analysis. They offer an exhaustive examination of various biological molecules on a comprehensive scale, delving deep into the molecular intricacies within cells, organisms, or biological systems. These methodologies, which include genomics, transcriptomics, proteomics, and metabolomics, collectively contribute to a holistic exploration of the molecular landscape, providing a nuanced understanding across multiple layers of biological information.

Genomics, as a fundamental branch of OMICS, is dedicated to the study of an organism's entire genome. This encompasses not only all genes but also non-coding DNA sequences, enabling the identification of genetic variations, mutations, and their intricate associations with various diseases. This in-depth exploration contributes to a deeper understanding of the genetic underpinnings of health and pathology, laying the foundation for advancements in precision medicine.

Transcriptomics, another pivotal component of OMICS, takes a comprehensive look at the complete set of RNA transcripts within a cell or tissue. This includes messenger RNA (mRNA), microRNA, and other RNA species, providing valuable insights into gene expression patterns and the intricate regulatory mechanisms governing cellular processes. Transcriptomics, therefore, serves as a critical tool in deciphering the dynamic interplay of genes and their regulatory elements, offering a comprehensive perspective on cellular responses and adaptations.

Proteomics, a key player in the OMICS suite, is dedicated to studying the entire complement of proteins in a biological sample. Through the identification, quantification, and characterisation of proteins, proteomics sheds light on their functions, interactions, and alterations in disease states. This branch offers a comprehensive understanding of the dynamic protein landscape, contributing significantly to our grasp of cellular processes and the molecular basis of health and disease.

Metabolomics, completing the quartet of OMICS disciplines, investigates the comprehensive set of small molecules, or metabolites, within a biological system. By offering a snapshot of cellular processes, including energy metabolism, signaling pathways, and responses to environmental stimuli, metabolomics plays a crucial role in uncovering the metabolic intricacies underlying physiological and pathological conditions.

The integration of these OMICS approaches goes beyond their individual capacities, providing a synergistic and holistic view of biological systems. This integrative approach facilitates a more comprehensive understanding of complex phenomena such as diseases, enabling researchers to unravel the molecular intricacies at different levels. By combining genomics, transcriptomics, proteomics, and metabolomics, scientists gain deeper insights into the interconnected networks governing cellular functions and dysregulations associated with various health conditions.

Moreover, the multifaceted contributions of OMICS technologies extend beyond fundamental research, significantly impacting fields such as personalised medicine, biomarker discovery, and the elucidation of disease mechanisms. The comprehensive data generated by these approaches allows for the identification of specific molecular signatures associated with diseases, paving the way for personalised therapeutic interventions tailored to individual patients. Additionally, OMICS technologies play a pivotal role in biomarker discovery, aiding in the identification of reliable indicators that can facilitate early disease detection, prognosis, and monitoring of treatment responses.

In essence, OMICS technologies, with their diverse branches and integrated capabilities, stand at the forefront of modern biological research. Their transformative impact extends to diverse applications, from unravelling the intricacies of genetic and molecular landscapes to advancing personalised medicine and enhancing our understanding of complex diseases. As these technologies continue to evolve, their contributions are poised to shape the future of biomedical research, paving the way for innovative approaches to healthcare and disease management on an even more sophisticated and personalised level.

16.6.2 Applications of OMICS in Understanding the Molecular Basis of Neurodegenerative Diseases

In the dynamic and ever-expanding landscape of neuroscience, the transformative impact of OMICS technologies has been nothing short of revolutionary. These cutting-edge methodologies, spanning genomics, transcriptomics, proteomics, and metabolomics, serve as powerful lenses, delving into the intricate molecular tapestry underlying neurodegenerative diseases such as Alzheimer's, Parkinson's, and Huntington's with unprecedented depth and precision.

Setting the pace in this discipline of medical exploration lies genomics, where researchers navigate the intricate code of life to unravel genetic risk factors and susceptibility genes associated with neurodegenerative diseases. The exploration begins with a meticulous investigation of both rare and common genetic variations, unlocking the secrets of how these nuances contribute to the onset and progression of these complex conditions. The power of genomics extends further, facilitating the stratification of patient populations based on their genetic profiles. This personalised approach holds immense promise in tailoring treatment strategies, ushering in an era where interventions can be finely tuned to align with an individual's unique genetic makeup.

Then we move toward the domain of transcriptomics, a discipline that surpasses the predefined models of genes to enter the dynamic system of gene expression. Profiling gene expression patterns becomes a pivotal tool for researchers seeking to elucidate the molecular changes unfolding during the progression of neurodegenerative diseases. The identification of dysregulated pathways and potential therapeutic targets hinges on the intricate dance of gene expression, offering a molecular roadmap for the development of intervention strategies. Transcriptomics extends its reach into the setting of non-coding RNAs, such as microRNAs, playing pivotal roles in orchestrating neurodegenerative processes and adding yet another layer to our understanding.

Moving further into the molecular symphony, we encounter the impactful domain of proteomics. Here, the focus shifts to the quantitative analysis of protein profiles, unravelling alterations in abundance and post-translational modifications that signify the molecular shifts accompanying neurodegeneration. The identification of protein biomarkers specific to various neurodegenerative diseases becomes a game-changer for diagnostics and therapeutic development. Moreover, proteomics illuminates the intricate choreography of protein–protein interactions and functional networks within affected tissues, providing a holistic view of the molecular ballet orchestrating neurodegeneration.

In the final movement of this alignment, metabolomics takes the stage, unveiling the metabolic underpinnings of neurodegenerative diseases. Profiling changes in metabolite levels provides insights into disrupted metabolic pathways, laying bare the biochemical fingerprints marking different stages of neurodegeneration. The identification of specific metabolomic signatures not only refines our understanding of disease progression but also sheds light on the impact of metabolic dysregulation on neuronal function and viability.

The true power of OMICS technologies emerges when these diverse data streams are seamlessly integrated, weaving a comprehensive map of the molecular events driving neurodegenerative diseases. Through this integrative approach, researchers gain a systems-level understanding, uncovering the intricate interactions between genes, proteins, and metabolites. The synergy of genomics, transcriptomics, proteomics, and metabolomics surmounts the sum of its parts, providing a meticulous comprehension of the molecular harmony playing out in the brains of those affected by neurodegenerative disorders.

These insights are not confined to the area of academic curiosity; they hold tangible promise for clinical applications. The multi-OMICS approach paves the way for the development of targeted therapies, where interventions can be tailored to address specific molecular aberrations identified in individual patients. Additionally, the discovery of biomarkers through these methodologies opens avenues for early diagnosis, allowing for proactive interventions and potentially altering the trajectory of disease progression. In the era of precision medicine, the integration of OMICS technologies stands as a beacon, guiding researchers and clinicians toward more effective strategies for understanding, diagnosing, and treating neurodegenerative disorders.

16.6.3 Integration of Multi-OMICS Data for a Comprehensive Understanding

In the relentless pursuit of unravelling the intricate molecular tapestry underlying neurodegenerative diseases, researchers increasingly recognise the paramount significance of integrating data from diverse OMICS platforms, including genomics, transcriptomics, proteomics, and metabolomics. This integrative approach goes beyond the limitations associated with studying individual molecular layers and provides a more comprehensive, nuanced understanding of the complex biological processes involved in debilitating conditions such as Alzheimer's, Parkinson's, and Huntington's disease.

By amalgamating information derived from genomics, which identifies genetic predispositions, with transcriptomics data revealing gene expression patterns, researchers gain profound insights into how specific genetic variations manifest at the functional level. Proteomics contributes by shedding light on the actual proteins synthesised, their modifications, and interactions, thereby adding a layer of granularity to the molecular narrative. Metabolomics, in turn, completes the picture by capturing the downstream impact on small molecules, offering a dynamic snapshot of the intricate metabolic landscape.

This integrative multi-OMICS strategy not only enhances our understanding of the molecular mechanisms propelling neurodegeneration but also provides a robust platform for identifying converging pathways and key molecular players. The synergy emanating from these data sets enables

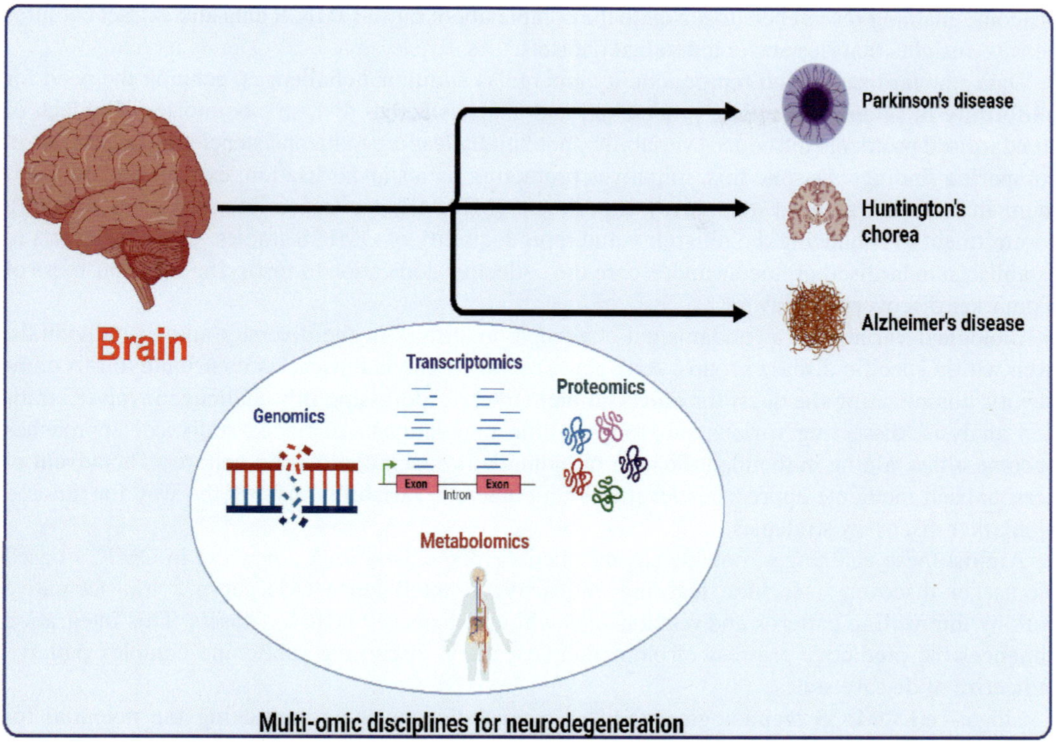

FIGURE 16.2 Multi-omics approaches used in neurodegenerative disorders. (Created with BioRender.com.)

the discovery of potential biomarkers crucial for early disease detection and facilitates the identification of therapeutic targets. Moreover, it fosters the development of more precise and personalised interventions by considering the interconnected nature of biological systems.

In summary, the integration of multi-OMICS data stands as an indispensable cornerstone in contemporary neurodegenerative disease research, paving the way for a comprehensive and systems-level understanding that is essential for advancing diagnostics, treatment strategies, and, ultimately, improving the quality of life for individuals grappling with these challenging disorders. This integrative approach not only broadens the scope of scientific inquiry but also holds tremendous potential for transformative breakthroughs in the diagnosis and treatment of neurodegenerative conditions (**Figure 16.2**).

16.6.4 Challenges and Advancements in OMICS-based Biomarker Discovery

The exploration of OMICS-based biomarker discovery unfolds as a compelling journey marked by both challenges and transformative advancements. Biomarkers, serving as measurable indicators of biological processes, assume a pivotal role in disease diagnosis, prognosis, and treatment response monitoring. This section delves into the intricacies of this dynamic field, illuminating the hurdles faced and the strides made in pushing the boundaries of biomarker discovery.

In the labyrinth of OMICS-based biomarker exploration, challenges emerge as formidable adversaries. The complexity arises notably from the amalgamation and interpretation of extensive datasets spanning genomics, transcriptomics, proteomics, and metabolomics. This intricate process demands sophisticated computational tools and methodologies, reflecting the sheer intricacy of biological systems. Nevertheless, the landscape is not devoid of hope, as advancements in bioinformatics and systems biology have birthed integrative analysis platforms. These platforms stand as beacons, enabling researchers to navigate the complexity of multi-OMICS data and extract comprehensive insights that rise above individual datasets.

Data standardisation and reproducibility emerge as additional challenges, echoing the need for uniformity in sample collection, processing, and analysis across diverse laboratories. The lack of standardised protocols introduces variability, potentially leading to inconsistencies and obstacles in comparing findings. Despite this, initiatives promoting data standardisation, exemplified by minimum information about a microarray experiment (MIAME) for transcriptomics, demonstrate a commitment to enhancing the reliability and reproducibility of OMICS studies. Ongoing efforts to establish standardised protocols underscore the collective dedication to fortifying the robustness of biomarker discovery endeavors.

Biological variability, a fundamental challenge, manifests in the diversity among individuals, even within specific disease groups. Age, genetics, and lifestyle intricacies contribute to this complexity, complicating the quest for universal biomarkers. Addressing this challenge involves stratified analysis, dissecting various subgroups within populations. Advanced statistical approaches become allies, aiding in the identification of biomarkers specific to certain cohorts. The advent of personalised medicine approaches leverages this inherent variability, paving the way for tailored biomarker discovery strategies.

Amidst these challenges, notable advancements emerge, heralding a new era in OMICS-based biomarker discovery. Machine learning and artificial intelligence (AI) carve a transformative path by unravelling patterns and relationships within expansive OMICS datasets. This integration enhances the predictive prowess of biomarker discovery, discerning subtle and complex patterns indicative of disease states.

Single-cell OMICS technologies stand as a revolutionary force, unlocking the potential for unprecedented insights into cellular diversity and disease-related changes. By dissecting individual cells, these technologies provide a high-resolution panorama, capturing nuances that traditional approaches may overlook.

Longitudinal studies, offering a dynamic temporal view of disease progression, become invaluable in identifying biomarkers associated with various disease stages. This approach transcends static snapshots, providing a comprehensive understanding of the temporal evolution of biomarkers.

Functional validation strategies, integrating OMICS data with techniques like CRISPR-Cas9 gene editing or pharmacological interventions, reinforce the reliability of identified biomarkers. Experimental validation ensures not just association but a causal link between biomarkers and the pathogenesis of the disease.

As the OMICS-based biomarker discovery field evolves, navigating these challenges and harnessing technological advancements becomes increasingly crucial. These advancements, collectively, hold the promise of uncovering clinically relevant biomarkers with heightened specificity and sensitivity. This progress propels us closer to the realisation of the full diagnostic and therapeutic potential inherent in personalised medicine, marking a paradigm shift in the landscape of healthcare and disease management.

16.7 TRANSLATIONAL NEURODEGENERATIVE MODELS

16.7.1 Importance of Translational Models in Bridging the Gap between Basic Research and Clinical Applications

Translational neurodegenerative models stand as indispensable conduits, navigating the complex trajectory from foundational laboratory discoveries to tangible clinical applications, thereby serving as crucial bridges between foundational research and clinical interventions. These models, encompassing a spectrum from simplified in vitro cultures to intricate in vivo systems, play a pivotal role in narrowing the divide between the foundational understanding of neurodegenerative diseases and the development of practical clinical interventions.

Within the intricate landscape of neurodegenerative diseases, these translational models emerge as pivotal tools, striving to mirror the inherent complexity of these conditions, encapsulating genetic, molecular, and environmental factors. Whether derived from patient-specific induced pluripotent stem cells (iPSCs) or intricate animal models, these systems are designed to faithfully replicate the intricate pathology observed in human patients, providing researchers with a platform that closely mimics the complex interplay of factors involved in disease initiation and progression.

A fundamental and far-reaching contribution of translational models lies in their unique capacity to validate potential therapeutic targets. Before advancing to the critical stages of clinical trials, researchers can systematically assess the safety and efficacy of interventions in a controlled experimental setting. This process provides invaluable guidance for the development of novel treatments, ensuring that only the most promising candidates progress toward human trials.

Beyond the validation of therapeutic targets, these models deepen our understanding of the underlying molecular and cellular mechanisms steering neurodegeneration. By closely mimicking human pathology, translational models enable researchers to dissect the intricate interplay of factors involved in disease initiation and progression, unravelling the complex mysteries that underlie these debilitating conditions.

In the expansive preclinical landscape, translational models serve as indispensable platforms for evaluating the safety and efficacy of a wide array of potential therapies. Whether it be drugs, gene therapies, or other interventions, these models offer critical insights that inform the decision-making process, allowing researchers to make informed choices before exposing human subjects to experimental treatments.

Furthermore, translational models significantly contribute to the identification and validation of biomarkers associated with neurodegenerative diseases. These biomarkers, essential for early disease detection, treatment response monitoring, and patient stratification, emerge from the controlled and precisely monitored environments of these models, providing valuable tools for clinical diagnosis and personalised treatment strategies.

Facilitating the paradigm shift toward personalised medicine, translational models, particularly those derived from patient-specific cells like iPSCs, possess the unique ability to capture individual genetic and molecular signatures. This allows for the development of tailored interventions that take into account the unique characteristics of each patient, ushering in an era of more precise and effective treatments tailored to the specific needs of individuals grappling with neurodegenerative diseases.

In practical terms, the strategic incorporation of translational models early in the drug development pipeline serves as a proactive approach. This not only helps identify potential challenges but also reduces the likelihood of costly and time-consuming failures in later-stage clinical trials, thus enhancing the overall efficiency of the drug development process. This proactive strategy holds promise for the development of more effective treatments for neurodegenerative diseases, ensuring that promising candidates are rigorously tested before advancing to human trials.

In essence, translational neurodegenerative models emerge as indispensable tools in the challenging journey toward clinical breakthroughs, offering a nuanced and comprehensive understanding of disease processes. Their multifaceted contributions, spanning validation of therapeutic targets, deepening molecular understanding, facilitating preclinical evaluations, and contributing to biomarker identification, position them as invaluable assets in the ongoing quest for meaningful advancements in the field of neurodegenerative disease research and treatment.

16.7.2 Overview of in Vitro and in Vivo Models for Neurodegenerative Diseases

Gaining profound insights into the intricate landscape of neurodegenerative diseases demands the adept utilisation of a broad spectrum of research models, where both in vitro and in vivo approaches play integral roles. In the field of in vitro models, meticulously conducted within controlled laboratory environments, a nuanced exploration of the cellular and molecular processes associated with

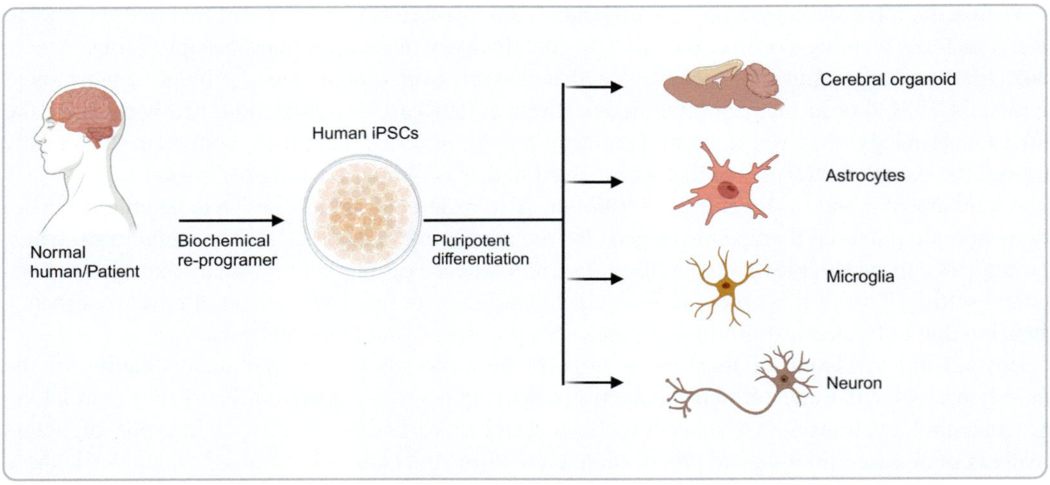

FIGURE 16.3 hiPSCs as a tool for developing translational model for mimicking neurological conditions; cerebral organoids as 3D model and different types of nervous cells (astrocytes, microglia & neuron) for 2D model development. (Created with BioRender.com.)

neurodegenerative diseases unfolds. Within this context, cell cultures, whether derived from neural tissues or induced pluripotent stem cells (iPSCs), stand out as versatile platforms. They offer researchers the ability to delve into intricate molecular mechanisms, conduct systematic drug screenings, and investigate early pathological changes. However, despite their undeniable utility, it is crucial to acknowledge that these in vitro models, with their simplified environments, may not entirely encapsulate the multifaceted intricacies inherent in the human brain.

Similarly, three-dimensional structures such as brain organoids, also derived from iPSCs, strive to emulate the structural complexity of the brain, providing a unique perspective for studying neuronal development and disease-related changes within a more organ-like context. Nevertheless, the inherent variability in the structure and function of these organoids poses a challenge to standardisation efforts, reflecting the intricacies of accurately replicating the diverse facets of the human brain within a controlled laboratory setting (**Figure 16.3**).

On the flip side, in vivo models involving studies conducted within living organisms contribute indispensable insights into the progression of neurodegenerative diseases, the efficacy of therapeutic interventions, and the behavioral alterations associated with these conditions. Rodent models, especially those involving genetically modified or induced mice and rats, serve as practical and widely employed tools for examining disease progression, testing potential interventions, and evaluating behavioral changes. Yet, despite their practicality, it's crucial to acknowledge that differences between rodent and human brains present a challenge in terms of translating findings from these models to the human context.

Moreover, non-human primates, such as rhesus macaques, emerge as valuable contributors to our understanding due to their closer evolutionary proximity to humans. These models offer a more accurate representation of higher-order brain functions, presenting a bridge between rodent models and the complex neurological landscape of humans. However, ethical concerns, high costs, and logistical challenges restrict their widespread use.

Additionally, models derived directly from patient samples, exemplified by patient-specific iPSCs, bring forth a personalised approach to studying disease mechanisms and testing potential therapeutic interventions. Despite their promise, these models come with their set of challenges, including variability among patient populations and the inherent complexity associated with faithfully replicating the intricate in vivo environment.

In summary, the strategic integration of these diverse models, encompassing both in vitro and in vivo approaches, empowers researchers to comprehensively investigate neurodegenerative diseases. This strategic approach facilitates a nuanced understanding that goes beyond the limitations of individual models, fostering the development of targeted and effective treatments for these complex and challenging conditions. As researchers navigate this intricate landscape, the synergy between various models becomes paramount, paving the way for breakthroughs in our understanding and treatment of neurodegenerative diseases.

16.8 INTERDISCIPLINARY APPROACHES AND COLLABORATIONS

16.8.1 The Need for Interdisciplinary Collaboration in Advancing Neurodegenerative Disease Research

Navigating the intricate and multifaceted terrain of neurodegenerative diseases demands the weaving of an elaborate interdisciplinary tapestry, interwoven with collaborative threads that unite experts from a diverse array of fields in a concerted and synergistic effort. The inherent complexities of disorders such as Alzheimer's, Parkinson's, and Huntington's surpass the boundaries of any singular discipline, underscoring the imperative for an integrated approach to make substantial strides in both research and therapeutic development.

Within this expansive interdisciplinary framework, the fusion of expertise from genomics, neurobiology, pharmacology, bioinformatics, and clinical sciences becomes imperative. This collaborative synergy fosters a nuanced and comprehensive exploration of disease mechanisms, unlocking the potential for the identification of novel therapeutic targets that may elude a more siloed approach. The collective intelligence of diverse disciplines enriches the understanding of the intricate interplay of genetic, molecular, and environmental factors contributing to neurodegenerative diseases.

Translating discoveries from the laboratory bench to clinical applications poses a considerable challenge, and the continuum from basic research to tangible treatments benefits immensely from interdisciplinary collaboration. This collaborative approach ensures that insights gained in controlled laboratory settings find practical applications that directly impact the well-being of patients. The seamless integration of diverse perspectives aids in the efficient translation of scientific knowledge into therapeutic interventions, bridging the gap between bench side discoveries and bedside applications.

Technological advancements, such as high-throughput sequencing, advanced imaging modalities, and sophisticated data analytics, underscore the need for a multidisciplinary approach in neurodegenerative disease research. Collaborative efforts empower researchers to harness these cutting-edge tools effectively, delving into the intricate details of neurodegenerative diseases with a level of sophistication that is the limitations of individual disciplines. The convergence of technologies and expertise from various fields enhances the precision and depth of research, paving the way for breakthroughs that would be challenging to achieve within the confines of a single discipline.

Patient-centricity becomes a cornerstone of therapeutic development through the incorporation of insights from clinicians, geneticists, and social scientists. This holistic approach not only considers the intricate biological aspects of the diseases but also delves into the broader implications for patients and their families. Collaborative strategies align therapeutic approaches with real-world needs, ensuring that interventions address the multifaceted challenges faced by individuals affected by neurodegenerative diseases.

The impact of neurodegenerative diseases extends beyond the confines of the brain, affecting overall health, mental well-being, and the quality of life. Collaborations among neuroscientists, psychologists, and public health experts provide a panoramic understanding of these broader implications, shaping comprehensive care strategies that go beyond the narrow confines of a singular discipline. Integrating mental health, social, and public health perspectives into the research agenda ensures a more holistic approach to neurodegenerative diseases.

Innovation often blossoms at the crossroads of different fields. Collaborative endeavors between neuroscientists and engineers, for instance, yield innovative technologies for early diagnosis or novel therapeutic delivery methods. This interdisciplinary innovation becomes a catalyst for pushing the boundaries of existing research paradigms and ushering in transformative breakthroughs that have the potential to revolutionise the diagnosis, treatment, and management of neurodegenerative diseases.

The vast scale of data generated in neurodegenerative disease research, particularly through OMICS technologies, necessitates a global collaborative approach. Interdisciplinary networks, spanning international boundaries, facilitate the pooling of resources, expertise, and data. This collaborative sharing accelerates progress, offering a comprehensive and diverse perspective essential for understanding the complex landscape of these diseases. Global collaboration is not just advantageous but becomes imperative for tackling the global burden of neurodegenerative diseases.

In essence, interdisciplinary collaboration stands not merely as an advantageous approach but as an imperative for advancing neurodegenerative disease research. As the scientific landscape evolves, and the challenges become more intricate, the tapestry of collaboration emerges as the linchpin for unravelling the mysteries of these diseases. Through this collaborative effort, researchers can translate discoveries into tangible improvements in patient care and outcomes, bringing us closer to effective treatments and, ultimately, a cure. The synergistic collaboration of diverse minds and expertise becomes the driving force for transformative advancements that have the potential to reshape the landscape of neurodegenerative disease understanding and treatment.

16.9 CURRENT CHALLENGES AND FUTURE DIRECTIONS

In the ever-evolving landscape of neurodegenerative disease research, the intersection of gene therapy and OMICS approaches amplifies the gravity of ethical considerations, necessitating even more meticulous examination and thoughtful navigation. The profound implications inherent in manipulating an individual's genetic makeup, a core aspect of gene therapy, call for an exhaustive exploration of concerns surrounding unintended consequences, potential off-target effects, and the enduring implications of genetic alterations. Consequently, fostering transparent communication between researchers, clinicians, and the public becomes not only imperative but increasingly complex. This transparency must be accompanied by the establishment of even more robust ethical guidelines, stringent informed consent procedures, and comprehensive data security measures. These measures are indispensable to uphold responsible and ethical research practices, ensuring not only scientific integrity but also mitigating potential risks that may arise from the intricacies of gene therapy and OMICS research.

Simultaneously, the vast data generated by OMICS research introduces its own set of ethical concerns, expanding beyond the boundaries of molecular intricacies to encompass issues of privacy, consent, and responsible data use. Striking a delicate balance between the myriad benefits of research and the protection of individual privacy demands an unwavering commitment to ethical considerations. Addressing these evolving challenges mandates the creation of even more stringent ethical frameworks, the intensified promotion of open communication, and the seamless integration of ethical principles into the very fabric of gene therapy and OMICS research design and execution.

Despite significant strides in preclinical research elucidating the intricate mechanisms of neurodegenerative diseases, the translation of these findings into clinically applicable interventions remains a formidable challenge, calling for an even more heightened focus on addressing intricate biological, physiological, and logistical hurdles in the transition from animal models to human subjects. While rodent models continue to provide valuable insights, acknowledging their limitations in fully replicating the complexity of human neurobiology underscores the pressing need for the development of increasingly accurate and predictive preclinical models to enhance the translational success of interventions.

The inherent heterogeneity of neurodegenerative diseases presents yet another formidable challenge. The variability in patient populations, disease subtypes, and progression stages complicates the identification of universal therapeutic targets. In response, the paradigm of personalised medicine approaches, tailoring interventions to individual characteristics, not only remains a promising avenue but calls for an even more nuanced and individualised approach to overcome this challenge, unlocking more precise and effective treatments tailored to the unique needs of each patient.

The design and execution of clinical trials, fundamental components in the journey from lab bench to bedside, introduce additional obstacles that warrant an even more strategic and collaborative approach. Optimising trial endpoints, refining patient selection criteria, and validating biomarkers become not only critical but increasingly complex components for success. Collaborative efforts that bring together academia, industry, and regulatory bodies are not only essential but necessitate an even more seamless integration of expertise to streamline the clinical translation process, expedite the development of effective therapies, and navigate the evolving complexities inherent in neurodegenerative disease research.

Looking toward the future, the landscape of neurodegenerative disease research holds even greater promise with the advent of emerging technologies and methodologies. CRISPR-based therapies, with their precise genome editing capabilities, not only open avenues for correcting or modifying disease-associated genetic mutations but also demand an even more careful consideration of ethical implications. RNA therapeutics, including antisense oligonucleotides and small interfering RNAs, present targeted approaches that call for an even more nuanced understanding of gene expression modulation.

Single-cell OMICS technologies, with their unprecedented resolution, not only enable the identification of subtle cellular heterogeneity within tissues but demand an even more sophisticated integration into the broader research framework. Advances in neuroimaging technologies, such as PET and fMRI, not only provide non-invasive visualisation of brain structures and functions but necessitate an even more intricate analysis of the obtained data. The integration of neuroimaging with other OMICS data, while offering a comprehensive view of molecular and structural changes associated with neurodegenerative diseases, calls for an even more sophisticated approach to data integration and interpretation.

Artificial intelligence and machine learning algorithms, while significantly contributing by analysing large-scale OMICS datasets, not only enhance the identification of patterns, biomarkers, and potential therapeutic targets but demand an even more rigorous validation and interpretation. These technologies play a pivotal role in the development of predictive models for disease progression, personalised treatment strategies, and the discovery of novel drug candidates, requiring an even more refined understanding of their applications in the context of neurodegenerative diseases.

In the context of gene therapy and OMICS, personalised medicine emerges not only as a linchpin but as an even more indispensable tool for tailoring interventions to individual characteristics. Addressing the challenges posed by disease heterogeneity becomes not just a promising avenue but a crucial imperative that demands even more precise and individualised strategies. By advancing ethical practices, overcoming translational challenges, embracing emerging technologies, and incorporating even more personalised approaches, the future of neurodegenerative disease research holds the promise of groundbreaking discoveries and transformative interventions that will significantly impact the lives of those grappling with these challenging conditions. The evolving complexities of ethical considerations, translational challenges, and technological advancements necessitate an even more interdisciplinary and forward-thinking approach to ensure the continued progress and success of neurodegenerative disease research.

16.10 CONCLUSION

In the exploration of gene therapy, OMICS approaches, and translational models in the context of neurodegenerative diseases, this chapter has navigated through a rich landscape teeming with scientific promise and profound challenges. Summarising the key points discussed unveils a multifaceted tapestry where ethical considerations intersect with technological advancements, and where the translation of preclinical insights converges with the emerging era of personalised medicine.

The ethical considerations surrounding gene therapy and OMICS research have been meticulously scrutinised, placing a spotlight on the critical need for transparency, the establishment of rigorous ethical guidelines, and robust data security measures. It underscores the delicate balance required to advance scientific progress while upholding ethical responsibility, especially as these transformative technologies venture into previously uncharted territories.

Addressing the challenges in translating preclinical findings to clinical applications necessitates the development of more accurate preclinical models, the integration of personalised medicine approaches to navigate disease heterogeneity, and collaborative efforts to optimise clinical trial design. The complexities inherent in neurodegenerative diseases demand a comprehensive and interdisciplinary approach, a recurring theme woven throughout the chapter.

Looking forward, emerging technologies such as CRISPR-based therapies, RNA therapeutics, single-cell OMICS, neuroimaging, and artificial intelligence offer unprecedented opportunities. These innovations stand at the front line of shaping the future of neurodegenerative disease research and treatment. CRISPR's precision in genome editing, the targeted modulations provided by RNA therapeutics, the high-resolution insights from single-cell OMICS, and the integrative power of artificial intelligence collectively hold the potential to revolutionise our understanding and management of these debilitating disorders.

The impact of gene therapy and OMICS approaches extends beyond the laboratory; it reaches into the branch of personalised medicine. Tailoring interventions to individual characteristics, understanding disease heterogeneity, and optimising therapeutic outcomes represent a paradigm shift in neurodegenerative disease treatment. The promise of personalised medicine beckons a new era where treatments are not only effective but also finely tuned to the unique genetic and molecular signatures of each patient.

As we reflect on the potential impact of these advancements, it becomes evident that the future of neurodegenerative disease treatment is marked by possibilities. The call to action is clear: continued research and collaboration are imperative. The journey from bench to bedside is ongoing, and the complexities of neurodegenerative diseases demand persistent efforts to unravel their mysteries.

In conclusion, this chapter has traversed the dimensions of gene therapy, OMICS approaches, and translational models, shedding light on the transformative potential embedded in their intricate landscapes. The fusion of ethical considerations, technological innovations, and personalised medicine heralds a future where neurodegenerative diseases may not only be understood at unprecedented depths but also treated with unparalleled precision. The path ahead is challenging, but it is also laden with opportunities to reshape the narrative of neurodegenerative diseases and enhance the lives of those affected. As we embark on this journey, the collaborative efforts of the scientific community will be paramount in unlocking the full potential of these advancements and translating them into tangible benefits for individuals grappling with neurodegenerative conditions.

BIBLIOGRAPHY

1. Ginn, S. L., Alexander, I. E., Edelstein, M. L., Abedi, M. R., & Wixon, J. (2018). Gene Therapy clinical trials worldwide to 2017: An update. Journal of Gene Medicine, 20(5), e3015. https://doi.org/10.1002/jgm.3015
2. Rinschen, M. M., Ivanisevic, J., Giera, M., & Siuzdak, G. (2019). Identification of bioactive metabolites using activity metabolomics. Nature Reviews Molecular Cell Biology, 20(6), 353–367. https://doi.org/10.1038/s41580-019-0121-0
3. Holtzman, D. M., Morris, J. C., & Goate, A. M. (2011). Alzheimer's disease: The challenge of the second century. Science Translational Medicine, 3(77), 77sr1. https://doi.org/10.1126/scitranslmed.3002369
4. Duan, D. (2018). Systemic AAV micro-dystrophin Gene therapy for duchenne muscular dystrophy. Molecular Therapy, 26(10), 2337–2356. https://doi.org/10.1016/j.ymthe.2018.08.005
5. Bezard, E., Yue, Z., Kirik, D., & Spillantini, M. G. (2013). Animal models of Parkinson's disease: Limits and relevance to neuroprotection studies. Movement Disorders, 28(1), 61–70. https://doi.org/10.1002/mds.23324

6. Nelson, C. E., & Gersbach, C. A. (2016). Engineering delivery vehicles for genome editing. Annual Review of Chemical and Biomolecular Engineering, 7, 637–662. https://doi.org/10.1146/annurev-chembioeng-080615-034454
7. Mastrokolias, A., Pool, R., Mina, E., Hettne, K. M., van Duijn, E., & van der Mast, R. C., et al. (2018). Integration of multi-omics data and deep phenotyping enables prediction of cytokine responses. Nature Immunology, 19(12), 1552–1554. https://doi.org/10.1038/s41590-018-0246-6
8. Takahashi, K., & Yamanaka, S. (2006). Induction of pluripotent stem cells from mouse embryonic and adult fibroblast cultures by defined factors. Cell, 126(4), 663–676. https://doi.org/10.1016/j.cell.2006.07.024
9. Ophinni, Y., Inoue, M., Kotaki, T., & Kameoka, M. (2018). CRISPR/Cas9 system targeting regulatory genes of HIV-1 inhibits viral replication in infected t-cell cultures. Scientific Reports, 8(1), 7784. https://doi.org/10.1038/s41598-018-26170-3
10. Zhu, X., Zhang, Y., Yang, X., Hao, C., & Duan, H. (2021). Gene Therapy for neurodegenerative disease: Clinical potential and directions. Frontiers in Molecular Neuroscience, 14, 618171. https://doi.org/10.3389/fnmol.2021.618171
11. Ayala, Y. M., & Nguyen, A. D. (2021). RNA-based therapies for neurodegenerative diseases. Missouri Medicine, 118(4), 340–345.
12. Kline, R. A., Lößlein, L., Kurian, D., Aguilar Martí, J., Eaton, S. L., Court, F. A., Gillingwater, T. H., & Wishart, T. M. (2022). An optimized comparative proteomic approach as a tool in neurodegenerative disease research. Cells, 11(17), 2653. https://doi.org/10.3390/cells11172653
13. La Cognata, V., Morello, G., & Cavallaro, S. (2021). Omics data and their integrative analysis to support stratified medicine in neurodegenerative diseases. International Journal of Molecular Sciences, 22(9), 4820. https://doi.org/10.3390/IJMS22094820
14. Wang, E., Lemos Duarte, M., Rothman, L. E., Cai, D., & Zhang, B. (2022). Non-coding RNAs and Alzheimer's disease: Perspectives from omics studies. Human Molecular Genetics, 31(R1), R54–R61. https://doi.org/10.1093/hmg/ddac202
15. Rana, M. S., Podder, N. K., Rana, H. K., Hasan, M. I., Azam, M. S., Rahim, M. A., & Saha, S. (2023). Identification of genomic associations between Parkinson's and neurodegenerative diseases using bioinformatics models. In *2023 International Conference on Electrical, Computer and Communication Engineering (ECCE)* (pp. 1–6). IEEE.
16. Chang, C. J., Cravatt, B. F., Johnson, D. S., & Lim, M. H. (2014). Molecular medicine and neurodegenerative diseases. Chemical Society Reviews, 43(19), 6668–6671. https://doi.org/10.1039/C4CS90065K
17. Suk, K. (2006). Proteomics-based discovery of biomarkers and therapeutic targets in neurodegenerative diseases: Perspective of microglia and neuroinflammation. Expert Opinion on Therapeutic Patents, 16(3), 237–247. https://doi.org/10.1517/13543776.16.3.237
18. Bang, L., Dansha, H., Xiao-Jiang, L., & Xiangrong, G. (2022). Modeling neurodegenerative diseases using non-human primates: Advances and challenges. Ageing and Neurodegenerative Diseases, 2, 12. https://doi.org/10.20517/and.2022.14
19. Nibedita, N. (2021). Molecular network for management of neurodegenerative diseases and their translational importance using animal biotechnology as a tool in preclinical studies. In Advances in Animal Genomics (pp. 219–235). Academic Press.
20. Stonebarger, G. A., Bimonte-Nelson, H. A., & Urbanski, H. F. (2021). The rhesus macaque as a translational model for neurodegeneration and Alzheimer's disease. Frontiers in Aging Neuroscience, 13, 734173. https://doi.org/10.3389/fnagi.2021.734173
21. Keeler, A. M., & Flotte, T. R. (2019). Recombinant adeno-associated virus Gene therapy in light of Luxturna (and Zolgensma and Glybera): Where are we, and how did we get here?. Annual Review of Virology, 6(1), 601–621. https://doi.org/10.1146/annurev-virology-092818-015530

17 Implication and Importance of Stem Cell Therapy in a Translational Model of Neurodegenerative Diseases

Shashank Kumar Ojha
School of Pharmacy, The Hebrew University of Jerusalem, Jerusalem, Israel

17.1 INTRODUCTION

Neurodegenerative diseases are characterised by chronic and progressive loss of structure and function of nerve cells in the brain or peripheral nervous system (PNS) [1]. The most common neurodegenerative disorders comprise Alzheimer's disease (AD), Parkinson's disease (PD), Huntington's disease (HD), amyotrophic lateral sclerosis (ALS), and multiple sclerosis (MS). These disorders result in a group of symptoms like memory impairment, motor dysfunction, cognitive deterioration, and behavioral changes [2, 3]. The exact causes of most neurodegenerative diseases are still not known and are multifactorial, involving an individual/combination effect of genetic, environmental, and age-related factors [4]. Currently, there are no cures for these diseases, and treatment mainly focuses on managing symptoms and improving patients' quality of life. Stem cell technology development and the use of alternative cell tissue sources are pointing to exciting frontiers in this area of research.

Stem cell therapy can be used in neurodegenerative diseases and involves the use of stem cells with the potential to differentiate into various types of nerve cells and restore or replace damaged or lost neurons in the central nervous system (CNS) or PNS [5, 6]. This promising approach may halt or slow disease progression and improve functional outcomes for patients with conditions like AD, PD, ALS, and other neurodegenerative disorders [7]. For example, Stem cell-derived cells can be differentiated in vitro in various stages of neuroblasts (immature neurons) and can be used for transplantation in the place of dysfunctional/lost neurons. Endogenous stem cells present in various parts of the brain can also be induced to differentiate into various types of neuron or glial cells [8, 9]. While stem cell therapy for neurodegenerative diseases is still in its early stages and faces several challenges, it holds great promise as a potential avenue for future treatments to combat these devastating conditions.

Each of the neurodegenerative diseases has specific clinical features, underlying mechanisms, and affected brain regions, but all share the common characteristic of progressive degeneration of nerve cells, resulting in significant impairments in motor function, cognition, and/or behavior [10]. Current treatments focus on managing symptoms and improving the patients' quality of life, as there are no definitive cures for these conditions [11]. Research continues to make a deeper understanding of the diseases and explore potential therapeutic interventions.

17.2 CURRENT TREATMENT APPROACHES AND THEIR LIMITATIONS

The causes of neurodegenerative diseases and the degree of disability in neurodegenerative diseases are very variable, which limits the treatment and available therapeutic options [12]. Some of the current treatment approaches and their limitations include the following:

 DOI: 10.1201/9781032620442-17

Approaches

1. *Pharmacological therapies*: The available medications alleviate some symptoms and may slow down disease progression; for example, levodopa is given to PD patients to manage motor symptoms, but sometimes it leads to side effects like dementia. Cholinesterase inhibitors are prescribed to Alzheimer's patients to improve cognitive function. However, the available drugs have limited effectiveness with symptoms, and they do not help in the underlying cause of neurodegeneration [13].
2. *Physical and occupational therapy*: Physical therapy helps in improving physical activity/motor function in the patient's life while occupational therapy helps in improving the daily life activities of patients. These therapies are beneficial, but they do not alter the course of the disease or halt neurodegeneration [14].
3. *Stem cell therapy*: Several studies from recent decades are exploring the potential of stem cells to replace damaged neurons or promote neuroprotection and regeneration. However, stem cell therapies are still in the early stages of development and study and need investigation for their long-term safety and efficacy [14, 15].
4. *Gene therapy*: Gene therapy is a promising medical approach and a powerful therapeutic strategy for various types of neurodegenerative disorders, including AD, PD, and HD. Although it is promising, it is complex, and its effectiveness depends on precise targeting. And the clinical trials of gene therapy in the past few years have not achieved satisfactory therapeutic effects [16].
5. *Supportive care*: Providing comprehensive care like counselling, education, and assistance to manage symptoms, can help patients to live the best quality of life [17]. However, supportive care does not cure the underlying disease.

Limitations

1. *Heterogeneity of diseases*: Since there are many different types of neurodegenerative disorders, each with a unique or different set of underlying mechanisms and affected brain areas, it is difficult to create generalised therapeutic strategies [18, 19].
2. *Diagnosis at advanced stages*: The diagnosis of neurodegenerative disease is frequently delayed until significant neuronal damage has already been done, which makes therapies less effective and preventive [18, 20].
3. *Side effects*: The available drugs are known to manage symptoms but with time start showing side effects that impact the overall well-being of patients [21].
4. *High cost*: Some treatments, particularly experimental medicines, and individualised methods, might be pricey and may not be available to all patients [22].

There is optimism to develop more effective and targeted treatment as our technology advances, research progresses, and our understanding of neurobiology deepens. Collaboration among researchers, doctors, and pharmaceutical corporations is necessary to create novel treatments for these difficult neurological conditions.

17.3 STEM CELLS

17.3.1 Introduction to Stem Cells

The exceptional property of stem cells is their capacity to differentiate into a variety of specialised cell types throughout the body [23]. They are essential for tissue repair, regeneration, and embryonic development [23]. Stem cells can do self-renewal, which enables them to produce exact copies of themselves, and they can differentiate into specialised cell types with defined tasks, in contrast to most other cells, which have a single function and a restricted capacity for replication [24].

Due to their potential use in several areas, such as regenerative medicine, tissue engineering, drug development, and disease modelling, stem cells have attracted a lot of interest from the scientific and medical community [24]. Their capacity to replace lost or injured cells gives hope for the treatment of diseases like diabetes, heart disorder, and spinal cord injuries [23, 24].

17.3.2 Characteristics and Properties of Stem Cells Related to Neurodegenerative Diseases

Stem cells have potential with different characteristics and properties to treat different neurodegenerative illnesses. These characteristics make stem cells a valuable source for developing new treatment methods to treat these disorders. Several of the essential traits and qualities include the following:

1. *Self-renewal*: Stem cells divide to make more stem cells like themselves and maintain their stem cell pool throughout life. This self-renewal capacity ensures a continuous supply of stem cells for potential therapies, making them valuable for long-term treatments [25].
2. *Differentiation*: Stem cells can differentiate into several different types of specialised cells. In the case of neurodegenerative illnesses, they can differentiate into various cells like neurons, astrocytes, and oligodendrocytes, which are essential for healing and regenerating damaged neural tissue [25].
3. *Neurotrophic factors*: These are molecules that support the survival, growth, and function of neurons. Stem cells (mesenchymal and neural) can secrete various neurotrophic factors like BDNF, NGF, GDNF, etc. These factors can promote neuronal growth, survival, and function. They also help in maintaining the microenvironment for neurons [26].
4. *Immunomodulatory properties*: Stem cells can interact with immune cells and may modulate the immune response. This modulation will prevent the non-specific attack of immune cell on healthy neurons, as in multiple sclerosis, for example [27].
5. *Migration and homing*: During injury/inflammation/neurodegenerative disorder in the brain, stem cell can migrate to the site and may release some therapeutic agents to prevent the progress of the disease [28].
6. *Personalised medicine*: The ability to produce induced pluripotent stem cells (iPSCs) particular to an individual enables the development of disease models that closely mimic a person's condition. These models from iPSCs can be utilised for drug testing and customised therapies [29].

17.3.3 Mechanisms of Action for Stem Cell Therapy in Neurodegenerative Diseases

Stem cell therapy has been shown to exert neuroprotective effects in neurodegenerative diseases. These effects involve various mechanisms that help preserve and protect neurons from degeneration and promote their survival. Some of the key neuroprotective mechanisms of stem cell therapy is given in **Figure 17.1**.

17.3.4 Stem Cell-Based Therapies in Neurodegenerative Diseases

Here is an overview of some important and common preclinical and clinical studies of neurodegenerative disorder.

Preclinical Studies

1. *Alzheimer's disease*: In preclinical studies of rodent AD models, stem cell treatment has shown significant results and successfully improved cognitive functions. They have used neural stem cells and iPSCs to study the disease mechanism and efficacy of stem cells in treatment of the disease [30].

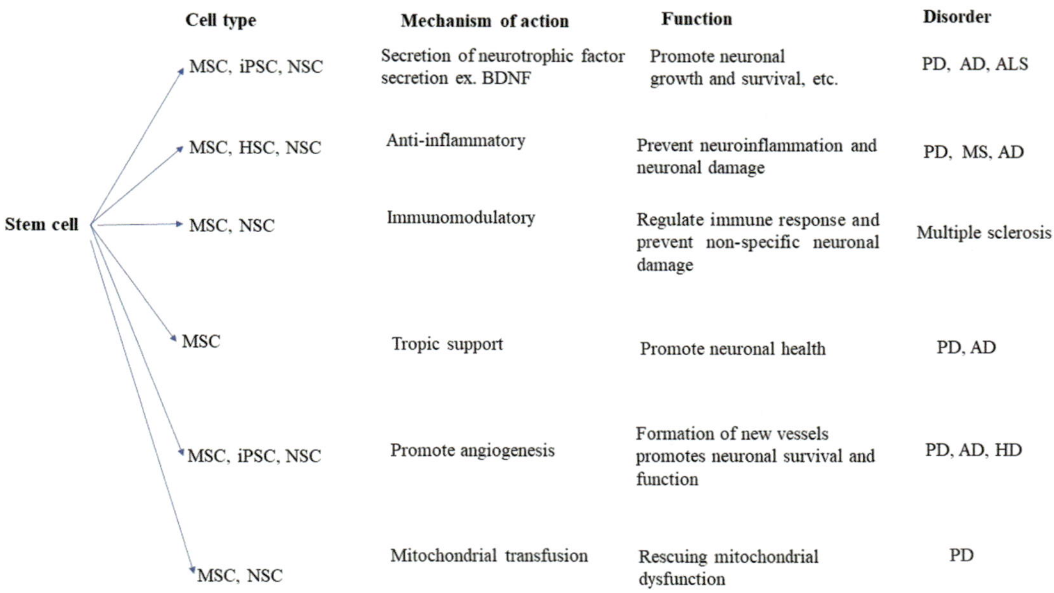

FIGURE 17.1 Use of stem cell type with different mechanisms of action and function in the treatment of different neurodegenerative disorders. *Abbreviations*: AD: Alzheimer's disease; ALS: Amylotropic lateral sclerosis; HD: Huntington's disease; HSC: haematopoetic stem cell; iPSC: induced pluripotent cell; MSC: mesenchymal stem cell; NSC: neural stem cell; PD: Parkinson's disease.

2. *Parkinson's disease*: In the preclinical model of PD, studies have shown that transplantation of iPSCs, embryonic stem cells, mesenchymal stem cells (MSCs) or neural stem cells helped in improving motor function and dopamine levels in the brain [31].
3. *Huntington's disease*: Several studies have shown that grafted striatal tissue can differentiate into desired cell types in rodents HD models and improve basic motor and cognitive functions [32]
4. *Amyotrophic lateral sclerosis*: Mesenchymal stem cells are multipotent cells having immunomodulatory, anti-inflammatory, and differentiation abilities, making them a good option for grafting for therapy. Results of preclinical trials have shown that treatment with MSCs could improve neuroprotection and reduce inflammation and neuro-regenerative properties that modulate biological functions [33].

Clinical Studies

1. *Alzheimer's disease*: Clinical trials of stem cells in AD are still in the early stages. The use of mesenchymal stem cells has been done in a few studies and it results in releasing neurotrophic factors and providing neuronal support [34].
2. *Parkinson's disease*: Transplantation of stem cell-derived dopaminergic neurons are in trial/use to treat patients of Parkinson's disease. Several trials are still in process to assess safety and efficacy and to optimise and address more potential side effects [35]. A few trials have shown positive results in motor impairment [35, 36].
3. *Multiple sclerosis*: Replacement with mesenchymal stem cells or neural stem cells has been studied in clinical trials for multiple sclerosis. Replacement with stem cell transplantation helps in replacing faulty immune systems and prevents neuronal damage [37, 38]. Some studies have shown promising results in halting disease progression and reducing disability in certain cases of multiple sclerosis [38].
4. *Amyotrophic lateral sclerosis*: Mesenchymal and neural stem cells are being used in the treatment of ALS [39]. The clinical trials of these cells help in slowing disease progression

and preserving motor function to some extent. These studies have primarily focused on safety and feasibility [40].

17.4 FUTURE DIRECTIONS AND CHALLENGES

New advancements in technology like DNA sequencing and its application in genomics studies provide a piece of novel information about the genetics of neurological disorders. Despite the advancements in such technology and advanced studies, understanding the cause/pathophysiology of neurodegenerative disorder is still a challenge. More studies on stem cell-related research will be focused in future to tackle the challenges in neurodegenerative disease.

1. *Personalised medicine*: Human-induced pluripotent stem cells (iPSCs) are a very promising model for disease modeling and drug discovery. The iPSCs provide the opportunity to develop a therapy based on the genetic and molecular basis of patients. iPSCs also help in delineating a molecular pathway of neurological disorders and can also help in exploring new targets for potential drugs. iPSC studies with genome editing approaches will help in pre-clinical trials in a personalised fashion [29, 41].
2. *Gene editing*: With the discovery of CRISPR-Cas9, gene editing has become a powerful tool to revolutionise the treatment of neurogenetic disorders and open the therapeutic window, with rapid advances made over the past few years. CRISPR-Cas9 can correct the mutation before stem cell transplantation in genetic neurodegenerative disorders which makes it a more specific, precise, and targeted therapy [42].
3. *Exosome-based therapies*: This approach will help in delivering bioactive molecules like proteins, DNA, etc. with stem cells to prevent damaged neurons and protect the progress of neurodegenerative disorders [43, 44].
4. *Combinatorial treatment*: Treatment with stem cells in addition to gene editing, immunomodulatory agents, and growth factors may have a synergistic effect in the treatment of neurodegenerative disorders [45].
5. *Brain organoids and organ-on-a-chip models*: Stem cell-based formation of brain organoids and brain-on-a-chip models provide better complexity and a more relevant platform to study neurodegenerative diseases and test available and potential therapies [46–48].
6. *Biomarker development*: Patient-specific or disease-specific derived neuron or stem cell-diseased models will aid in the discovery of disease-specific biomarkers and will facilitate early diagnosis and monitoring of disease progression in neurodegenerative disorders [49].

17.5 CONCLUSION

Over the past few years research on stem cell therapy has significantly increased. Stem cell technological improvements help in doing better research and opening new therapeutic windows for the treatment of several neurodegenerative diseases. The neuroprotection provided by stem cells occurs by slowing down disease progression, reducing symptoms, and improving quality of life with neurological conditions. Further research on stem cells will be crucial to understanding the mechanisms of action, and that will be more effective for developing treatment for various neurodegenerative diseases.

REFERENCES

1. Gao HM, Hong JS. Why neurodegenerative diseases are progressive: uncontrolled inflammation drives disease progression. *Trends in Immunology*. 2008;29(8):357–365.
2. Jeon I, Lee N, Li JY, Park IH, Park KS, Moon J, Shim SH, Choi C, Chang DJ, Kwon J, Oh SH. Neuronal properties, in vivo effects, and pathology of a Huntington's disease patient-derived induced pluripotent stem cells. *Stem Cells*. 2012;30(9):2054–2062.

3. Stephenson J, Nutma E, van der Valk P, Amor S. Inflammation in CNS neurodegenerative diseases. *Immunology*. 2018;154(2):204–219.
4. Migliore L, Coppedè F. Genetics, environmental factors and the emerging role of epigenetics in neurodegenerative diseases. *Mutation Research/Fundamental and Molecular Mechanisms of Mutagenesis*. 2009;667(1-2):82–97.
5. Sivandzade F, Cucullo L. Regenerative stem cell therapy for neurodegenerative diseases: an overview. *International Journal of Molecular Sciences*. 2021;22(4):2153.
6. Bonaventura G, et al. Stem cells: Innovative therapeutic options for neurodegenerative diseases? *Cells*. 2021;10(8): 1992.
7. Kim SU, De Vellis J. Stem cell-based cell therapy in neurological diseases: A review. *Journal of Neuroscience Research*. 2009;87(10):2183–2200.
8. Joyce N, et al. Mesenchymal stem cells for the treatment of neurodegenerative disease. *Regenerative Medicine*. 2010;5(6):933–946.
9. Lindvall O, Kokaia Z. Stem cells in human neurodegenerative disorders—Time for clinical translation? *The Journal of Clinical Investigation*. 2010;120(1):29–40.
10 Roberson, Erik D. Treatment of central nervous system degenerative disorders. In *Goodman & Gilaman's the Pharmacological Basis of Therapeutics* 13e Eds. Laurence L. Brunton, et al. McGraw-Hill Education, 2017.
11. Maresova P, Hruska J, Klimova B, Barakovic S, Krejcar O. Activities of daily living and associated costs in the most widespread neurodegenerative diseases: A systematic review. *Clinical Interventions in Aging*. 2020;15:1841–1862.
12. Ford E, et al. Human pluripotent stem cells-based therapies for neurodegenerative diseases: Current status and challenges. Cells. 2020;9(11): 2517.
13. O'brien JT, Burns P. Clinical practice with anti-dementia drugs: A revised (second) consensus statement from the British Association for Psychopharmacology. 2011;25(8):997–1019.
14. Vaddadi KS, et al. A randomised, placebo-controlled, double blind study of treatment of Huntington's disease with unsaturated fatty acids. *Neuroreport*. 2002;13(1):29–33.
15. Heris RM, et al. The potential use of mesenchymal stem cells and their exosomes in Parkinson's disease treatment. *Stem Cell Research & Therapy*. 2022;13(1):371.
16. Parambi DG, Alharbi KS, Kumar R, Harilal S, Batiha GE, Cruz-Martins N, Magdy O, Musa A, Panda DS, Mathew B. Gene therapy approach with an emphasis on growth factors: Theoretical and clinical outcomes in neurodegenerative diseases. *Molecular Neurobiology*. 2022;1:1–43.
17. Nance MA. Comprehensive care in Huntington's disease: A physician's perspective. *Brain Research Bulletin*. 2007;72(2-3):175–178.
18. Olanow CW, et al., Why have we failed to achieve neuroprotection in Parkinson's disease? *Annals of Neurology: Official Journal of the American Neurological Association and the Child Neurology Society*. 2008;64(S2):S101–S110.
19. Soto C, Pritzkow S. Protein misfolding, aggregation, and conformational strains in neurodegenerative diseases. *Nature Neuroscience*. 2018;21(10):1332–1340.
20. DeKosky, S.T., and K.J.S. Marek, Looking backward to move forward: Early detection of neurodegenerative disorders. *Science*. 2003;302(5646):830–834. doi: 10.1126/science.1090349. PMID: 14593169.
21. Kargiotis, O., et al., Quality of life in multiple sclerosis: Effects of current treatment options. *International Review of Psychiatry*. 2010;22(1):67–82.
22. Seifirad S, Haghpanah V. Inappropriate modeling of chronic and complex disorders: How to reconsider the approach in the context of predictive, preventive and personalized medicine, and translational medicine. *EPMA Journal*. 2019;10(3):195–209.
23. Donovan PJ, Gearhart JJN. The end of the beginning for pluripotent stem cells. *Nature*. 2001;414(6859): 92–97.
24. Fuchs E, Segre JA. Stem cells: A new lease on life. *Cell*. 2000;100(1):143–155.
25. Wobus AM, Boheler KR. Embryonic stem cells: Prospects for developmental biology and cell therapy. *Physiological Reviews*. 2005;85(2):635–678.
26. Lladó J, et al., Neural stem cells protect against glutamate-induced excitotoxicity and promote survival of injured motor neurons through the secretion of neurotrophic factors. *Molecular and Cellular Neuroscience*. 2004;27(3):322–331.
27. Xu C, et al., Mesenchymal stem cells maintain the microenvironment of central nervous system by regulating the polarization of macrophages/microglia after traumatic brain injury. *International Journal of Neuroscience*. 2017;127(12):1124–1135.
28. Müller FJ, Snyder EY, Loring JF. Gene therapy: Can neural stem cells deliver?. *Nature Reviews Neuroscience*. 2006;7(1):75–84

29. Stoddard-Bennett T, Reijo Pera R. Treatment of Parkinson's disease through personalized medicine and induced pluripotent stem cells. *Cells*. 2019;8(1):26.
30. Barak M, et al., Human iPSC-derived neural models for studying Alzheimer's disease: From neural stem cells to cerebral organoids. *Stem Cell Reviews and Reports*. 2022;18(2):792–820.
31. Isacson O, et al., Toward full restoration of synaptic and terminal function of the dopaminergic system in Parkinson's disease by stem cells. *Annals of Neurology: Official Journal of the American Neurological and the Child Neurology Society*. 2003;53(S3):S135–S148.
32. Nakao N, Itakura T. Fetal tissue transplants in animal models of Huntington's disease: The effects on damaged neuronal circuitry and behavioral deficits. *Progress in Neurobiology*. 2000;61(3):313–338.
33. Lewis CM, Suzuki M. Therapeutic applications of mesenchymal stem cells for amyotrophic lateral sclerosis. *Stem Cell Research & Therapy*. 2014;5:1-0.
34. Pacheco-Herrero M, et al. Current status and challenges of stem cell treatment for Alzheimer's disease. *Journal of Alzheimer's Disease*. 2021;84(3):917–935.
35. Takahashi J. iPS cell-based therapy for Parkinson's disease: A Kyoto trial. *Regenerative Therapy*. 2020;13:18–22.
36. Barker RA, et al. Human trials of stem cell-derived dopamine neurons for Parkinson's disease: Dawn of a new era. *Cell Stem Cell*. 2017;21(5):569–573.
37. Karussis D, et al. Hematopoietic stem cell transplantation in multiple sclerosis. *Expert Review of Neurotherapeutics*. 2013;13(5):567–578.
38. Holloman JP, et al. The development of hematopoietic and mesenchymal stem cell transplantation as an effective treatment for multiple sclerosis. *American Journal of Stem Cells*. 2013;2(2):95.
39. Trounson A, et al. Clinical trials for stem cell therapies. *BMC Medicine*. 2011;**9**(1):1–7.
40. Trounson A, McDonald C. Stem cell therapies in clinical trials: Progress and challenges. *Cell Stem Cell*. 2015;17(1):11–22.
41. Corti S, et al. Human pluripotent stem cells as tools for neurodegenerative and neurodevelopmental disease modeling and drug discovery. *Expert Opinion on Drug Discovery*. 2015;10(6):615–629.
42. Kolli N, et al. Application of the gene editing tool, CRISPR-Cas9, for treating neurodegenerative diseases. *Neurochemistry International*. 2018;112:187–196.
43. Xiong Y. Mahmood A, Chopp M, Emerging potential of exosomes for treatment of traumatic brain injury. *Neural Regeneration Research*. 2017;12(1):19.
44. Joo HS, et al. Current knowledge and future perspectives on mesenchymal stem cell-derived exosomes as a new therapeutic agent. *International Journal of Molecular Sciences*. 2020;21(3):727.
45. Caprnda M, et al. Immunomodulatory effects of stem cells: Therapeutic option for neurodegenerative disorders. *Biomedicine and Pharmacotherapy*. 2017;91:60–69.
46. Fanizza F, et al. Induced pluripotent stem cell-based organ-on-a-chip as personalized drug screening tools: A focus on neurodegenerative disorders. *Journal of Tissue Engineering*. 2022;13:20417314221095339.
47. Bang S, et al. Emerging brain-pathophysiology-mimetic platforms for studying neurodegenerative diseases: Brain organoids and Brains-on-a-chip. *Advanced Healthcare Materials*. 2021;10(12):2002119.
48. Jusop AS, et al. Development of brain organoid technology derived from IPSC for the neurodegenerative disease modelling: A glance through. *Frontiers in Molecular Neuroscience*. 2023. Aug 3;16:1173433
49. Silva MC, Haggarty SJ. Human pluripotent stem cell–derived models and drug screening in CNS precision medicine. *Annals of the New York Academy of Sciences*. 2020;1471(1):18–56.

Index

Note: *Italic* and **Bold** page numbers refer to *figures* and **tables.**

D

E

F

G

For Product Safety Concerns and Information please contact our
EU representative GPSR@taylorandfrancis.com Taylor & Francis
Verlag GmbH, Kaufingerstraße 24, 80331 München, Germany